# Mathematics for Physical Chemistry

## Second Edition

### Robert G. Mortimer

*Department of Chemistry*
*Rhodes College*
*Memphis, Tennessee*

HARCOURT
ACADEMIC
PRESS

This book is printed on acid-free paper. ∞

Copyright © 1999, 1981 by ACADEMIC PRESS

All Rights Reserved.
No part of this publication may be reproduced or transmitted in any form or by any
means, electronic or mechanical, including photocopy, recording, or any information
storage and retrieval system, without permission in writing from the publisher.

Academic Press
*a division of Harcourt Brace & Company*
525 B Street, Suite 1900, San Diego, California 92101-4495, USA
http://www.apnet.com

Academic Press
24-28 Oval Road, London NW1 7DX, UK
http://www.hbuk.co.uk/ap/

Library of Congress Catalog Card Number: 98-89310

International Standard Book Number: 0-12-508340-8

PRINTED IN THE UNITED STATES OF AMERICA
99  00  01  02  03  04  QW  9  8  7  6  5  4  3  2  1

# ◼ CONTENTS

# 3 Mathematical Functions and Differential Calculus

# 4 Integral Calculus

# 5 Calculus with Several Independent Variables

# 6 Mathematical Series and Transforms

# 7 Differential Equations

# 8 Operators, Matrices, and Group Theory

# 9 The Solution of Algebraic Equations

# 10 The Treatment of Experimental Data

# 11 Using Computers in Physical Chemistry

# ■ PREFACE

This book provides a survey of the mathematics needed for physical chemistry courses at the undergraduate level. Although several kinds of elementary physical chemistry courses exist, all have several mathematics courses as prerequisites. However, in three decades of teaching physical chemistry, I have found that many students have not been introduced to all the mathematical topics needed in the course and that most need some practice in applying their mathematical knowledge to the problems of physical chemistry.

I have tried to write all parts of this book so that they can be used for self-study by someone not familiar with the material, although any book such as this cannot be a substitute for the traditional training offered in mathematics courses. Solved examples and problems for the reader are interspersed throughout the presentations, and these form an important part of the presentations. As you study any topic in the book, you should follow the solution to each example and work each problem as you come to it.

The first nine chapters of the book are constructed around a sequence of mathematical topics. Chapter 10 is a discussion of mathematical topics needed in the analysis of experimental data, and Chapter 11 is a brief introduction to the use of computers in physical chemistry, including computer programming in the BASIC language. Most of the material in at least the first four chapters should be a review for nearly all readers of the book. I have tried to write all of the chapters after the first four so that they can be studied in any sequence, or piecemeal as the need arises.

This edition is a revision of a first edition published by Macmillan in 1981. I have reviewed every paragraph and have made those changes that were necessary to improve the clarity and correctness of the presentations. Additional information on

Fourier transforms has been added, and a brief discussion of Laplace transforms has been added to Chapter 5. A discussion of the use of Laplace transforms in solving differential equations has been added to Chapter 7. The material in Chapter 11 on word processors, spreadsheets, graphics software, and complete mathematics packages is new. I have added chapter summaries, a new preview of each chapter, and a list of important facts and ideas as the beginning of each chapter. I have continued to give a list of objectives for each chapter.

This book serves three functions: (1) a review of topics already studied and an introduction to new topics for those preparing for a course in physical chemistry, (2) a supplementary text to be used during a physical chemistry course, and (3) a reference book for graduate students and practicing chemists.

I am pleased to acknowledge the cooperation and help of David Phanco and Garrett Brown of Academic Press and of Rama Viswanathan and Kim Salt, who reviewed all or part of the manuscript. It is also a pleasure to acknowledge the assistance of all those who helped with the first edition of this book.

*Robert G. Mortimer*

# NUMBERS, MEASUREMENTS, MATHEMATICS, AND PROBLEM SOLVING

## Preview

Mathematics is one of the most important tools of physical chemistry. It deals with various physical quantities that have numerical values. In this chapter, we introduce the correct use of numerical values to represent measured physical quantities. Such a value generally consists of a number and a unit of measurement, and both parts of the value must be manipulated correctly. We introduce the use of significant digits to communicate the probable accuracy of the measured value. We also review the factor-label method, which is a routine method of expressing a measured quantity in terms of a different unit of measurement. Finally, we make some elementary comments on problem-solving techniques.

## Principal Facts and Ideas

1. Specification of a measured quantity consists of a number and a unit.
2. A unit of measurement is an arbitrarily defined quantity.
3. The SI units have been officially adopted by international organizations of physicists and chemists.
4. Reported values of all quantities should be rounded so that insignificant digits are not reported.
5. Consistent units must be used in any calculation.
6. The factor-label method can be used to convert from one unit of measurement to another.
7. The principal initial task in problem-solving is to plan a method of processing the given information to obtain the desired answer.

## Objectives

After you have studied the chapter, you should be able to:

1. use numbers and units correctly to express measured quantities;
2. understand the relationship of uncertainties in measurements to the use of significant digits;
3. use consistent units, especially the SI units, in equations and formulas;
4. use the factor-label method to convert from one unit of measurement to another;
5. analyze a typical physical chemistry problem and plan a method of problem-solving.

## SECTION 1.1. NUMBERS AND MEASUREMENTS

The most common use that chemists make of numbers is to report values for measured quantities. Specification of a measured quantity generally includes a number and a unit of measurement. For example, a length might be given as 12.00 inches, or 30.48 centimeters, or 0.3048 meters, etc. Specification of the quantity is not complete until the unit of measurement is specified, since 30.48 centimeters is definitely not the same as 30.48 inches. We discuss numbers in this section of the chapter, and will use some common units of measurement. We discuss units in the next section.

### Numbers

There are several sets into which we can classify numbers. Those which can represent physical quantities are called *real numbers*. They can range from positive numbers of indefinitely large magnitude to negative numbers of indefinitely large magnitude. Among the real numbers are the *integers* $0, \pm 1, \pm 2, \pm 3$, etc., which are part of the set of *rational numbers*. Other members of the set of rational numbers are quotients of two integers, such as $\frac{2}{3}, \frac{7}{9}, \frac{37}{53}$, etc.

Still other real numbers are called *algebraic irrational numbers*. They include square roots of rational numbers, cube roots of rational numbers, etc., which are not themselves rational numbers. All of the rest of the real numbers are called *transcendental irrational numbers*. Two commonly encountered transcendental irrational numbers are the ratio of the circumference of a circle to its diameter, called $\pi$ and given by $3.141592654\ldots$, and the base of natural logarithms, called $e$ and given by $2.718281828\ldots$. The three dots (an "ellipsis") that follow the given digits indicate that more digits follow. Irrational numbers have the property that if you have some means of finding what the correct digits are, you will never reach a point beyond which all of the remaining digits are zero, or beyond which the digits form some other repeating pattern. However, with a rational number, one or the other of these two things will always happen.[1]

In addition to real numbers, mathematicians have defined imaginary numbers into existence. The *imaginary unit*, $i$, is defined to equal $\sqrt{-1}$. An *imaginary number* is equal to a real number times $i$, and a *complex number* is equal to a real number

---

[1] It has been said that early in the twentieth century the legislature of the state of Indiana, in an effort to simplify things, passed a resolution that henceforth in that state, $\pi$ should be exactly equal to 3.

plus an imaginary number. If $x$ and $y$ are real numbers, then the quantity $z = x + iy$ is a complex number. $x$ is called the *real part* of $z$, and the real number $y$ is called the *imaginary part* of $z$. We will discuss complex numbers in a later chapter.

---

### Problem 1.1

Take a few simple fractions, such as $\frac{2}{3}$, $\frac{4}{9}$, or $\frac{3}{7}$, and express them as decimal numbers, finding either all of the nonzero digits or the repeating pattern of digits.

---

## Measurements and Significant Digits

A measured quantity can almost never be known with complete exactness. It is therefore a good idea to communicate the probable accuracy of a reported measurement. For example, let us consider the specification of the length of an object, say a piece of glass tubing, which we have measured with a meter stick. Assume that our measured value is 387.8 millimeters (mm) and that our experimental error is no greater than 0.6 millimeter. The best way to specify the length of the glass tubing is

$$\text{length} = 387.8\,\text{mm} \pm 0.6\,\text{mm}.$$

If for some reason we cannot include a statement of the probable error, we should at least avoid including digits that are probably wrong. We must sometimes round off some of the digits of the value. In this case, our error is somewhat less than 1 mm, so the correct number is probably closer to 388 mm than to either 387 mm or 389 mm. If we do not want to report the expected experimental error, we report the length as 388 mm and assert that the three digits given are *significant digits*. This means that we want to communicate that the given digits are correct. If we had reported the length as 387.8 mm, the last digit is *insignificant*. That is, it is probably wrong.

You should always avoid reporting digits that are not significant. When you carry out calculations involving measured quantities, you should always determine how many significant digits your answer can have and round off your result to that number of digits. When values of physical quantities are given in a physical chemistry textbook or in this book, you can assume that all digits specified are significant. In most cases, counting the number of significant digits is simple. You just count the digits. However, a problem arises in counting significant digits if the number contains one or more zeros. Any zero which occurs between nonzero digits counts as a significant digit. However, any zeros which are given only to specify the location of a decimal point do *not* represent significant digits. For example, the number 0.0000345 contains three significant digits. The number 76,000 contains only two significant digits. However, the number 0.000034500 contains five significant digits. The zeros at the left are present only to locate the decimal point, but the final two zeros are not needed to locate the decimal point, and therefore must have been included because the number is known with sufficient accuracy to require statement of these digits. You should apply these rules when reading a number written by someone else and should conform your values to them when you write numbers.

A problem arises when zeros that appear to occur only to locate the decimal point are actually significant. For example, if a mass is known to be closer to 3500 grams than to 3499 grams or to 3501 grams, there are four significant digits. If one simply wrote 3500 grams, persons with training in significant digits would assume that the zeros are not significant and that there are two significant digits. Some people communicate the fact that there are four significant digits by writing 3500. grams. The explicit decimal point communicates the fact that the zeros are significant digits. Others put a bar over the zeros, writing 3500̄ to indicate that there are four significant digits.

## Scientific Notation

The communication difficulty involving trailing zeros can be avoided by the use of *scientific notation*, in which a number is expressed as the product of two factors, one of which is a number lying between 1 and 10, and the other is 10 raised to some integer power. The mass mentioned above would thus be written as $3.500 \times 10^3$ grams, and there are clearly four significant digits indicated, since the trailing zeros are not required to locate a decimal point. If the mass were known to only two significant digits, it would be written as $3.5 \times 10^3$ grams.

Scientific notation is also convenient if extremely small or extremely large numbers must be written. For example, Avogadro's constant, the number of molecules or other formula units per mole, is easier to write as $6.02214 \times 10^{23}$ mol$^{-1}$ than as 602,214,000,000,000,000,000,000 mol$^{-1}$, and the charge on an electron is easier to write as $1.60217 \times 10^{-19}$ coulomb than as 0.000000000000000000160217 coulomb.

---

### Problem 1.2

Convert the following numbers to scientific notation, using the correct number of significant digits:

    a. 0.000645
    b. 67,342,000
    c. 0.000002
    d. 6432.

---

## Determining the Number of Significant Digits in a Calculated Quantity

When you calculate a numerical value that depends on a set of numerical values substituted into a formula, etc., the accuracy of the result depends on the accuracy of the first set of values. Therefore, the number of significant digits in the result depends on the numbers of significant digits in the first set of values. Any result containing insignificant digits must be rounded to the proper number of digits.

The process of rounding is straightforward in most cases. The calculated number is simply replaced by that number containing the proper number of digits that is closer to the calculated value than any other number containing this many digits. Thus, if there are three significant digits, 4.567 is rounded to 4.57, and 4.564 is rounded to 4.56. However, if your only insignificant digit is a 5, your calculated number is midway between two rounded numbers, and you must decide whether to round up or to round down. It is best to have a rule that will "round down" half of the time and "round up" half of the time. One widely used rule is to round to the even digit, since there is a 50% chance that any digit will be even. For example, 2.5 would be rounded to 2, and

3.5 would be rounded to 4. An equally valid procedure that is apparently not generally used would be to toss a coin and round up if the coin comes up "heads" and to round down if it comes up "tails."

There are several useful rules of thumb that allow you to determine the proper number of significant digits in the result of a calculation. For multiplication of two or more factors, the rule is that the product will have the same number of significant digits as the factor with the fewest significant digits. The same rule holds for division.

### Example 1.1

What is the area of a rectangular object whose length is given as 7.78 meter (m), whose width is given as 3.486 m, and whose height is 1.367 m?

### Solution

The volume is the product of the three dimensions:

$$V = (7.78\,\text{m})(3.486\,\text{m})(1.367\,\text{m}) = 37.07451636\,\text{m}^3 \approx 37.1\,\text{m}^3.$$

We round the volume to 37.1 m$^3$, since the factor with the fewest digits has three digits.

### Example 1.2

Compute the smallest and largest values that the volume in Example 1.1 might have and determine whether the answer given in Example 1.1 is correctly stated.

### Solution

The smallest value that the length might have, assuming the given value to have only significant digits, is 7.775 m, and the largest value that it might have is 7.785 m. The smallest possible value for the width is 3.4855 m and the largest value is 3.4865 m. The smallest possible value for the height is 1.3665 m and the largest value is 1.3675. The minimum value for the volume is

$$V_{min} = (7.775\,\text{m})(3.4855\,\text{m})(1.3665\,\text{m}) = 37.0318254562\,\text{m}^3.$$

The maximum value is

$$V_{max} = (7.785\,\text{m})(3.4865\,\text{m})(1.3675\,\text{m}) = 37.1172354188\,\text{m}^3.$$

Obviously, all of the digits beyond the first three are insignificant. The rounded result of 37.1 m$^3$ in Example 1.1 contains all of the digits that can justifiably be given.

The result of Example 1.2 is fairly typical. There is some chance that 37.0 m$^3$ might be closer to the actual volume than is 37.1 m$^3$. This is often the case: the last digit said to be significant might be incorrect by $\pm 1$. We still consider it to be a significant digit. The rule of thumb for significant digits in addition or subtraction is slightly different: For a digit to be significant, it must arise from a significant digit in every term of the sum or difference.

### Example 1.3

Determine the combined length of two objects, one of length 0.783 m and one of length 17.3184 m.

**Solution**

We add

$$
\begin{array}{r}
0.783 \text{ m} \\
17.3184 \text{ m} \\
\hline
18.1014 \text{ m} \approx 18.101 \text{ m.}
\end{array}
$$

The final 4 is not a significant digit, because the fourth digit after the decimal point in the top number is unknown, and fourth digit after the decimal point in the sum could be significant only if that digit were significant in every term of the sum. The 4 in the sum could be significant only if the fourth digit in the first number were known to be zero. We must round the answer to 18.101 m. Even after this rounding, we have obtained a number with five significant digits while one of our terms has only three significant digits.

---

In a calculation with several steps, some people round off the insignificant digits at each step. However, this can lead to accumulation of round-off error. A reasonable policy is to carry along at least one insignificant digit during the calculation, and then to round off the insignificant digits at the final answer. When using an electronic calculator, it is easy to use all of the digits carried by the calculator, and then to round off at the end of the calculation.

If you are carrying out operations other than additions, subtractions, multiplications, and divisions, you may have trouble in determining which digits are significant. If you must take sines, cosines, logarithms, etc., it may be necessary to do the operation with the smallest and the largest values that the number on which you must operate can have (incrementing and decrementing the number). For numbers that are not too large or too small, you can usually keep the same number of digits in a logarithm, sine, cosine, etc., that occur in its argument. However, more accurate rules of thumb can be found.[2]

---

### Example 1.4

Calculate the following. Determine the correct number of significant digits by incrementing or decrementing.

a. $\sin(372.15°)$
b. $\ln(567.812)$
c. $e^{-9.813}$.

---

**Solution**

a. By use of a calculator,

$$
\sin(372.155°) = 0.210557
$$
$$
\sin(372.145°) = 0.210386.
$$

---

[2] Donald E. Jones, "Significant Digits in Logarithm Antilogarithm Interconversions," *J. Chem. Educ.* **49**, 753 (1972).

Therefore,

$$\sin(372.15°) = 0.210_5,$$

where we introduce the notation that a number that is not quite significant is written as a subscript. Note that even though the argument of the sine had five significant digits, the sine has only three significant digits (almost four).

b.  By use of a calculator,

$$\ln(567.8125) = 6.341791259$$
$$\ln(567.8115) = 6.341789497.$$

Therefore,

$$\ln(567.812) = 6.34179.$$

In this case, the logarithm has the same number of significant digits as its argument. If the argument of a logarithm is very large, the logarithm can actually have many more significant digits than its argument, since the logarithm of a large number is a slowly varying function of its argument.

c.  By the use of a calculator,

$$e^{-9.8135} = 0.00005470803$$
$$e^{-9.8125} = 0.00005476277.$$

Therefore,

$$e^{-9.813} = 0.00054_7.$$

Although the argument of the exponential had four significant digits, the exponential has not quite three significant digits.

---

### Problem 1.3

Calculate the following to the proper numbers of significant digits.

a.  $(37.815 + 0.00435)(17.01 + 3.713)$
b.  $625[e^{12.1} + \sin(30.0°)]$
c.  $65.718 \times 14.3$
d.  $17.13 + 14.7651 + 3.123 + 7.654 - 8.123.$

---

## SECTION 1.2. UNITS OF MEASUREMENT

In Section 1.1, we mentioned measuring the length of an object with a meter stick. Such a measurement would be impossible without a standard definition of the meter (or other unit of length), and for many years science and commerce were hampered by the lack of accurately defined units of measurement. This problem has been largely overcome by accurate measurements and international agreements.

The internationally accepted system of units of measurements is called the *Systéme International d'Unités*, abbreviated *SI*. This is an *MKS system*, which means that length is measured in meters, mass in kilograms, and time in seconds. In 1960 the international chemical community agreed to use SI units, which had been in use

**TABLE 1.1   SI Units**

**SI base units (units with independent definitions)**

| Physical quantity | Name of unit | Symbol | Definition |
|---|---|---|---|
| Length | meter | m | Length such that the speed of light is exactly 299,792,458 $\text{m s}^{-1}$. |
| Mass | kilogram | kg | The mass of a platinum-iridium cylinder kept at the International Bureau of Weights and Measures. |
| Time | second | s | The duration of 9,192,631,770 cycles of the radiation of a certain emission of the cesium atom. |
| Electric | ampere | A | The magnitude of current which, when flowing in each of two long parallel wires 1 m apart in free space, results in a force of $2 \times 10^{-7}$ N per meter of length. |
| Temperature | kelvin | K | Absolute zero is 0 K, triple point of water is 273.16 K. |
| Luminous intensity | candela | cd | The luminous intensity, in the perpendicular direction, of a surface of $1/600,000 \, \text{m}^2$ of a black body at temperature of freezing platinum at a pressure of 101,325 $\text{N/m}^2$. |
| Amount of substance | mole | mol | Amount of substance which contains as many elementary units as there are carbon atoms in exactly 0.012 kg of the carbon-12 isotope. |

**Other SI units ("derived units")**

| Physical quantity | Name of unit | Physical dimensions | Symbol | Definition |
|---|---|---|---|---|
| Force | newton | $\text{kg m s}^{-2}$ | N | $1 \, \text{N} = 1 \, \text{kg m s}^{-2}$ |
| Energy | joule | $\text{kg m}^2 \text{s}^{-2}$ | J | $1 \, \text{J} = 1 \, \text{kg m}^2 \text{s}^{-2}$ |
| Electrical charge | coulomb | A s | C | $1 \, \text{C} = 1 \, \text{A s}$ |
| Pressure | pascal | $\text{N m}^{-2}$ | Pa | $1 \, \text{Pa} = 1 \, \text{N m}^{-2}$ |
| Magnetic field | tesla | $\text{kg s}^{-2} \text{A}^{-1}$ | T | $1 \, \text{T} = 1 \, \text{kg s}^{-2} \text{A}^{-1}$ $= 1 \, \text{weber m}^{-2}$ |
| Luminous flux | lumen | cd sr | lm | $1 \, \text{lm} = 1 \, \text{cd sr}$ (sr = steradian) |

by physicists for some time.[3] The seven base units given in Table 1.1 form the heart of the system. Included in the table are also some "derived" units, which owe their definitions to the definitions of the seven base units.

Some non-SI units continue to be used, such as the *atmosphere* (atm), which is a pressure defined to equal $101,325 \, \text{N m}^{-2}$ (101,325 Pa), the liter (L), which is exactly $0.001 \, \text{m}^3$, and the *torr*, which is a pressure such that exactly 760 torr equal 1

[3] See "Policy for NBS Usage of SI Units," *J. Chem. Educ.* **48**, 569 (1971).

**TABLE 1.2  Prefixes for Multiple and Submultiple Units**

| Multiple | Prefix | Abbreviation |
|----------|--------|--------------|
| $10^{12}$ | tera- | T |
| $10^9$ | giga- | G |
| $10^6$ | mega- | M |
| $10^3$ | kilo- | k |
| 1 | — | — |
| $10^{-1}$ | deci-* | d |
| $10^{-2}$ | centi-* | c |
| $10^{-3}$ | milli- | m |
| $10^{-6}$ | micro- | $\mu$ |
| $10^{-9}$ | nano- | n |
| $10^{-12}$ | pico- | p |
| $10^{-15}$ | femto- | f |
| $10^{-18}$ | atto- | a |

*The use of the prefixes for $10^{-1}$ and $10^{-2}$ is being discouraged, but centimeters will probably not be abandoned for many years to come.

atmosphere. The Celsius temperature scale also remains in common use among chemists. The definition of 0°C is 273.15 K, and the degree Celsius is the same size as the kelvin.

In the United States of America, English units of measurement are still in common use. The inch (in) has been redefined to equal exactly 0.0254 meter. The pound (lb) is equal to 0.4536 kg (not an exact definition; good to four significant digits).

Multiples and submultiples of SI units are commonly used.[4] Examples are the millimeter and kilometer. These multiples and submultiples are denoted by standard prefixes attached to the name of the unit, as listed in Table 1.2. The abbreviation for a multiple or submultiple is obtained by attaching the prefix abbreviation to the unit abbreviation, as in Gm (gigameter) or ns (nanosecond). Note that since the base unit of length is the kilogram, the table would imply the use of things such as the mega kilogram. Double prefixes are not used. We use gigagram instead of megakilogram.

Any measured quantity is not completely specified until its units are given. If $a$ is a length, one must say

$$a = 10.345\,\text{m} \tag{1.1}$$

not just

$$a = 10.345 \qquad (not\ correct).$$

---

[4]There is a somewhat apocryphal story about Robert A. Millikan, a Nobel-prize-winning physicist who was not noted for false modesty. A rival is supposed to have told Millikan that he had defined a new unit for the quantitative measure of conceit and had named the new unit the kan. However, 1 kan was an exceedingly large amount of conceit, so that for most purposes the practical unit was to be the millikan.

It is permissible to write

$$a/m = 10.345$$

which means that the length $a$ divided by 1 meter (m) is 10.345, a dimensionless number. When constructing a table of values, it is convenient to label the columns or rows with such dimensionless quantities.

When you make numerical calculations, you should make certain that you use consistent units for all quantities. Otherwise, you will likely get the wrong answer. This means that (1) you must convert all multiple and submultiple units to the base unit, and (2) you cannot mix different systems of units. For example, you cannot substitute a length in inches into a formula in which the other quantities are in SI units without converting. It is a good idea to write the unit as well as the number, as in Eq. (1.1), even for scratch calculations. This will help you avoid some kinds of mistakes by inspecting any equation and making sure that both sides are measured in the same units.

## The Factor-Label Method

This is an elementary method for the routine conversion of a quantity measured in one unit to the same quantity measured in another unit. The method consists of multiplying the quantity by a fraction that is equal to unity in a physical sense, with the numerator and denominator equal to the same quantity expressed in different units. This does not change the quantity physically, but numerically expresses it in another unit, and so changes the number expressing the value of the quantity.

For example, to express 3.00 km in terms of meters, one writes

$$(3.00\,\text{km})\left(\frac{1000\,\text{m}}{1\,\text{km}}\right) = 3000\,\text{m} = 3.00 \times 10^3\,\text{m}. \tag{1.2}$$

You can check the units by considering a given unit to "cancel" if it occurs in both the numerator and denominator. Thus, the left-hand side of Eq. (1.2) has units of meters, because the km on the top cancels the km on the bottom. In applying the method, you should write out the factors explicitly, including the units. You should carefully check that the unwanted units "cancel." Only then should you proceed to the numerical calculation.

---

### Example 1.5

Convert the speed of light, $2.9979 \times 10^8\,\text{m s}^{-1}$, to the same quantity in miles per hour. Use the definition of the inch, $1\,\text{in} = 0.0254\,\text{m}$ (exactly).

---

### Solution

$$(2.9979 \times 10^8\,\text{m s}^{-1})\left(\frac{1\,\text{in}}{0.0254\,\text{m}}\right)\left(\frac{1\,\text{ft}}{12\,\text{in}}\right)\left(\frac{1\,\text{mi}}{5280\,\text{ft}}\right)\left(\frac{60\,\text{s}}{1\,\text{min}}\right)\left(\frac{60\,\text{min}}{1\text{h}}\right)$$
$$= 6.7061 \times 10^8\,\text{mi h}^{-1}.$$

The conversion factors which correspond to exact definitions do not limit the number of significant digits. For example, 12 inches equal exactly 1 foot, so that our answer has five significant digits.

---

### Problem 1.4

Express the following in terms of SI base units.

a. 26.17 mi
b. 55 mi h$^{-1}$
c. 7.5 nm ps$^{-1}$
d. 13.6 eV (electron volts).

The electron volt, a unit of energy, equals $1.6022 \times 10^{-19}$ J.

---

## SECTION 1.3. PROBLEM SOLVING

In solving problems in physical chemistry, you will probably learn more than you do from reading your textbook or listening to lectures. A typical physical chemistry problem is similar to what was once called a "story problem" or a "word problem" in elementary school. You are given some factual information (or asked to find some), together with a verbal statement of what answer is required, but you must find your own method of obtaining the answer from the given information. In a simple problem, this may consist only of substituting numerical values into a formula, but in a more complicated problem you might have to derive your own mathematical formula, or you might have to draw a graph and draw conclusions from it or write a computer program. The method, or *algorithm*, must be developed for each problem.

---

### Example 1.6

Calculate the volume occupied by 1.278 mol of an ideal gas if the pressure is 6.341 atm and the temperature is 298.15 K.

---

### Solution

The ideal gas equation is

$$PV = nRT, \tag{1.3}$$

where $V$ is the volume, $n$ is the amount of gas in moles, $T$ is the temperature, $P$ is the pressure, and $R$ is the gas constant, equal to 8.3145 J K$^{-1}$ mol$^{-1}$. Since there are four variables, we can calculate the value of one of them if the values of the other three are given. We solve the equation for $V$ by dividing both sides of the equation by $P$:

$$V = \frac{nRT}{P}. \tag{1.4}$$

We calculate the volume by substitution into Eq. (1.4) and conversion of the pressure from atmospheres to pascals by use of the factor label method:

$$V = \frac{(1.278\,\text{mol})(8.3145\,\text{J K}^{-1}\,\text{mol}^{-1})(298.15\,\text{K})}{6.341\,\text{atm}} \left(\frac{1\,\text{atm}}{101325\,\text{Pa}}\right)$$

$$= 4.931 \times 10^{-3}\,\text{J Pa}^{-1}$$

$$= 4.931 \times 10^{-3}\,\text{J Pa}^{-1} \left(\frac{1\,\text{Pa}}{1\,\text{N m}^{-2}}\right)\left(\frac{1\,\text{N m}}{1\,\text{J}}\right) = 4.931 \times 10^{-3}\,\text{m}^3.$$

We have made additional conversions to make the units as simple as possible. The answer is given to four significant digits, because both the pressure and the amount of gas are specified to four significant digits.

---

Let's summarize the procedure of Example 1.6. We first determined that enough information was given and that the ideal gas equation of state was sufficient to work the problem. This equation was solved algebraically for the volume to give a working formula. The given values of quantities were substituted into the formula, and the necessary unit conversion was made. After all factors were written out and the units checked, the multiplications and divisions were carried out. This was a simple problem. In working a more complicated problem, it might be useful to map out on a piece of paper how you are going to get from the given information to the desired answer.[5]

---

### Problem 1.5

Calculate the temperature of 10.65 mol of an ideal gas if the volume is 37.6 L and the pressure is 532.6 torr.

In the remaining chapters of this book, you will see a number of examples worked out, and you will be asked to work a number of problems. In most of these, a method must be found and applied that will lead from the given information to the desired answer.

In some problems there will be a choice of methods. Perhaps you must choose between a graphical procedure and a numerical procedure, or between an approximate formula and an exact formula. In some of these cases, it would be foolish to carry out a more difficult solution, because an approximate solution will give you an answer that will be sufficient for the purpose at hand. In other cases, you will need to carry out a more nearly exact solution. You will need to learn how to distinguish between these two cases.

---

## SUMMARY OF THE CHAPTER

In this chapter, we introduced the use of numerical values in physical chemistry. In order to use such values correctly, one must handle the units of measurement in which they are expressed. Techniques for doing this, including the factor-label method, were introduced. One must also recognize the uncertainties in experimentally measured

---

[5]Some unkind soul has defined a mathematician as a person capable of designing a mathematically precise path from an unwarranted assumption to a foregone conclusion.

quantities. In order to avoid implying a greater accuracy than actually exists, one must express calculated quantities with the proper number of significant digits. Basic rules for significant digits were presented.

## ADDITIONAL READING

Arthur W. Adamson, *Understanding Physical Chemistry*, Benjamin, New York, 1969. This is a book of problems in physical chemistry, designed to supplement any physical chemistry textbook. There is a brief section in each chapter that summarizes the relevant theory. Solutions are given to the problems.

Leonard C. Labowitz and John S. Arents, *Physical Chemistry Problems and Solutions*, Academic Press, New York, 1969. This is another book of problems, with solutions given to all problems. In each section, there are three categories of problems, arranged by difficulty.

M. L. McGlashan, *Physico-chemical Quantities and Units*, Royal Inst. Chem. Publ. No. 15, London, 1968. This is a description of the SI.

C. R. Metz, *2000 Solved Problems in Physical Chemistry*, McGraw–Hill, New York, 1990. This is a good source of problems to be used for practicing problem-solving techniques in physical chemistry.

M. A. Paul, "The International System of Units (SI): Development and Progress," *J. Chem. Doc.* **11**, 3(1971). This is another description and explanation of the SI.

G. Polya, *How to Solve It—A New Aspect of Mathematical Method*, Princeton Univ. Press, Princeton, NJ, 1945. This small book, which should be in every college or university library, contains a detailed discussion of general methods of solving problems. The techniques apply to physical chemistry problems just as well as to mathematical problems.

## ADDITIONAL PROBLEMS

### 1.6

a. Find the number of inches in a meter. How many significant digits could be given?
b. Find the number of meters in 1 mile and the number of miles in 1 kilometer. How many significant digits could be given?

### 1.7

A furlong is one-eighth of a mile and a fortnight is 2 weeks. Find the speed of light in furlongs per fortnight, specifying the correct number of significant digits.

### 1.8

The Rankine temperature scale is defined so that the Rankine degree is the same size as the Fahrenheit degree, and $0°R$ is the same as 0 K.

a. Find the Rankine temperature at $0.00°C$.
b. Find the Rankine temperature at $0.00°F$.

### 1.9

Calculate the mass of AgCl that can be precipitated from 10.00 mL of a solution of NaCl containing 0.345 mol $L^{-1}$. Report your answer to the correct number of digits.

### 1.10

The volume of a right circular cylinder is given by

$$V = \pi r^2 h,$$

where $V$ is the volume, $r$ the radius, and $h$ the height. If a certain right circular cylinder has a radius given as 0.134 m and a height given as 0.318 m, find its volume, specifying it with the correct number of digits. Calculate the smallest and largest volumes that the cylinder might have and check your first answer for the volume.

### 1.11

The value of a certain angle is given as 31°. Using a table or a calculator, find the smallest and largest values that its sine and cosine might have and specify the sine and cosine to the appropriate number of digits.

### 1.12

    a. Some elementary chemistry textbooks give the value of $R$, the ideal gas constant, as 0.0821 L atm $K^{-1}$ $mol^{-1}$. Using the SI value, 8.3145 J $K^{-1}$ $mol^{-1}$, obtain the value in L atm $K^{-1}$ $mol^{-1}$ to five significant digits.

    b. Calculate the pressure in atmospheres and in N $m^{-2}$ (pascal) of a sample of an ideal gas containing 0.13678 mol, confined in a volume of 1.000 L at a temperature of 298.15 K, using the value of $R$ in SI units.

    c. Calculate the pressure in part b in atmospheres and in N $m^{-2}$ (Pa) using the value of $R$ in L atm $K^{-1}$ $mol^{-1}$.

### 1.13

The van der Waals equation of state gives better accuracy than the ideal gas equation of state. It is

$$\left(P + \frac{a}{\bar{V}^2}\right)(\bar{V} - b) = RT$$

where $a$ and $b$ are parameters that have different values for different gases and where $\bar{V} = V/n$, the molar volume. For carbon dioxide, $a = 0.3640$ Pa $m^6$ $mol^{-2}$, $b = 4.267 \times 10^{-5}$ $m^3$ $mol^{-1}$. Calculate the pressure of carbon dioxide, assuming the conditions in part b of Problem 1.12.

### 1.14

The specific heat capacity (specific heat) of a substance is crudely defined as the amount of heat required to raise the temperature of unit mass of the substance by 1 degree. The specific heat capacity of water is 4.18 kJ °$C^{-1}$ $kg^{-1}$. Find the rise in temperature if 100.0 J of heat is transferred to 1.000 kg of water.

# 2 MATHEMATICAL VARIABLES AND OPERATIONS

## Preview

In this chapter, we discuss elementary algebraic operations on mathematical variables. We include the simple algebraic operations, trigonometric functions, logarithms and exponentials of real scalar variables, algebraic operations on real vector variables, and algebraic operations on complex scalar variables.

## Principal Facts and Ideas

1. Algebra is a branch of mathematics in which operations are performed symbolically instead of numerically, according to a well-defined set of rules.
2. Trigonometric functions are examples of mathematical functions: To a given value of an angle there corresponds a value of the sine function, etc.
3. There is a set of useful trigonometric identities.
4. A vector is a quantity with magnitude and direction.
5. Vector algebra is an extension of ordinary algebra with its own rules and defined operations.
6. A complex number has a real part and an imaginary part that is proportional to $i$, defined to equal $\sqrt{-1}$.
7. The algebra of complex numbers is an extension of ordinary algebra with its own rules and defined operations.

## Objectives

After you have studied the chapter, you should be able to:

1. manipulate variables algebraically to simplify complicated algebraic expressions;
2. manipulate trigonometric functions correctly;
3. work correctly with logarithms and exponentials;
4. calculate correctly the sum, difference, scalar product, and vector product of any two vectors, whether constant or variable;
5. perform elementary algebraic operations on complex numbers; form the complex conjugate of any complex number and separate the real and imaginary parts of any complex expression.

## SECTION 2.1. ALGEBRAIC OPERATIONS ON REAL SCALAR VARIABLES

Algebra is a branch of mathematics that was invented by Greek mathematicians and developed by Hindu, Arab, and medieval European mathematicians. The great utility of algebra comes from the fact that letters are used to represent constants and variables, and that operations are indicated by symbols such as $+, -, \times, /, \sqrt{\ }$, etc. Operations can be carried out symbolically instead of numerically, so that formulas and equations can be modified and simplified before numerical calculations are carried out. This ability allows calculations to be carried out that arithmetic cannot handle.

The numbers and variables on which we operate in this section are called *real numbers* and *real variables*, because they do not include imaginary numbers such as the square root of $-1$, which we discuss in Section 2.6. They are also called *scalars*, to distinguish them from *vectors*, which have direction as well as magnitude, and which we discuss in Section 2.5. Real scalar numbers have *magnitude*, a specification of the size of the number, and *sign*, which can be positive or negative.

### The Four Elementary Algebraic Operations

The rules for operations on numbers with sign can be simply stated:

a. The product of two factors of the same sign is positive, and the product of two factors of different signs is negative.
b. The quotient of two factors of the same sign is positive, and the quotient of two factors of different signs is negative.
c. The difference of two numbers is the same as the sum of the first number and the negative of the second.
d. Multiplication is *commutative*, which means that[1]

$$\boxed{a \times b = b \times a}.$$   (2.1)

---

[1] We enclose equations that you will likely use frequently in a box.

e. Multiplication is *associative*, which means that

$$a \times (b \times c) = (a \times b) \times c \;.$$  (2.2)

f. Multiplication and addition are *distributive*, which means that

$$a \times (b + c) = a \times b + a \times c \;.$$  (2.3)

## Algebraic Manipulations

Algebra involves symbolic operations, that is, you manipulate symbols instead of carrying out numerical operations. For example, you can "divide" an expression by writing a symbol in a denominator and can then cancel the symbol in the denominator against the same symbol in the numerator of the same fraction. Or you can factor a polynomial expression and possibly cancel one or more of the factors against the same factors in a denominator. Or you can solve an equation by symbolically carrying out some set of operations on both sides of an equation, eventually isolating one of the symbols on one side of the equation.

Remember that if one side of an equation is operated on by anything that changes its value, the same operation must be applied to the other side of the equation to maintain the validity of the equality. Operations that do not change the value of an expression, such as factoring an expression, multiplying out factors, multiplying the numerator and denominator of a fraction by the same thing, etc., can be done to one side of an equation without destroying the validity of the equation.

### Example 2.1

Write the following expression in a simpler form:

$$A = \frac{(2x + 5)(x + 3) - 2x(x + 5) - 14}{x^2 + 2x + 1}.$$

### Solution

We multiply out the factors in the numerator and combine terms, factor the denominator, and cancel a common factor:

$$A = \frac{2x^2 + 11x + 15 - 2x^2 - 10x - 14}{(x + 1)(x + 1)} = \frac{x + 1}{(x + 1)^2} = \frac{1}{x + 1}.$$

### Problem 2.1

Write the following expression in a simpler form:

$$B = \frac{(x^2 + 2x)^2 - x^2(x - 2)^2 + 12x^4}{6x^3 + 12x^4}.$$

### Problem 2.2

The van der Waals equation of state provides a more nearly exact description of real gases than does the ideal gas equation. It is

$$\left( P + \frac{n^2 a}{V^2} \right)(V - nb) = nRT,$$

where $P$ is the pressure, $V$ is the volume, $n$ is the amount of gas in moles, $T$ is the absolute temperature, and $R$ is the gas constant (the same constant as in the ideal gas equation). The symbols $a$ and $b$ represent "parameters," which means that they are constants for a particular gas, but have different values for different gases.

    a. Manipulate the equation so that $\bar{V}$, defined as $V/n$, occurs instead of $V$ and $n$ occurring separately.
    b. Manipulate the equation into an expression for $P$ in terms of $T$ and $\bar{V}$.

---

## Additional Algebraic Operations

In addition to the four elementary algebraic operations, there are some other important algebraic operations. The *magnitude*, or *absolute value*, of a scalar quantity is denoted by placing vertical bars before and after the symbol for the quantity. This operation means

$$|x| = \begin{cases} x & \text{if } x \geq 0 \\ -x & \text{if } x < 0. \end{cases} \tag{2.4}$$

For example,

$$|4.5| = 4.5$$
$$|-3| = 3.$$

The magnitude of a number is always nonnegative (positive or zero).

Another important set of algebraic operations is the taking of *powers* and *roots*. If a quantity $x$ is multiplied by itself $n - 1$ times, so that there are $n$ factors, we represent this by the symbol $x^n$, representing $x$ to the $n$th power. For example,

$$x^2 = x \times x, \qquad x^3 = x \times x \times x, \qquad x^n = x \times x \times x \times x \times \cdots \times x \ (n \text{ factors}). \tag{2.5}$$

The number $n$ in the expression $x^n$ is called the *exponent* of $x$. If the exponent is not an integer, we can still define $x^n$, but we will discuss this in Section 2.3, when we discuss logarithms. An exponent that is a negative number indicates the reciprocal of the quantity with a positive exponent:

$$x^{-1} = \frac{1}{x}, \qquad x^{-3} = \frac{1}{x^3}. \tag{2.6}$$

*Roots* of real numbers are defined in an inverse way. For example, the *square root* of $x$ is denoted by $\sqrt{x}$ and is defined as the number that yields $x$ when squared:

$$(\sqrt{x})^2 = x. \tag{2.7a}$$

The *cube root* of $x$ is denoted by $\sqrt[3]{x}$, and is defined as the number that when cubed (raised to the third power) yields $x$:

$$(\sqrt[3]{x})^3 = x. \tag{2.7b}$$

Fourth roots, fifth roots, etc., are defined in similar ways.

---

**Problem 2.3**

Find the value of the expression

$$\frac{3(2+4)^2 - 6(7+|-17|)^3 + (\sqrt{37-|-1|})^3}{(1+2^2)^4 - (|-7|+6^3)^2 + \sqrt{12+|-4|}}.$$

---

The operation of taking a root is the same as raising a number to a fractional exponent. For example,

$$\sqrt[3]{x} = x^{1/3}.$$

Equation (2.7b) is the same as

$$(\sqrt[3]{x})^3 = (x^{1/3})^3 = x = (x^3)^{1/3} = \sqrt[3]{x^3}.$$

This equation illustrates the fact that the order of taking a root and raising to a power can be reversed without changing the result.

There are two numbers that when squared will yield a given positive real number, for example, $2^2 = 4$ and $(-2)^2 = 4$. We can say that both 2 and $-2$ are square roots of 4 if we wish, but when the symbol $\sqrt{4}$ is used, only the positive square root, 2, is meant. To specify the negative square root of $x$, we write $-\sqrt{x}$. If we confine ourselves to real numbers, there is no square root, fourth root, sixth root, etc., of a negative number. In Section 2.6, we define imaginary numbers, which can be square roots of negative quantities. Both positive and negative numbers can have cube roots, fifth roots, etc., since an odd number of negative factors yields a negative product.

The square roots, cube roots, etc., of integers and other rational numbers are either rational numbers or algebraic irrational numbers. The square root of 2 is an example of an *algebraic irrational number*. When written as a decimal number, it does not exhibit any pattern of repeating digits. An algebraic irrational number produces a rational number when raised to the proper integral power. An irrational number that does not produce a rational number when raised to any integral power is a *transcendental irrational number*. Examples are $e$, the base of natural logarithms, and $\pi$, the ratio of a circle's circumference to its diameter.

## SECTION 2.2. TRIGONOMETRIC FUNCTIONS

The ordinary *trigonometric functions* include the sine, the cosine, the tangent, the cotangent, the secant, and the cosecant. These are sometimes called the *circular trigonometric functions* to distinguish them from the hyperbolic trigonometric functions discussed briefly in Section 2.4.

The trigonometric functions can be defined geometrically as in Fig. 2.1, which shows two angles, $\alpha_1$ and $\alpha_2$. Along the horizontal reference line drawn from the

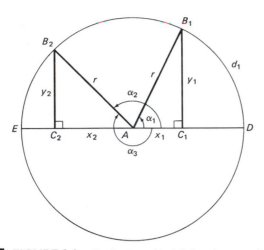

**FIGURE 2.1**    The figure used in defining trigonometric functions.

point $E$ to the point $D$, the points $C_1$ and $C_2$ are chosen so that the triangles are right triangles. In the right triangle $AB_1C_1$, the radius $r$ is called the hypotenuse, the vertical side of length $y_1$ is called the opposite side, and the horizontal side of length $x_1$ is called the adjacent side. We define

$$\sin(\alpha_1) = \frac{y_1}{r} \qquad \text{(opposite side over hypotenuse)} \tag{2.8}$$

$$\cos(\alpha_1) = \frac{x_1}{r} \qquad \text{(adjacent side over hypotenuse)} \tag{2.9}$$

$$\tan(\alpha_1) = \frac{y_1}{x_1} \qquad \text{(opposite side over adjacent side)} \tag{2.10}$$

$$\cot(\alpha_1) = \frac{x_1}{y_1} \qquad \text{(adjacent side over opposite side)} \tag{2.11}$$

$$\sec(\alpha_1) = \frac{r}{x_1} \qquad \text{(hypotenuse over adjacent side)} \tag{2.12}$$

$$\csc(\alpha_1) = \frac{r}{y_1} \qquad \text{(hypotenuse over opposite side)} \ . \tag{2.13}$$

The trigonometric functions of the angle $\alpha_2$ are defined in the same way, except that as drawn in Fig. 2.1, the distance $x_2$ must be counted as negative, because the point $B_2$ is to the left of $A$. If the point $B_2$ were below $A$, then $y_2$ would be counted as negative.

There are three common ways to specify the size of an angle (the "measure" of an angle). *Degrees* are defined so that a right angle is $90°$ (90 degrees), and a full circle contains $360°$. The *grad* is defined so that 100 grad is a right angle and a full circle contains 400 grad. For most mathematical purposes, the best way to specify the size of an angle is with *radians*. The measure of an angle in radians is defined to be the length of the arc subtending the angle divided by the radius of the circle. In Fig. 2.1, the arc $DB_1$ subtends the angle $\alpha_1$, so that in radians

$$\alpha_1 = \frac{d_1}{r}, \tag{2.14}$$

where $d_1$ is the length of the arc $DB_1$. The full circle contains $2\pi$ radians, and 1 radian corresponds to $360°/(2\pi) = 57.2957795\ldots°$. The right angle, $90°$, is $\pi/2$ radians $= 1.5707963\ldots$radians.

The trigonometric functions are examples of *mathematical functions*. The angle for which we evaluate the functions is called the *independent variable*, or the *argument* of the function. If we choose a value for the independent variable, the function provides a corresponding value for another variable, which we call the *dependent variable*. For example, if we write

$$f = \sin(\alpha) \tag{2.15}$$

then $f$ is the dependent variable and $\alpha$ the independent variable. The trigonometric functions are *single-valued*: for each value of the angle $\alpha$, there is one and only one value of the sine, one and only one value of the cosine, etc. Mathematicians usually use the name "function" to apply only to single-valued functions. We will discuss mathematical functions in more detail in Chapter 3.

## Trigonometric Identities

There are a number of relations between trigonometric functions that are valid for all values of the given angles. Such relations are said to be *identically true*, or to be *identities*. For example, from Eqs. (2.8) through (2.13), we can write

$$\boxed{\cot(\alpha) = \frac{1}{\tan(\alpha)}} \tag{2.16}$$

$$\boxed{\sec(\alpha) = \frac{1}{\cos(\alpha)}} \tag{2.17}$$

$$\boxed{\csc(\alpha) = \frac{1}{\sin(\alpha)}}. \tag{2.18}$$

The negative angle $\alpha_3$ in Fig. 2.1 has the same triangle, and therefore the same trigonometric functions as the positive angle $\alpha_2$. (Negative angles are measured in a clockwise direction, whereas positive angles are measured in a counterclockwise direction.) Since $\alpha_3$ is equal to $-(2\pi - \alpha_2)$ if the angles are measured in radians, we can write

$$\sin[-(2\pi - \alpha)] = \sin(\alpha - 2\pi) = \sin(\alpha) \tag{2.19}$$

with similar equations for the other trigonometric functions.

If an angle is increased by $2\pi$ radians (360°), the new angle corresponds to the same triangle as does the old angle, and we can write

$$\sin(\alpha + 2\pi) = \sin(\alpha) \tag{2.20}$$
$$\cos(\alpha + 2\pi) = \sin(\alpha) \tag{2.21}$$

with similar equations for the other trigonometric functions. The trigonometric functions are *periodic functions* with period $2\pi$. That is, if any multiple of $2\pi$ is added to the argument, the value of the function is unchanged.

Other identities can be obtained geometrically. For example,

$$\boxed{\sin(\alpha) = -\sin(-\alpha)} \tag{2.22}$$

$$\boxed{\cos(\alpha) = \cos(-\alpha)} \tag{2.23}$$

$$\boxed{\tan(\alpha) = -\tan(-\alpha)}. \tag{2.24}$$

Equations (2.22) and (2.24) express the fact that the sine and the tangent are *odd functions*, and Eq. (2.23) expresses the fact that the cosine is an *even function*. Additional trigonometric identities are listed in Appendix 5. Equation (14) of this appendix is very useful and can be obtained from the famous *theorem of Pythagorus*. Pythagorus drew a figure with three squares such that one side of each square formed a side of the same right triangle. He then proved by geometry that the area of the square on the hypotenuse was equal to the sum of the areas of the squares on the other two sides. In terms of the quantities in Fig. 2.1

$$\boxed{x^2 + y^2 = r^2}. \tag{2.25}$$

We divide both sides of this equation by $r^2$ and use Eqs. (2.8) and (2.9) to obtain

$$\boxed{[\sin(\alpha)]^2 + [\cos(\alpha)]^2 = \sin^2(\alpha) + \cos^2(\alpha) = 1}. \tag{2.26}$$

Notice the common notation for a power of a trigonometric function: the exponent is written after the symbol for the trigonometric function and before the parentheses enclosing the argument.

## A Useful Approximation: Mathematical Limits

Comparison of Eqs. (2.8) and (2.14) shows that for a fairly small angle, the sine of an angle and the measure of the angle in radians are approximately equal, since the sine differs from the angle only by having the opposite side in place of the arc length, which is approximately the same size. In fact,

$$\lim_{\alpha \to 0} \frac{\sin(\alpha)}{\alpha} = 1 \qquad (\alpha \text{ must be measured in radians}). \tag{2.27a}$$

The symbol on the left stands for a *mathematical limit*. In this case, the equation means that if we let the value of $\alpha$ become smaller and smaller until it becomes more and more nearly equal to zero, the ratio of $\sin(\alpha)$ to $\alpha$ becomes more and more nearly equal to unity. In some cases, there is a distinction between letting the variable draw closer in value to a constant value from the positive side or from the negative side. To indicate that $\alpha$ approaches zero from the positive side (takes on positive values closer and closer to zero), we would write

$$\lim_{\alpha \to 0^+} \frac{\sin(\alpha)}{\alpha} = 1. \tag{2.27b}$$

To indicate that $\alpha$ approaches zero from the negative side, we would write

$$\lim_{\alpha \to 0^-} \frac{\sin(\alpha)}{\alpha} = 1. \tag{2.27c}$$

In the present case, the limits in Eqs. (2.27b) and (2.27c) are the same, and there is no need to specify which one is meant.

For fairly small angles, we write as an approximation

$$\alpha \approx \sin(\alpha) \qquad (\alpha \text{ must be measured in radians}). \tag{2.28}$$

The symbol $\approx$ means "approximately equal to." Since the adjacent side is nearly equal to the hypotenuse for small angles, we can also write

$$\alpha \approx \tan(\alpha) \approx \sin(\alpha) \qquad \begin{array}{l}(\alpha \text{ must be measured in radians for} \\ \text{the first equality}).\end{array} \tag{2.29}$$

Equations (2.28) and (2.29) are valid for both positive and negative values of $\alpha$. If you are satisfied with an accuracy of about 1%, you can use Eq. (2.29) for angles with magnitude up to about 0.2 radians (approximately 11°).

---

**Problem 2.4**

For an angle that is nearly as large as $\pi/2$, find an approximate equality similar to Eq. (2.29) involving $(\pi/2) - \alpha$, $\cos(\alpha)$, and $\cot(\alpha)$.

---

## General Properties of Trigonometric Functions

To use trigonometric functions easily, you must have a clear mental picture of the way in which the sine, cosine, and tangent depend on their arguments. Figures 2.2, 2.3, and 2.4 show these functions.

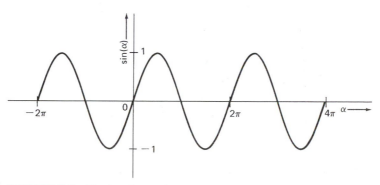

**FIGURE 2.2**  The sine of an angle $\alpha$.

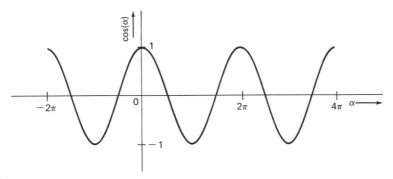

**FIGURE 2.3**  The cosine of an angle $\alpha$.

The tangent has a complicated behavior, becoming larger without bound as its argument approaches $\pi/2$ from the left, and becoming more negative without bound as its argument approaches the same value from the right. We can write

$$\lim_{\alpha \to \pi/2^+} [\tan(\alpha)] = -\infty$$

$$\lim_{\alpha \to \pi/2^-} [\tan(\alpha)] = \infty.$$

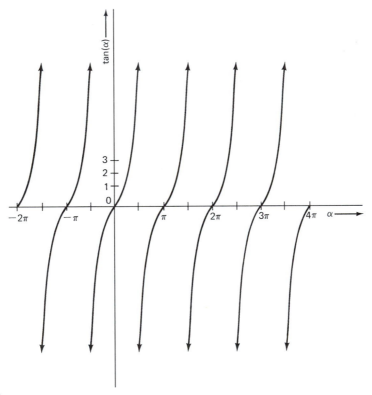

**FIGURE 2.4**  The tangent of an angle $\alpha$.

In these equations, the superscript $+$ on the $\pi/2$ means that the value of $\alpha$ approaches $\pi/2$ from the right. That is, $\alpha$ is greater than $\pi/2$ as it becomes more and more nearly equal to $\pi/2$. The $-$ superscript means that $\alpha$ approaches $\pi/2$ from the left. The symbol $\infty$ stands for *infinity*, which is larger than any number that you or anyone else can name. This quantity is sometimes called "undefined."

## Inverse Trigonometric Functions

It is possible to think of trigonometric functions as defining a mathematical function in an inverse way. For example, if

$$y = \sin(x) \tag{2.30}$$

we can define a function to give a value for $x$ as a dependent variable for each value of $y$. We write

$$x = \arcsin(y). \tag{2.31}$$

This can be read as "$x$ is the angle whose sine is $y$." The *arcsine* function is also called the *inverse sine*, and another notation is also common:

$$x = \sin^{-1}(y). \tag{2.32}$$

The $-1$ superscript indicates an inverse function. It is not an exponent, even though exponents are written in the same position.

From Fig. 2.2, you can see that there are many angles that have the same value of the sine. In order to make Eq. (2.31) or (2.32) into a single valued function, we must restrict the set of values of $x$ which we consider. With the arcsine function, this set of values is taken from $-\pi/2$ to $+\pi/2$ and is said to include the *principal values* of the arcsine function. The set of principal values of the arctangent and arccosecant functions is from $-\pi/2$ to $+\pi/2$, the same as with the arcsine. The principal values of the arccosine, arccotangent, and arcsecant are taken from 0 to $\pi$.

---

### Problem 2.5

Sketch graphs of the arcsine function, the arccosine function, and the arctangent function. Include only the principal values.

---

## SECTION 2.3. LOGARITHMS

We have discussed the operation of raising a number to an integral power. The expression $a^2$ means $a \times a$, $a^3$ means $a \times a \times a$, etc. In addition, you can have exponents that are not integers. If we write $a^x$, $x$ can be a variable that takes on any real value. We assume $a$ to be positive, so that $a^x$ is positive. The *logarithm to the base a* is defined by the relation that if

$$y = a^x, \tag{2.33a}$$

then x is called the logarithm of $y$ to the base a and denoted by

$$x = \log_a(y). \tag{2.33b}$$

If $a = 10$, the logarithms are called *common logarithms*: If $10^x = y$, then $x = \log_{10}(y)$, the common logarithm of $y$.

For integral values of $x$, it is easy to generate the following short table of common logarithms:

| $y$ | $x = \log_{10}(y)$ |
|---|---|
| 1 | 0 |
| 10 | 1 |
| 100 | 2 |
| 1000 | 3 |
| 0.1 | $-1$ |
| 0.01 | $-2$ |
| 0.001 | $-3$ |
| etc. | |

In order to understand logarithms that are not integers, we need to understand exponents that are not integers.

---

### Example 2.2

Find the logarithm of $\sqrt{10}$.

---

### Solution

The square root of 10 is the number that yields 10 when multiplied by itself:

$$(\sqrt{10})^2 = 10.$$

We use the fact about exponents

$$(a^x)^z = a^{xz}. \tag{2.34}$$

Since 10 is the same thing as $10^1$,

$$\sqrt{10} = 10^{1/2}. \tag{2.35}$$

Therefore

$$\log_{10}(\sqrt{10}) = \log_{10}(3.162277\ldots) = \frac{1}{2}.$$

---

Equation (2.34) and some other relations governing exponents can be used to generate other logarithms, as in the following problem.

---

### Problem 2.6

Use Eq. (2.34) and the fact that $10^{-n} = 1/(10^n)$ to generate the negative logarithms in the short table of logarithms.

---

Table 2.1 lists some properties of exponents and logarithms. An additional fact is that negative numbers do not possess real logarithms.

We will not discuss further how the logarithms of various numbers are computed, but extensive tables of logarithms are available, with up to seven or eight significant

■ **TABLE 2.1** **Properties of Exponents and Logarithms**

$$\text{If } a^x = y, x = \log_a(y)$$

| Exponent fact | Logarithm fact | Equation number |
|---|---|---|
| $a^0 = 1$ | $\log_a(1) = 0$ | (2.36) |
| $a^{1/2} = \sqrt{a}$ | $\log_a(\sqrt{a}) = \frac{1}{2}$ | (2.37) |
| $a^1 = a$ | $\log_a(a) = 1$ | (2.38) |
| $a^{x_1} a^{x_2} = a^{x_1 + x_2}$ | $\log_a(y_1 y_2) = \log_a(y_1) + \log_a(y_2)$ | (2.39) |
| $a^{-x} = \dfrac{1}{a^x}$ | $\log_a(1/y) = -\log_a(y)$ | (2.40) |
| $a^{x_1}/a^{x_2} = a^{x_1 - x_2}$ | $\log_a(y_1/y_2) = \log_a(y_1) - \log_a(y_2)$ | (2.41) |
| $(a^x)^z = a^{xz}$ | $\log_a(y^z) = z \log_a(y)$ | (2.42) |
| $a^\infty = \infty$ | $\log_a(\infty) = \infty$ | (2.43) |
| $a^{-\infty} = 0$ | $\log_a(0) = -\infty$ | (2.44) |

digits, and most electronic calculators provide values of logarithms with as many as 10 or 11 significant digits. Before the invention of inexpensive electronic calculators, tables of logarithms were used when a calculation required more significant digits than a slide rule could provide. For example, to multiply two numbers together, one would look up the logarithms of the two numbers, add the logarithms manually, and then look up the antilogarithm of the sum (the number possessing the sum as its logarithm).

## Natural Logarithms

Besides 10, there is another commonly used base of logarithms. This is a transcendental irrational number called $e$ and equal to $2.7182818\ldots$.

$$\boxed{\text{If } e^y = x \quad \text{then } y = \log_e(x) = \ln(x)}. \tag{2.45}$$

The base of natural logarithms, $e$, is named after Leonhard Euler, 1707–1783, a great Swiss mathematician. The definition of $e$ is

$$e = \lim_{n \to \infty} \left(1 + \frac{1}{n}\right)^n = 2.7182818\ldots . \tag{2.46}$$

The notation $\ln(x)$ is more common than $\log_e(x)$, and logarithms to the base $e$ are called *natural logarithms*. They are also occasionally called *Naperian logarithms*, after John Napier, 1550–1617, a Scottish landowner, theologian, and mathematician, who was one of the inventors of logarithms. Unfortunately, some mathematicians use the symbol $\log(y)$ without a subscript for natural logarithms. Chemists use the symbol $\log(y)$ without a subscript for common logarithms and the symbol $\ln(y)$ for natural logarithms. Natural logarithms occur in many formulas of physical chemistry, often arising in solutions of differential equations, etc.

If the common logarithm of a number is known, its natural logarithm can be computed as

$$e^{\ln(y)} = 10^{\log_{10}(y)} = \left(e^{\ln(10)}\right)^{\log_{10}(y)} = e^{\ln(10)\log_{10}(y)}. \tag{2.47}$$

The natural logarithm of 10 is equal to $2.302585\ldots$, so we can write

$$\boxed{\ln(y) = \ln(10)\log_{10}(y) = (2.302585\ldots)\log_{10}(y)}\,. \tag{2.48}$$

In order to remember Eq. (2.48) correctly, keep the fact in mind that since $e$ is smaller than 10, the natural logarithm is larger than the common logarithm.

---

**Problem 2.7**

Without using a calculator or a table of logarithms, find the following:

a. $\ln(100.000)$
b. $\ln(0.0010000)$
c. $\log_{10}(e)$.

---

## SECTION 2.4. THE EXPONENTIAL FUNCTION: HYPERBOLIC TRIGONOMETRIC FUNCTIONS

The *exponential function* is encountered frequently in physical chemistry. This function is the same as raising $e$ to a power and is denoted either by the usual notation for a power, or by the notation $\exp(\ldots)$,

$$y = ae^{bx} \equiv a\exp(bx), \tag{2.49}$$

where $a$ and $b$ are constants. Figure 2.5 shows a graph of this function for $b > 0$.

The graph in Fig. 2.5 exhibits an important behavior of the exponential function. For $b > 0$, it doubles each time the independent variable increases by a fixed amount whose value depends on the value of $b$. If $b < 0$, the function decreases to half its value each time the independent variable increases by a fixed amount.

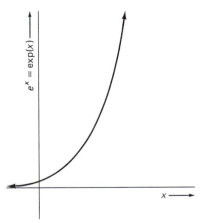

**FIGURE 2.5**    The exponential function.

An example of the exponential function is in the decay of radioactive isotopes. If $N_o$ is the number of atoms of the isotope at time $t = 0$, the number at any other time, $t$, is given by

$$N(t) = N_o e^{-t/\tau}, \tag{2.50}$$

where $\tau$ is called the *relaxation time*. The time that is required for the number of atoms to drop to half its original value is called the *half-time* or *half-life*.

---

**Example 2.3**

Show that the half-life, $t_{1/2}$, is equal to $\tau \ln(2)$.

---

**Solution**

If $t_{1/2}$ is the half-life, then

$$e^{-t_{1/2}/\tau} = \frac{1}{2}.$$

Thus

$$\frac{t_{1/2}}{\tau} = -\ln\left(\frac{1}{2}\right) = \ln(2). \tag{2.51}$$

---

**Problem 2.8**

A certain population is growing exponentially, so that it doubles in size each 30 years.

   a. If the population has a size of $4.00 \times 10^6$ individuals at $t = 0$ write the formula giving the population after a number of years equal to $t$.
   b. Find the size of the population at $t = 150$ years.

**Problem 2.9**

A reactant in a first-order chemical reaction without back reaction has a concentration governed by the same formula as radioactive decay,

$$[A]_{t'} = [A]_0 e^{-kt'},$$

where $[A]_0$ is the concentration at time $t = 0$, $[A]_{t'}$ is the concentration at time $t = t'$, and $k$ is a function of temperature called the rate constant. If $k = 0.123$ s$^{-1}$, find the time required for the concentration to drop to 21.0% of its initial value.

---

## Hyperbolic Trigonometric Functions

These functions are closely related to the exponential function. The *hyperbolic sine* of $x$ is denoted by $\sinh(x)$, and defined by

$$\sinh(x) = \frac{1}{2}(e^x - e^{-x}). \tag{2.52}$$

The *hyperbolic cosine* is denoted by $\cosh(x)$, and defined by

$$\cosh(x) = \frac{1}{2}(e^x + e^{-x}). \tag{2.53}$$

The other hyperbolic trigonometric functions are the *hyperbolic tangent*, denoted by $\tanh(x)$; the *hyperbolic cotangent*, denoted by $\coth(x)$; the *hyperbolic secant*, denoted by $\operatorname{sech}(x)$; and the *hyperbolic cosecant*, denoted by $\operatorname{csch}(x)$. These functions are given by the equations

$$\tanh(x) = \frac{\sinh(x)}{\cosh(x)} \tag{2.54}$$

$$\coth(x) = \frac{1}{\tanh(x)} \tag{2.55}$$

$$\operatorname{sech}(x) = \frac{1}{\cosh(x)} \tag{2.56}$$

$$\operatorname{csch}(x) = \frac{1}{\sinh(x)}. \tag{2.57}$$

---

### Problem 2.10

Find the value of each of the hyperbolic trigonometric functions for $x = 0$ and $x = \pi/2$. Compare these values with the values of the ordinary (circular) trigonometric functions for the same values of the independent variable.

---

## SECTION 2.5. VECTORS: COORDINATE SYSTEMS

Quantities that have both magnitude and direction are called *vectors*. For example, the position of an object can be represented by a vector, since the position can be specified by giving the distance and the direction from a reference point (an origin). A force is also a vector, since it is not completely specified until its magnitude and direction are both given. Some other vectors that are important in physical chemistry are the dipole moments of molecules, magnetic and electric fields, angular momenta, and magnetic dipoles.

We will use a letter set in boldface type to represent a vector. For example, the force on an object is denoted by **F**. When you are writing by hand, there is no easy way to write boldface letters, so you can use a letter with an arrow over it (e.g., $\vec{F}$) or you can use a wavy underscore (e.g., F̰), which is the typesetter's symbol for boldface type.

### Vectors in Two Dimensions

Let us begin with a restricted case, in which all vectors are required to lie in a plane. This case includes position vectors for objects that remain on a flat surface. We represent this physical surface by a mathematical plane, which is a map of the surface, so that to each location in the physical surface there corresponds a point of the mathematical plane. We choose some reference point as an origin and pick

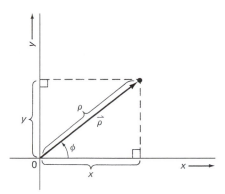

**FIGURE 2.6**   A position vector, $\rho$, in a plane, with plane polar coordinates and cartesian coordinates.

some convenient line passing through the origin as our *x axis*. One end of this axis is designated as the positive end. The line passing through the origin perpendicular to the *x* axis is designated as our *y axis*, and the end that is counterclockwise 90° from the positive end of the *x* axis is designated as its positive end. These axes are shown in Fig. 2.6. In this figure, the origin is labeled as point $O$, and the location of some object is labeled as point $P$.

The directed line segment beginning at $O$ and ending at $P$ is the *position vector* for the object. We denote the position vector in two dimensions by $\rho$. In the figure, we draw an arrowhead on the directed line segment to make its direction clear. The negative of a vector is a vector of the same length directed in the opposite direction. For example, the vector going from $O$ to $P$ is the negative of the vector going from $P$ to $O$.

One way to specify the location of $P$ is to give the magnitude of $\rho$ and the value of the angle $\phi$ between the positive end of the *x* axis and $\rho$, measured counterclockwise from the axis. We will denote the magnitude of any vector by the same letter in ordinary type, or by the symbol for the vector with vertical bars before it and after it, as $|\rho| = \rho$. The magnitude of a vector is the length of the line segment from $O$ to $P$, and it is a nonnegative scalar. A vector and its negative have the same magnitude, as do all the vectors of the same length pointing in any other directions. The variables $\rho$ and $\phi$ are called the *plane polar coordinates* of the point $P$. If we allow $\rho$ to range from zero to $\infty$ and allow $\phi$ to range from zero to $2\pi$ radians, we can specify the location of any point in the entire plane.

There is another common way to specify the location of $P$. We draw two line segments from $P$ perpendicular to the axes, as shown in Fig. 2.6. The distance from the origin to the intersection on the *x* axis is called $x$ and is considered to be positive if the intersection is on the positive half of the axis, and negative if the intersection is on the negative half of the axis. The distance from the origin to the intersection on the *y* axis is called $y$, and its sign is assigned in a similar way. The variables $x$ and $y$ are the *cartesian coordinates* of $P$. The point $P$ can be designated by its cartesian coordinates within parentheses, as $(x, y)$. The values of $x$ and $y$ are also called the *cartesian components* of the position vector, and the position vector $\rho$ is sometimes denoted by the same symbol as the point at which its head lies, as $(x, y)$. Cartesian coordinates are named for René duPerron Descarte, 1596–1650, French mathematician, philosopher, and natural scientist, who is famous for his statement, "I think, therefore I am."

Changing from polar coordinates to cartesian coordinates is an example of *transformation of coordinates*, and can be done using the equations

$$x = \rho \cos(\phi) \qquad (2.58)$$
$$y = \rho \sin(\phi), \qquad (2.59)$$

where $\rho$ is the magnitude of the vector $\boldsymbol{\rho}$.

---

### Problem 2.11

Show that Eqs. (2.58) and (2.59) are correct.

---

The coordinate transformation in the other direction is also possible. From the theorem of Pythagorus, Eq. (2.25),

$$\rho = \sqrt{x^2 + y^2}. \qquad (2.60)$$

From the definition of the tangent function, Eq. (2.10),

$$\phi = \arctan\left(\frac{y}{x}\right). \qquad (2.61)$$

However, since we want $\phi$ to range from 0 to $2\pi$ radians, we must specify this range for the inverse tangent function, instead of using the principal value. We must decide in advance which quadrant $\phi$ lies in.

---

### Problem 2.12

    a. Find $x$ and $y$ if $\rho = 10$ and $\phi = \pi/6$ radians.
    b. Find $\rho$ and $\phi$ if $x = -5$ and $y = 10$.

---

A position vector is only one example of a vector. Anything, such as a force, a velocity, or an acceleration, which has magnitude and direction, is a vector. Figure 2.6 is a map of physical space, and a distance in such a diagram is measured in units of length, such as meters. Other kinds of vectors can also be represented on vector diagrams by directed line segments. However, such a diagram is not a map of physical space, and the length of a line segment representing a vector will represent the magnitude of a force, or the magnitude of a velocity, or something else.

Position vectors ordinarily remain with their tails at the origin, but since other vector diagrams do not necessarily represent a physical (geographical) space, we will consider a vector to be unchanged if it is moved from one place in a vector diagram to another, as long as its length and its direction do not change.

### Vector Algebra in Two Dimensions

Figure 2.7 is a vector diagram in which two vectors, **A** and **B**, are shown. The *sum of the two vectors* is obtained as follows: (1) Move the second vector so that its tail

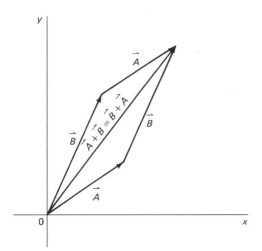

**FIGURE 2.7**   Two vectors and their sum.

coincides with the head of the first. (2) Draw the sum vector from the tail of the first
vector to the head of the second. The addition of vectors is commutative: $\mathbf{A} + \mathbf{B}$ is
the same as $\mathbf{B} + \mathbf{A}$.

The components of A and B are defined in the same way as the components
of the position vector in Fig. 2.6. The $x$ components are called $A_x$ and $B_x$, and the
y components are called $A_y$ and $B_y$. Vector addition can be performed using the
components of the vectors. If the sum of $\mathbf{A}$ and $\mathbf{B}$ is called $\mathbf{C}$,

$$C_x = A_x + B_x \tag{2.62}$$
$$C_y = A_y + B_y. \tag{2.63}$$

The *difference of two vectors* is the sum of the first vector and the negative of the
second. The negative of $\mathbf{B}$ is denoted by $-\mathbf{B}$ and is the vector with components $-B_x$
and $-B_y$. If the vector $\mathbf{A} - \mathbf{B}$ is called $\mathbf{D}$,

$$D_x = A_x - B_x \tag{2.64}$$
$$D_y = A_y - B_y. \tag{2.65}$$

The vector $\mathbf{A} - \mathbf{B}$ has its head at the head of $\mathbf{A}$ and its tail at the head of $\mathbf{B}$ if both $\mathbf{A}$
and $\mathbf{B}$ have their tails at the same place.

If $\mathbf{A}$ is a vector and $a$ is a scalar, the *product of the scalar and the vector* $a\mathbf{A}$ has
the components

$$(aA)_x = aA_x \tag{2.66}$$
$$(aA)_y = aA_y. \tag{2.67}$$

If $a$ is a positive scalar, the vector $a\mathbf{A}$ points in the same direction as $\mathbf{A}$, and if $a$ is a
negative scalar, the vector $a\mathbf{A}$ points in the opposite direction. The magnitude of $a\mathbf{A}$
is equal to $|a||\mathbf{A}|$.

The magnitude of any other vector in two dimensions is obtained in the same
manner as the magnitude of a position vector, as given in Eq. (2.60). Thus

$$A = |\mathbf{A}| = \sqrt{A_x^2 = A_y^2}. \tag{2.68}$$

**Problem 2.13**

The vector **A** has the components $A_x = 2$, $A_y = 3$. The vector **B** has the components $B_x = 3$, $B_y = 4$.

    a. Find the components and the magnitude of $\mathbf{A} + \mathbf{B}$.
    b. Find the components and the magnitude of $\mathbf{A} - \mathbf{B}$.
    c. Find the components and the magnitude of $2\mathbf{A} - 3\mathbf{B}$.

We next define the *scalar product* of two vectors, which is also called the *dot product* because of the use of a dot to represent the operation. If **A** and **B** are two vectors, and $\alpha$ is the angle between them, their scalar product is denoted by $\mathbf{A} \cdot \mathbf{B}$ and given by

$$\mathbf{A} \cdot \mathbf{B} = |\mathbf{A}||\mathbf{B}|\cos(\alpha). \tag{2.69}$$

The result is a scalar, as the name implies. The following are properties of the scalar product:

    1. If **A** and **B** are parallel, $\mathbf{A} \cdot \mathbf{B}$ is the product of the magnitudes of **A** and **B**.
    2. The scalar product of **A** with itself is the square of the magnitude of **A**:

$$\mathbf{A} \cdot \mathbf{A} = |\mathbf{A}|^2 = A^2 = A^2 = A_x^2 + A_y^2. \tag{2.70}$$

    3. If **A** and **B** are perpendicular to each other, $\mathbf{A} \cdot \mathbf{B} = 0$. Such vectors are said to be *orthogonal* to each other.
    4. If **A** and **B** point in opposite directions (are antiparallel), $\mathbf{A} \cdot \mathbf{B}$ is the negative of the product of the magnitudes of **A** and **B**.

These properties also hold for scalar products of vectors in three dimension.

There is another common product of two vectors, called the *vector product*. Since it is a vector that is perpendicular to the plane containing the two vectors, we defer its definition until we discuss three-dimensional vectors.

Another way to represent vectors is by using *unit vectors*. We define **i** to be a vector of unit length pointing in the direction of the positive end of the $x$ axis, and **j** to be a vector of unit length pointing in the direction of the positive end of the $y$ axis. These are shown in Fig. 2.8. A vector **A** is represented as

$$\mathbf{A} = \mathbf{i}A_x + \mathbf{j}A_y. \tag{2.71}$$

The first term on the right-hand side of this equation is a product of a scalar $A_x$ and a vector **i**, so it is a vector of length $A_x$ pointing along the x axis, as shown in Fig. 2.8. The other term is similarly a vector pointing along the y axis. The sum in Fig. 2.8 is shown in the figure as a vector sum.

A similar equation can be written for another vector, **B**:

$$\mathbf{B} = \mathbf{i}B_x + \mathbf{j}B_y. \tag{2.72}$$

The scalar product $\mathbf{A} \cdot \mathbf{B}$ can be written

$$\begin{aligned}\mathbf{A} \cdot \mathbf{B} &= (\mathbf{i}A_x + \mathbf{j}A_y) \cdot (\mathbf{i}B_x + \mathbf{j}B_y) \\ &= \mathbf{i} \cdot \mathbf{i}A_x B_x + \mathbf{i} \cdot \mathbf{j}A_x B_y + \mathbf{j} \cdot \mathbf{i}A_y B_x + \mathbf{j} \cdot \mathbf{j}A_y B_y.\end{aligned}$$

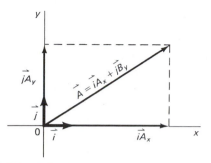

**FIGURE 2.8**   A vector in terms of the unit vectors **i** and **j**.

From the definitions of **i** and **j**,

$$\mathbf{i} \cdot \mathbf{i} = \mathbf{j} \cdot \mathbf{j} = 1 \tag{2.73}$$
$$\mathbf{i} \cdot \mathbf{j} = \mathbf{j} \cdot \mathbf{i} = 0 \tag{2.74}$$

so that

$$\mathbf{A} \cdot \mathbf{B} = A_x B_x + A_y B_y. \tag{2.75}$$

---

### Example 2.4

Consider the following vectors: $\mathbf{A} = 2.5\mathbf{i} + 4\mathbf{j}$ and $\mathbf{B} = 3\mathbf{i} - 5\mathbf{j}$.

a. Find $\mathbf{A} \cdot \mathbf{B}$.

b. Find $|\mathbf{A}|$ and $|\mathbf{B}|$ and use them to find the angle between $\mathbf{A}$ and $\mathbf{B}$.

---

### Solution

a. $\mathbf{A} \cdot \mathbf{B} = (2.5)(3) + (4)(-5) = 7.5 - 20 = -12.5$.

b. $|\mathbf{A}| = (6.25 + 16)^{1/2} = (22.25)^{1/2} = 4.717\ldots$

  $|\mathbf{B}| = (9 + 25)^{1/2} = (34)^{1/2} = 5.8309\ldots$

$$\cos(\alpha) = \frac{\mathbf{A} \cdot \mathbf{B}}{|\mathbf{A}||\mathbf{B}|} = \frac{-12.5}{(4.717)(5.831)} = -0.4545$$
$$\alpha = 2.043 \text{ rad} = 117.0° = \arccos(-0.4545).$$

---

### Problem 2.14

Consider two forces $\mathbf{A} = (3.00)\mathbf{i} - (4.00)\mathbf{j}$ and $\mathbf{B} = -(1.00)\mathbf{i} + (2.00)\mathbf{j}$.

a. Find $\mathbf{A} \cdot \mathbf{B}$ and $(2\mathbf{A}) \cdot (3\mathbf{B})$.

b. Find the angle between $\mathbf{A}$ and $\mathbf{B}$. Use the principal value of the arccosine, so that an angle of less than $\pi$ radians (180°) results.

---

## Vectors and Coordinate Systems in Three Dimensions

Figure 2.9 shows the three-dimensional version of cartesian coordinates. The axes are viewed from the first octant, in which $x$, $y$, and $z$ are all positive. A coordinate system such as that shown is called a *right-handed coordinate system*. For such a system, the thumb, index finger, and middle finger of the right hand can be aligned with the positive ends of the $x$, $y$, and $z$ axes, respectively. If the left hand must be used for such an alignment, the coordinate system is called a *left-handed coordinate system*.

The location of the point $P$ is specified by $x$, $y$, and $z$, which are called the *cartesian coordinates of the point*. These are the distances from the origin to the points on the axes reached by moving perpendicularly from $P$ to each axis. These coordinates can be positive or negative. The point $P$ is sometimes denoted by its coordinates, as $(x, y, z)$. The directed line segment from the origin to $P$ is the *position vector* of $P$, and $x$, $y$, and $z$ are the cartesian components of this vector. We denote this vector by **r**. The position vector is also denoted by the symbol $(x, y, z)$.

We can represent the position vector, or any other vector, by use of unit vectors, much as in two dimensions. In addition to the unit vectors **i** and **j**, we define **k**, a vector of unit length pointing in the direction of the positive end of the $z$ axis.

Figure 2.10 shows these unit vectors and the position vector written as

$$\mathbf{r} = \mathbf{i}x + \mathbf{j}y + \mathbf{k}z. \tag{2.76}$$

The magnitude of **r** can be obtained from the theorem of Pythagorus. In Fig. 2.10 you can see that $r$ is the hypotenuse of a right triangle with sides $\rho$ and $z$, so that

$$r^2 = \rho^2 + z^2 = x^2 + y^2 + z^2 \tag{2.77}$$

or

$$r = |\mathbf{r}| = \sqrt{x^2 + y^2 + z^2}. \tag{2.78}$$

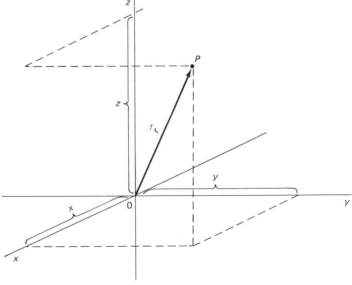

**FIGURE 2.9**   Cartesian coordinates in three dimensions.

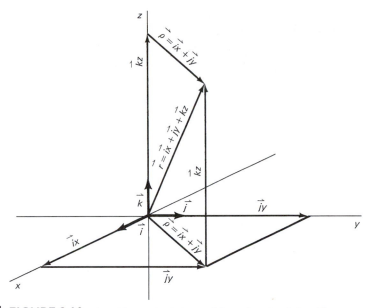

**FIGURE 2.10** A position vector in terms of the unit vectors **i**, **j**, and **k**.

Figure 2.11 shows the way in which *spherical polar coordinates* are used to specify the location of the point $P$ and the vector **r** from the origin to $P$. The vector $\rho$ in the x-y plane is also shown. It is called the projection of **r** in the x-y plane. Its head is reached from the head of **r** by moving to the x-y plane in a direction perpendicular to the plane. The three spherical polar coordinates are $r$, $\theta$, and $\phi$. $\theta$ is the angle between the positive $z$ axis and the position vector. $\phi$ is the angle between the positive $x$ axis

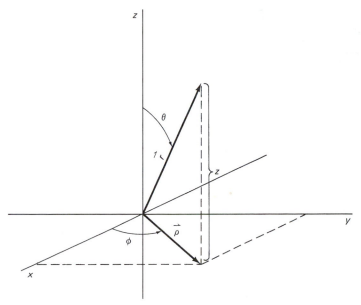

**FIGURE 2.11** Spherical polar coordinates.

and $\rho$, just as in our two-dimensional polar coordinates. The angle $\theta$ is allowed to range from 0 to $\pi$ and the angle $\phi$ is allowed to range from 0 to $2\pi$. The distance $r$ is allowed to range from 0 to $\infty$, and these ranges allow the location of every point in the entire three-dimensional space to be given.

The following equations and Eq. (2.78) can be used to transform from cartesian coordinates to spherical polar coordinates:

$$\theta = \arccos\left(\frac{z}{r}\right) \tag{2.79}$$

and

$$\phi = \arctan\left(\frac{y}{x}\right). \tag{2.80}$$

Equation (2.80) is the same as Eq. (2.61).

The following equations can be used to transform from spherical polar coordinates to cartesian coordinates:

$$x = r\sin(\theta)\cos(\phi) \tag{2.81}$$

$$y = r\sin(\theta)\sin(\phi) \tag{2.82}$$

$$z = r\cos(\theta). \tag{2.83}$$

### Example 2.5

Find the spherical polar coordinates of the point whose cartesian coordinates are (1.000, 1.000, 1.000).

### Solution

$$r = \sqrt{(1.000^2 + 1.000^2 + 1.000^2)} = \sqrt{3.000} = 1.732$$
$$\phi = \arctan\left(\frac{1.000}{1.000}\right) = \frac{\pi}{4}\ \text{radians} = 45.0°$$
$$\theta = \arccos\left(\frac{1.000}{1.732}\right) = 0.955\ \text{radians} = 54.7°.$$

### Problem 2.15

Find the spherical polar coordinates of the point whose cartesian coordinates are (2, −3, 4).

The *cylindrical polar coordinate system* is another three-dimensional coordinate system that is in common use. It uses the variables $\rho$, $\phi$, and $z$, already defined and shown in Fig. 2.11. The equations needed to transform from cartesian coordinates to cylindrical polar coordinates are Eqs. (2.60) and (2.61). The third coordinate, $z$, is the same in both cartesian and cylindrical polar coordinates. Equations (2.58) and (2.59) are used for the reverse transformation.

---

### Example 2.6

Find the cylindrical polar coordinates of the point whose cartesian coordinates are $(1.000, -4.000, -2.000)$.

---

### Solution

$$\rho = \sqrt{1.000^2 + 4.000^2} = \sqrt{17.000} = 4.123$$
$$\phi = \arctan\left(\frac{-4.000}{1.000}\right) = 4.957 \, \text{radians} = 284°$$
$$z = -2.000.$$

---

### Problem 2.16

a. Find the cylindrical polar coordinates of the point whose cartesian coordinates are $(-2.000, -2.000, 3.000)$.
b. Find the cartesian coordinates of the point whose cylindrical polar coordinates are $\rho = 25.00$, $\phi = 60.0°$, $z = 17.50$.

---

### Vector Algebra in Three Dimensions

The sum and scalar product of two vectors are very similar to those quantities in two dimensions. Let $\mathbf{A}$ and $\mathbf{B}$ be two vectors, represented in terms of their components and the unit vectors $\mathbf{i}$, $\mathbf{j}$, and $\mathbf{k}$ by

$$\mathbf{A} = \mathbf{i}A_x + \mathbf{j}A_y + \mathbf{k}A_z \tag{2.84a}$$

$$\mathbf{B} = \mathbf{i}B_x + \mathbf{j}B_y + \mathbf{k}B_z. \tag{2.84b}$$

The sum is still obtained by placing the tail of the second vector at the head of the first and drawing the sum vector from the tail of the first to the head of the second. If $\mathbf{C} = \mathbf{A} + \mathbf{B}$, then

$$\begin{aligned} C_x &= A_x + B_x \\ C_y &= A_y + B_y \\ C_z &= A_z + B_z \end{aligned} \tag{2.85}$$

The product of a vector $\mathbf{A}$ and a scalar $a$ is

$$\mathbf{C} = a\mathbf{A} = \mathbf{i}aA_x + \mathbf{j}aA_y + \mathbf{k}aA_z. \tag{2.86}$$

The magnitude of a vector is

$$|\mathbf{A}| = A = \left(A_x^2 + A_y^2 + A_z^2\right)^{1/2}. \tag{2.87}$$

The scalar product of two vectors is still given by

$$\mathbf{A} \cdot \mathbf{B} = |\mathbf{A}||\mathbf{B}| \cos(\alpha). \tag{2.88}$$

where $\alpha$ is the angle between the vectors.

Analogous to Eq. (2.75), we have

$$\mathbf{A} \cdot \mathbf{B} = A_x B_x + A_y B_y + A_z B_z. \tag{2.89}$$

---

### Example 2.7

Let $\mathbf{A} = 2\mathbf{i} + 3\mathbf{j} + 7\mathbf{k}$ and $\mathbf{B} = 7\mathbf{i} + 2\mathbf{j} + 3\mathbf{k}$. Find $\mathbf{A} \cdot \mathbf{B}$ and the angle between $\mathbf{A}$ and $\mathbf{B}$. Find $(3\mathbf{A}) \cdot \mathbf{B}$.

---

### Solution

$$\mathbf{A} \cdot \mathbf{B} = 14 + 6 + 21 = 41.$$

Let $\alpha$ be the angle between $\mathbf{A}$ and $\mathbf{B}$.

$$\alpha = \arccos\left(\frac{\mathbf{A} \cdot \mathbf{B}}{|A||B|}\right)$$

$$= \arccos\left(\frac{41}{\sqrt{(62)}\sqrt{(62)}}\right) = \arccos(0.6613) = 0.848 \, \text{rad} = 48.6°$$

$$(3\mathbf{A}) \cdot \mathbf{B} = 3 \times 14 + 3 \times 6 + 3 \times 21 = 123.$$

Notice that $(3\mathbf{A}) \cdot \mathbf{B} = 3(\mathbf{A} \cdot \mathbf{B})$.

---

### Problem 2.17

a. Find the position vector of the point whose spherical polar coordinates are $r = 2, \theta = 90°, \phi = 0°$. Call this vector $\mathbf{A}$.
b. Find the scalar product of the vector A from part a and the vector $\mathbf{B}$ whose components are $(1, 2, 3)$.
c. Find the angle between these two vectors.

---

We now introduce another kind of a product between two vectors, called the *vector product*, or *cross product*, and denoted by $\mathbf{A} \times \mathbf{B}$. If $\mathbf{C} = \mathbf{A} \times \mathbf{B}$, then $\mathbf{C}$ is defined to be perpendicular to the plane containing $\mathbf{A}$ and $\mathbf{B}$ and to have the magnitude

$$C = |\mathbf{C}| = |\mathbf{A}||\mathbf{B}| \sin(\alpha), \tag{2.90}$$

where $\alpha$ is the angle between $\mathbf{A}$ and $\mathbf{B}$, measured so that it is less than or equal to 180°. The direction of the cross product $\mathbf{A} \times \mathbf{B}$ is defined as follows: If the first vector listed, **A** in this case, is rotated through the angle $\alpha$ so that its direction coincides with that of **B**, then **C** points in the direction that an ordinary (right-handed) screw thread would move with this rotation. Another rule to obtain the direction is a "right-hand rule." If the thumb of the right hand points in the direction of **A** and the index finger points in the direction of **B**, the middle finger points in the direction of the cross product.

---

**Problem 2.18**

From the geometrical definition just given, show that

$$\boxed{\mathbf{A} \times \mathbf{B} = -\mathbf{B} \times \mathbf{A}}. \tag{2.91}$$

---

In Problem 2.18, you have shown that the vector product of two vectors is not commutative, which means that you get a different result if you switch the order of the two factors.

From Eq. (2.90),

$$\boxed{\mathbf{A} \times \mathbf{A} = \mathbf{0}}, \tag{2.92}$$

where **0** is the null vector, which has zero magnitude and no particular direction. To express the cross product in terms of components, we can write

$$\mathbf{i} \times \mathbf{j} = \mathbf{j} \times \mathbf{j} = \mathbf{k} \times \mathbf{k} = 0 \tag{2.93}$$

and

$$\mathbf{i} \times \mathbf{j} = \mathbf{k} \tag{2.94a}$$
$$\mathbf{j} \times \mathbf{i} = -\mathbf{k} \tag{2.94b}$$
$$\mathbf{i} \times \mathbf{k} = -\mathbf{j} \tag{2.94c}$$
$$\mathbf{k} \times \mathbf{i} = \mathbf{j} \tag{2.94d}$$
$$\mathbf{j} \times \mathbf{k} = \mathbf{i} \tag{2.94e}$$
$$\mathbf{k} \times \mathbf{j} = -\mathbf{i}. \tag{2.94f}$$

By use of these relations, we obtain

$$\boxed{\mathbf{C} = \mathbf{A} \times \mathbf{B} = \mathbf{i}(A_y B_z - A_z B_y) + \mathbf{j}(A_z B_x - A_x B_z) + \mathbf{k}(A_x B_y - A_y B_x)}. \tag{2.95}$$

---

**Problem 2.19**

Show that Eq. (2.95) follows from Eq. (2.94).

---

**Example 2.8**

Find the cross product $\mathbf{A} \times \mathbf{B}$, where $\mathbf{A} = (1, 2, 3)$ and $\mathbf{B} = (1, 1, 1)$.

**Solution**

Let $\mathbf{C} = \mathbf{A} \times \mathbf{B}$.

$$\mathbf{C} = \mathbf{i}(2-3) + \mathbf{j}(3-1) + \mathbf{k}(1-2)$$
$$= -\mathbf{i} + 2\mathbf{j} - \mathbf{k}.$$

**Example 2.9**

Show that the vector $\mathbf{C}$ obtained in Example 2.8 is perpendicular to $\mathbf{A}$.

**Solution**

We do this by showing that $\mathbf{A} \cdot \mathbf{C} = 0$.

$$\mathbf{A} \cdot \mathbf{C} = A_x C_x + A_y C_y + A_z C_z$$
$$= -1 + 4 - 3 = 0.$$

**Problem 2.20**

Show that the vector $\mathbf{C}$ obtained in Example 2.9 is perpendicular to $\mathbf{B}$, and that Eq. (2.90) is satisfied. Do this by finding the angle between $\mathbf{A}$ and $\mathbf{B}$ through calculation of $\mathbf{A} \cdot \mathbf{B}$.

An example of a vector product is the force on a moving charged particle due to a magnetic field. If $q$ is the charge on the particle measured in coulombs, $\mathbf{v}$ is the velocity of the particle in meters per second, and $\mathbf{B}$ is the magnetic induction (often called the "magnetic field") measured in tesla, the force in newtons is given by

$$\mathbf{F} = q\mathbf{v} \times \mathbf{B}. \tag{2.96}$$

Since this force is perpendicular to the velocity, it causes the trajectory of the particle to curve, rather than changing the speed of the particle.

**Example 2.10**

Find the force on an electron in a magnetic field if $\mathbf{v} = \mathbf{i}(1.000 \times 10^5 \, \mathrm{m\,s}^{-1})$ and $\mathbf{B} = \mathbf{j}(1.000 \times 10^{-4} T)$.

**Solution**

The value of $q$ is $-1.602 \times 10^{-19} \, \mathrm{C}$ (note the negative sign).

$$\mathbf{F} = (\mathbf{i} \times \mathbf{j})(1.602 \times 10^{-19} \, \mathrm{C})(1.000 \times 10^5 \, \mathrm{m\,s}^{-1})(1.000 \times 10^{-4} \, \mathrm{T})$$
$$= -\mathbf{k}(1.602 \times 10^{-18} \, \mathrm{A\,s\,m\,s}^{-1}\mathrm{kg\,s}^{-2}\mathrm{A}^{-1})$$
$$= \mathbf{k}(-1.602 \times 10^{-18} \, \mathrm{kg\,m\,s}^{-2}) = \mathbf{k}(-1.602 \times 10^{-18} \, \mathrm{N}).$$

The force on a charged particle due to an electric field is

$$\mathbf{F} = q\mathbf{E}, \tag{2.97}$$

where $\mathbf{E}$ is the electric field. If the charge is measured in coulombs and the field in volts per meter, the force is in newtons.

---

### Problem 2.21

Find the direction and the magnitude of the electric field necessary to provide a force on the electron in Example 2.10 that is equal in magnitude to the force due to the magnetic field but opposite in direction. If both these forces act on the particle, what will be their effect?

---

## SECTION 2.6. IMAGINARY AND COMPLEX NUMBERS

The *imaginary unit* is called $i$ (not to be confused with the unit vector $\mathbf{i}$) and is defined to be the square root of $-1$:

$$\boxed{i = \sqrt{-1}}. \tag{2.98}$$

If $b$ is a real number, the quantity $ib$ is said to be *pure imaginary*, and if $a$ is also real, the quantity

$$c = a + ib \tag{2.99}$$

is said to be *complex*. The real number $a$ is called the *real part* of $c$ and is denoted by

$$a = R(c). \tag{2.100}$$

The real number $b$ is called the *imaginary part* of $c$ and is denoted by

$$b = I(c). \tag{2.101}$$

All of the rules of ordinary arithmetic apply with complex numbers. Some of the rules follow.

Addition and multiplication are both associative. If $A$, $B$, and $C$ are complex numbers

$$A + (B + C) = (A + B) + C \tag{2.102}$$
$$A(BC) = (AB)C. \tag{2.103}$$

Addition and multiplication are distributive.

$$A(B + C) = AB + AC. \tag{2.104}$$

Addition and multiplication are both commutative.

$$A + B = B + A \tag{2.105}$$
$$AB = BA. \tag{2.106}$$

Subtraction is simply the addition of a number whose real and imaginary parts are the negatives of the number to be subtracted, and division is the multiplication by the reciprocal of a number. If

$$z = x + iy \tag{2.107}$$

then the reciprocal of $z$, called $z^{-1}$, is given by

$$\boxed{z^{-1} = \frac{x}{x^2 + y^2} - i\frac{y}{x^2 + y^2}}.$$  (2.108)

---

### Example 2.11

Show that $z(z^{-1}) = 1$.

---

### Solution

$$z(z^{-1}) = (x + iy)\left(\frac{x}{x^2 + y^2} - i\frac{y}{x^2 + y^2}\right)$$
$$= \frac{1}{x^2 + y^2}(x^2 + ixy - ixy - i^2y^2) = \frac{1}{x^2 + y^2}(x^2 - i^2y^2)$$
$$= 1.$$

---

### Problem 2.22

Show that

$$(a + ib)(c + id) = ac - bd + i(bc + ad).$$  (2.109)

### Problem 2.23

Show that

$$\frac{a + ib}{c + id} = \frac{(ac + bd - iad + ibc)}{c^2 + d^2}.$$  (2.110)

### Problem 2.24

Find the value of

$$(4 + 6i)(3 + 2i) + 4i - \frac{1 + i}{3 - 2i}.$$

---

Specifying a complex number is equivalent to specifying two real numbers. This is equivalent to specifying the coordinates of a point in a plane or the two components of a two-dimensional vector. We can therefore represent a complex number by the location of a point in a plane, as shown in Fig. 2.12. This kind of a figure is called an *Argand diagram*, and the plane of the figure is called the *Argand plane* or the *complex plane*. The horizontal coordinate represents the real part of the number and the vertical coordinate represents the imaginary part. The horizontal axis, labeled $R$, is called the *real axis*, and the vertical axis, labeled $I$, is called the *imaginary axis*.

The location of the point in the Argand plane can be given by polar coordinates, just as in Section 2.5. We now use the symbol $r$ for the distance from the origin to the point, and the symbol $\phi$ for the angle in radians between the positive real axis and the

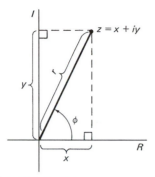

**FIGURE 2.12**   Representation of the complex number $z = x + iy$ in the Argand diagram.

line segment joining the origin and the point. The quantity $r$ is the *magnitude*, or *absolute value*, or *modulus*, and $\phi$ is the *argument*, or *phase*. From Eqs. (2.58) and (2.59),

$$x + iy = r\cos(\phi) + ir\sin(\phi). \tag{2.111}$$

There is a theorem, known as *Euler's formula*, which allows a complex number to be written as an exponential with an imaginary exponent,

$$\boxed{re^{i\phi} = r\cos(\phi) + ir\sin(\phi) = x + iy}, \tag{2.112}$$

where $e$ is the base of natural logarithms, $e = 2.7182818\ldots$, and where $r$ and $\phi$ are the magnitude and phase of the number. This form is called the *polar representation*. In this formula, $\phi$ must be measured in radians.

In the polar representation, the product of two complex numbers, say $z_1 = r_1 e^{i\phi_1}$ and $z_2 = r_2 e^{i\phi_2}$, is given in a convenient form:

$$\boxed{z_1 z_2 = r_1 r_2 e^{i(\phi_1 + \phi_2)}}. \tag{2.113}$$

The quotient $z_1 z_2$ is given by

$$\boxed{\frac{z_1}{z_2} = \left(\frac{r_1}{r_2}\right) e^{i(\phi_1 - \phi_2)}}. \tag{2.114}$$

*DeMoivre's formula* gives the result of raising a complex number to a given power:

$$\boxed{(re^{i\phi})^n = r^n e^{in\phi} = r^n[\cos(n\phi) + i\sin(n\phi)]}. \tag{2.115}$$

---

### Example 2.12

Evaluate the following.

a. $(4e^{i\pi})(3e^{2i\pi})$
b. $(8e^{2i\pi})(2e^{i\pi/2})$
c. $(8e^{4i})^2$.

**Solution**

a. $(4e^{i\pi})(3e^{2i\pi}) = 12e^{3i\pi} = 12e^{i\pi}$
b. $(8e^{2i\pi})(2e^{i\pi/2}) = 4e^{5i\pi/2}$
c. $(8e^{4i})^2 = 64e^{8i}$.

In part a of Example 2.12, we have used the fact that an angle of $2\pi$ radians gives the same point in the complex plane as an angle of 0, so that

$$\boxed{e^{2\pi i} = 1}.$$

(2.116a)

Similarly,

$$\boxed{e^{\pi i} = -1}.$$

(2.116b)

Since an angle is unchanged if any multiple of $2\pi$ is added or subtracted from it, we can write, for example,

$$e^{3\pi i} = e^{\pi i}.$$

(2.117)

If a number is given in the form $z = x + iy$, we can find the magnitude and the phase as

$$r = \sqrt{x^2 + y^2}$$

(2.118)

$$\phi = \arctan\left(\frac{y}{x}\right).$$

(2.119)

Just as with a transformation from cartesian coordinates to polar coordinates, we do not necessarily use the principal value of the arctangent function, but must obtain an angle in the proper quadrant, with $\phi$ ranging from 0 to $2\pi$.

**Problem 2.25**

Express the following complex numbers in the form $re^{i\phi}$:

a. $4 + 4i$
b. $-1$
c. $1$
d. $1 - i$.

Express the following complex numbers in the form $x + iy$:

a. $e^{i\pi}$
b. $3e^{\pi i/2}$
c. $e^{3\pi i/2}$.

The *complex conjugate* of a number is defined as the number that has the same real part and an imaginary part that is the negative of that of the original number. The complex conjugate is denoted by an asterisk or by a bar over the letter for the number:

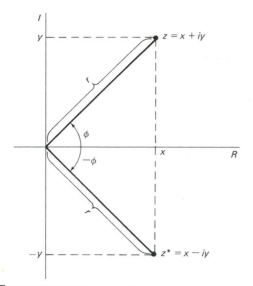

**FIGURE 2.13**  A complex number, $z = x + iy$, and its complex conjugate, $z^* = x - iy$, in the Argand plane.

If $z = x + iy$,

$$\bar{z} = z^* = (x + iy)^* = x - iy$$  (2.120)

Figure 2.13 shows the location of a complex number and of its complex conjugate in the Argand plane. The phase of the complex conjugate is $-\phi$ if the phase of the original number is $\phi$. The magnitude is the same, so

$$(re^{i\phi})^* = re^{-i\phi}$$  (2.121)

Although we do not prove it, the following fact is useful in obtaining the complex conjugate of a complex quantity: *The complex conjugate of any expression is obtained by changing the sign in front of every i that occurs in the expression.*

---

**Example 2.13**

Find the complex conjugates of the following, where $a, b, c$, and $d$ are real quantities:

    a.  $A = (1 + 2i)^{3/2} + \exp(3 + 4i)$
    b.  $B = a(b + ci)^2 + 4(c - id)^{-1}$.

---

**Solution**

    a.  $A^* = (1 - 2i)^{3/2} + \exp(3 - 4i)$
    b.  $B^* = a(b + ci)^2 + 4(c + id)^{-1}$.

**Problem 2.26**

Find the complex conjugates of

a. $A = (x + iy)^2 - 4e^{ixy}$
b. $B = (3 + 7i)^3 - (7i)^2$.

Once we have an expression for the complex conjugate of a quantity, we can use it to express the real and imaginary parts separately:

$$R(z) = \frac{z + z^*}{2} \tag{2.122}$$

$$I(z) = \frac{z - z^*}{2i}. \tag{2.123}$$

**Problem 2.27**

a. Use Eq. (2.120) to show that Eqs. (2.122) and (2.123) are correct.
b. Obtain the famous formulas

$$\boxed{\cos(\phi) = \frac{e^{i\phi} + e^{-i\phi}}{2} = R(e^{i\phi})} \tag{2.124}$$

$$\boxed{\sin(\phi) = \frac{e^{i\phi} - e^{-i\phi}}{2i} = I(e^{i\phi})}. \tag{2.125}$$

The magnitude of an expression can also be obtained by using the complex conjugate. We find that

$$\boxed{zz^* = (re^{i\phi})(re^{-i\phi}) = r^2} \tag{2.126}$$

so that

$$\boxed{r = \sqrt{zz^*}}, \tag{2.127}$$

where the positive square root is to be taken. The product of any complex number and its complex conjugate is always real and nonnegative.

**Problem 2.28**

Write a complex number in the form $x + iy$ and show that the product of the number with its complex conjugate is real and nonnegative.

**Example 2.14**

If $z = 4e^{3i} + 6i$, find $R(z)$, $I(z)$, $r$, and $\phi$.

---

**Solution**

$$R(z) = \frac{z + z^*}{2} = \frac{4e^{3i} + 6i + 4e^{-3i} - 6i}{2}$$
$$= 2(e^{3i} + e^{-3i}) = 4\cos(3) = -3.960$$
$$I(z) = \frac{z - z^*}{2i} = \frac{4e^{3i} + 6i - 4e^{-3i} + 6i}{2i}$$
$$= \frac{4(e^{3i} - e^{-3i})}{2i} + 6 = 4\sin(3) + 6 = 4\sin(171.89°) + 6$$
$$r = (zz^*)^{1/2} = (x^2 + y^2)^{1/2} = [(-3.960)^2 + (6.5645)^2]^{1/2}$$
$$= 7.666$$
$$\phi = \arctan\left(\frac{I}{R}\right) = \arctan\left(\frac{-6.5645}{3.960}\right) = \arctan(-1.6577).$$

The principal value of this arctangent is $-58.90°$. However, since $R(z)$ is negative and $I(z)$ is positive, we require an angle in the second quadrant.

$$\phi = 180° - 58.90° = 121.10° = 2.114\,\text{rad}.$$

---

**Problem 2.29**

If $z = \left(\dfrac{3 + 2i}{4 + 5i}\right)^2$, find $R(z)$, $I(z)$, $r$, and $\phi$.

---

The *square root of a complex number* is a number, which when multiplied by itself will yield the first number. Just as with real numbers, there are two square roots of a complex number. If $z = re^{i\phi}$, one of the square roots is given by

$$\sqrt{re^{i\phi}} = \sqrt{r}e^{i\phi/2}. \tag{2.128}$$

The other square root is obtained by realizing that if $\phi$ is increased by $2\pi$, the same point in the Argand plane is represented. Therefore, the square root of $re^i(2\pi + \phi)$ is the same as the other square root of $re^{i\phi}$.

$$\sqrt{re^{i\phi}} = \sqrt{re^{i(2\pi+\phi)}} = \sqrt{r}e^{i(\pi+\phi/2)}. \tag{2.129}$$

---

**Example 2.15**

Find the square roots of $3e^{i\pi/2}$.

---

**Solution**

One square root is, from Eq. (2.128),

$$\sqrt{3e^{i\pi/2}} = \sqrt{3}e^{i\pi/4}.$$

The other square root is, from Eq. (2.129),

$$\sqrt{3}e^{i(\pi+\pi/4)} = \sqrt{3}e^{i5\pi/4}.$$

**Problem 2.30**

Find the square roots of $4 + 4i$. Sketch an Argand diagram and locate the roots on it.

---

There are three *cube roots of a complex number*. These can be found by looking for the numbers that when cubed yield $re^{i\phi}$, $re^{i(2\pi+\phi)}$, and $re^{i(4\pi+\phi)}$. These numbers are

$$\sqrt[3]{re^{i\phi}} = \sqrt[3]{r}e^{i\phi/3}, \sqrt[3]{r}e^{i(2\pi+\phi)/3}, \sqrt[3]{r}e^{i(4\pi+\phi)/3}.$$

Higher roots are obtained similarly.

## SUMMARY OF THE CHAPTER

In this chapter we have presented the algebraic tools needed to manipulate expressions containing real scalar variables, real vector variables, and complex scalar variables. We have also introduced ordinary and hyperbolic trigonometric functions, exponentials, and logarithms.

## ADDITIONAL READING

James Stewart, *Calculus*, 2nd ed., Brooks/Cole, Pacific Grove, CA, 1991. This is a calculus textbook that uses some examples from physics in its discussions. You can read about coordinate systems, vectors, and complex numbers in almost any calculus textbook, including this one.

Herbert B. Dwight, *Tables of Integrals and Other Mathematical Data*, 4th ed., Macmillan Co., New York, 1962. This book is a very useful small compilation of formulas, including trigonometric identities, derivatives, definite and indefinite integrals, and infinite series.

D. D. Fitts, *Nonequilibrium Thermodynamics*, McGraw–Hill, New York, 1962. After discussing equilibrium thermodynamics in your physical chemistry course, you might be interested in glancing through this book, which presents the nonequilibrium version of the subject. There is also a brief discussion of dyadics, a kind of product of vectors that we did not discuss in this chapter.

Sherman K. Stein, *Calculus with Analytic Geometry*, 4th ed., McGraw–Hill, New York, 1987. This is another calculus textbook with a number of applications to physical problems.

## ADDITIONAL PROBLEMS

### 2.31

A Boy Scout finds a tall tree while hiking and wants to estimate its height. He walks away from the tree and finds that when he is 100 m from the tree, he must look upward at an angle of 35° to look at the top of the tree. His eye is 1.50 m from the ground, which is perfectly level. How tall is the tree?

### 2.32

The equation $x^2 + y^2 + z^2 = c^2$, where $c$ is a constant, represents a surface in three dimensions. Express the equation in spherical polar coordinates. What is the shape of the surface?

### 2.33

Express the equation $y = b$, where $b$ is a constant, in plane polar coordinates.

### 2.34

Express the equation $y = mx + b$, where $m$ and $b$ are constants, in plane polar coordinates.

### 2.35

Find the values of the plane polar coordinates that correspond to $x = -2$, $y = 4$.

### 2.36

Find the values of the cartesian coordinates that correspond to $r = 10$, $\theta = 45°$, $\phi = 135°$.

### 2.37

Find $\mathbf{A} - \mathbf{B}$ if $\mathbf{A} = 2\mathbf{i} + 3\mathbf{j}$ and $\mathbf{B} = \mathbf{i} + 3\mathbf{j} - \mathbf{k}$.

### 2.38

Find $\mathbf{A} \cdot \mathbf{B}$ if $\mathbf{A} = (0, 2)$ and $\mathbf{B} = (2, 0)$.

### 2.39

Find $|\mathbf{A}|$ if $\mathbf{A} = 3\mathbf{i} + 4\mathbf{j} - \mathbf{k}$.

### 2.40

Find $\mathbf{A} \times \mathbf{B}$ if $\mathbf{A} = (0, 1, 2)$ and $\mathbf{B} = (2, 1, 0)$.

### 2.41

Find the angle between $\mathbf{A}$ and $\mathbf{B}$ if $\mathbf{A} = \mathbf{i} + 2\mathbf{j} + \mathbf{k}$ and $\mathbf{B} = \mathbf{i} + \mathbf{j} + \mathbf{k}$.

### 2.42

A spherical object falling in a fluid has three forces acting upon it: (1) The gravitational force, whose magnitude is $F_g = mg$, where $m$ is the mass of the object. (2) The buoyant force, whose magnitude is $F_b = m_f g$, where $m_f$ is the mass of the displaced fluid, and whose direction is upward. (3) The frictional force, given by $\mathbf{F}_f = -6\pi \eta r \mathbf{v}$, where $r$ is the radius of the object, $\mathbf{v}$ its velocity, and $\eta$ the coefficient of viscosity of the fluid. If the object is falling at a constant speed of $0.500$ m s$^{-1}$, has a radius of $0.15$ m, a mass of $0.0600$ kg, and displaces a mass of fluid equal to $0.01257$ kg, find the value of $\eta$ and its units.

### 2.43

The solutions to the Schrödinger equation for the electron in a hydrogen atom have three quantum numbers associated with them, called $n$, $l$, and $m$, and these solutions are often denoted by $\psi_{nlm}$. One of the solutions is

$$\psi_{211} = \frac{1}{8\sqrt{\pi}}\left(\frac{1}{a_0}\right)^{3/2}\frac{r}{a_0}e^{-r/2a_0}\sin(\theta)e^{i\phi},$$

where $a_0$ is a distance equal to $0.529 \times 10^{-10}$ m, called the Bohr radius.

   a. Write this function in terms of cartesian coordinates.
   b. Write an expression for the magnitude of this complex function.
   c. This function is sometimes called $\psi_{2p1}$. Write expressions for the real and imaginary parts of the function, which are proportional to the related functions called $\psi_{2px}$ and $\psi_{2py}$.

**2.44**

Find the sum of $4e^{3i}$ and $5e^{2i}$.

**2.45**

Find the three cube roots of $3 - 2i$.

**2.46**

Find the real and imaginary parts of

$$\sqrt{3 + 2i} + (7 + 5i)^2.$$

Obtain a separate answer for each of the two square roots of the first term. Write the complex conjugate of each answer.

**2.47**

An object has a force on it given by $(4.75\,\text{N})\mathbf{i} + (7.00\,\text{N})\mathbf{j} + (3.50\,\text{N})\mathbf{k}$.

   a. Find the magnitude of the force.
   b. Find the projection of the force in the $x$-$y$ plane. That is, find the vector in the $x - y$ plane whose head is reached from the head of the force vector by moving in a direction perpendicular to the $x$-$y$ plane.

**2.48**

An object of mass 12.000 kg is moving in the $x$ direction. It has a gravitational force acting on it equal to $-mg\mathbf{k}$, where $m$ is the mass of the object and $g$ is the acceleration due to gravity, equal to $9.80\,\text{m s}^{-2}$. There is a frictional force equal to $-(0.240\,\text{N})\mathbf{i}$. What is the magnitude and direction of the resultant force (the vector sum of the forces on the object)?

**2.49**

The potential energy of a magnetic dipole in a magnetic field is given by

$$\mathcal{V} = -\mu \cdot \mathbf{B},$$

where $\mathbf{B}$ is the magnetic induction (magnetic field) and $\mu$ is the magnetic dipole. Make a graph of $\mathcal{V}/(|\mu||\mathbf{B}|)$ as a function of the angle between $\mu$ and $\mathbf{B}$.

# 3

# MATHEMATICAL FUNCTIONS AND DIFFERENTIAL CALCULUS

## Preview

In this chapter we present the concept of a mathematical function and discuss its relationship to the behavior of physical variables in actual systems. We define the derivative of a function of one independent variable and discuss its geometric interpretation. We discuss the use of derivatives in approximate calculations of changes in dependent variables and describe their use in finding minimum and maximum values of functions.

## Principal Facts and Ideas

1. A mathematical function of one variable is a rule for giving a value of a dependent variable for any value of an independent variable.
2. The derivative of a function is a measure of how rapidly the dependent variable changes with changes in the value of the independent variable. If $\Delta y$ is the change in the dependent variable produced by a change $\Delta x$ in the independent variable, then the derivative $dy/dx$ is defined by

$$\frac{dy}{dx} = \lim_{\Delta x \to 0} \frac{\Delta y}{\Delta x}.$$

3. The derivatives of many simple functions can be obtained by applying a few simple rules, either separately or in combination. For example,

$$\frac{d(x^n)}{dx} = nx^{n-1}.$$

4. A finite increment in a dependent variable, $\Delta y$, can sometimes be calculated approximately by use of the formula

$$\Delta y \approx \frac{dy}{dx} \Delta x.$$

5. Differential calculus can be used to find maximum and minimum values of a function. A relative minimum or maximum value of a variable $y$ which depends on $x$ is found at a point where $dy/dx = 0$.

## Objectives

After studying this chapter, you should:

1. understand the concept of a mathematical function;
2. be able to obtain a formula for the derivative of any fairly simple function without consulting a table;
3. be able to draw a rough graph of any fairly simple function and locate important features on the graph;
4. be able to find maximum and minimum values of a function of one variable.

## SECTION 3.1. MATHEMATICAL FUNCTIONS

One definition of a mathematical function is that it is a set of ordered pairs of numbers. This means, for example, that you have a table with two columns of numbers in it. Each number in the first column is associated with the number on the same line in the other column. The choice of a number in the first column delivers a unique value from the second column. For example, a physical chemistry student might measure the vapor pressure of liquid ethanol at ten different temperatures and present the results in such a table. In the first column, the student puts the values of the temperature, and on the same line in the second column he or she puts the observed vapor pressure for that temperature.

One variable, say the temperature, is chosen to be the *independent variable*. The vapor pressure is then the *dependent variable*. This means that if we choose a value of the temperature, the function provides the corresponding value of the vapor pressure. This is the important property of a function. It is as though the function says, "You give me a value for the independent variable, and I'll give you the corresponding value of the dependent variable." In some cases, we can reverse the roles of the dependent and independent variables. We discuss that later.

A two-column table of numerical values is of course not the only way to represent a mathematical function. Any rule that delivers a value of a dependent variable when a value of an independent variable is specified is a mathematical function. Two common representations of mathematical functions are mathematical formulas and graphs.

There are also functions with more than one independent variable. In this case, a value must be specified for each of the independent variables in order for the function to deliver a value of the dependent variable. We discuss this kind of function in Chapter 6.

## Properties of Functions Representing Physical Variables

Mathematical functions are useful in physics and chemistry because the physical variables describing the state of a physical system behave like mathematical functions. For example, the physical chemistry student in the laboratory measuring the vapor pressure of ethanol might choose a value of the temperature by setting the control on a thermoregulator. The system assumes a value of the vapor pressure that is determined by the temperature.

We make the following assumptions about the actual behavior of physical systems:

1. Of the macroscopic variables such as temperature, pressure, volume, density, entropy, energy, etc., only a certain number (depending on circumstances) can be independent variables. The others are dependent variables, governed by mathematical functions.

2. These mathematical functions are single-valued, except possibly at isolated points. This means that only one value of the dependent variable occurs for a given value of the independent variable.

3. These mathematical functions are continuous and differentiable, except possibly at isolated points. We will discuss later what this statement means.

Although a list of ten temperatures and the corresponding ten values of the vapor pressure qualifies as a mathematical function, such a function is not very useful, because it can be used only for these temperatures. We need a function which will provide a value of the vapor pressure for other temperatures in order to have a useful function. If we have an exact way to do this, the rule used to generate new values is an exact *representation of the function*. Approximate representations of a function must generally be used in physical chemistry. Approximate representations include interpolation between values in a table, a graph on which a curve is drawn passing through or near data points, and a mathematical formula chosen to represent the data points. In all these approximate representations of a function, we hope to make our rule for generating new values give nearly the same values as the "correct" function, or at least close enough for some purposes.

We assume that the exact representation of a function describing a property of a physical system is nearly always *single-valued*. A single-valued function delivers one and only one value of the dependent variable for any given value of the dependent variable. In discussing real functions of real variables, mathematicians will usually not call something a function unless it is single-valued.

We also assume that our functions are nearly always *continuous*. If a function is continuous, the dependent variable does not change abruptly if the independent variable changes gradually. If you are drawing a graph of a continuous function, you will not have to draw a vertical step in your curve or take your pencil away from the paper. This can be expressed mathematically by saying that the function

$$y = f(x) \tag{3.1}$$

is continuous at $x = a$ if

$$\lim_{x \to a^+} f(x) = \lim_{x \to a^-} f(x) = f(a). \tag{3.2}$$

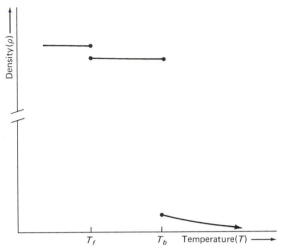

**FIGURE 3.1**    The density of a pure substance as a function of temperature (schematic).

As $x$ draws close to $a$ from either direction, $f(x)$ smoothly draws close to $f(a)$, the value that $y$ has at $x = a$.

In some cases, the functions that represent physical variables are continuous over their entire range of values. In other cases, they are *piecewise continuous*. That is, they are continuous except at a finite number of isolated points, at which finite jumps in the value of the function are permitted. Finite jump discontinuities are called *ordinary discontinuities*. Figure 3.1 shows schematically the density of a pure substance as a function of temperature at fixed pressure. The vertical axis is compressed in one place, and there is a large discontinuity in the density at the boiling temperature, $T_b$, and a smaller discontinuity at the freezing temperature, $T_f$. The density is piecewise continuous; it is continuous except at the freezing temperature $T_f$ and the boiling temperature $T_b$. If $T$ is made to approach $T_f$ from above, it smoothly approaches one value, the density of the liquid at $T_f$. If $T$ is made to approach $T_f$ from below, the density smoothly approaches another value, the density of the solid at this temperature.

The system can exist either as a solid or as a liquid at the freezing temperature, or it can be partly solid and partly liquid. The density appears to be *double-valued* at $T = T_f$ and also at $T = T_b$. At these temperatures, two phases can coexist, each having a different value of the density.

In Eq. (3.1), we have used the notation of a mathematician, using the letter $y$ to represent the dependent variable and the letter $f$ to represent the function that provides values of $y$. A physicist or chemist uses a different policy, writing for example for the density

$$\rho = \rho(T), \tag{3.3}$$

where the letter $\rho$ stands both for the density and for the function that provides values of the density. There are two reason for this policy: First, we have a lot of variables to discuss, and only a limited supply of letters; second, using two letters for one variable would double the number of letters to keep in mind during any discussion of a physical phenomenon.

## Graphical Representations of Functions

One way to communicate quickly the general behavior of a function is with a graph. A rough graph can quickly show the general trend of the dependence of a dependent variable on an independent variable. An accurate graph can be read to provide good approximate values of the dependent variable for specific values of the independent variable. A graph containing data points and a curve drawn through or near the points can also reveal the presence of an inacurrate data point.

---

### Problem 3.1

The following is a set of data for the vapor pressure of ethanol. Plot these points on a graph, with the temperature on the horizontal axis (the abscissa) and the vapor pressure on the vertical axis (the ordinate). Decide if there are any bad data points. Draw a smooth curve nearly through the points, disregarding any bad points. If you have access to a graphing program such as CricketGraph or Kaleida-Graph, or a spreadsheet program such as Excel, use the curve-fitting procedure in the software.

| Temperature/°C | Vapor pressure/torr |
|---|---|
| 25.00 | 55.9 |
| 30.00 | 70.0 |
| 35.00 | 97.0 |
| 40.00 | 117.5 |
| 45.00 | 154.1 |
| 50.00 | 190.7 |
| 55.00 | 241.9 |

---

## Important Families of Functions

There are a number of important families of functions which occur frequently in physical chemistry. A *family of functions* is a set of related functions. It is sometimes represented by a single formula that contains other symbols besides those of the independent variable. The choice of a set of values for these quantities specifies which member of the family of functions is meant. The following formula represents one such family. A member of this family is a *linear function* or *first-degree polynomial*:

$$y = mx + b. \tag{3.4}$$

In this family, we have a different function for each set of values of the parameters $m$ and $b$. The graph of each such function is a straight line, so Eq. (3.4) represents a family of different straight lines. The constant $b$ is called the *intercept*. It equals the value of the function for $x = 0$. The constant $m$ is called the *slope*. It gives the

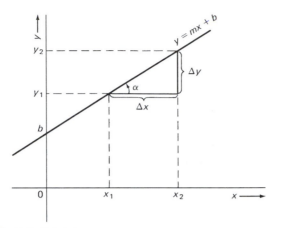

**FIGURE 3.2**    The graph of the linear function $y = mx + b$.

steepness of the line, or the relative rate at which the dependent variable changes as the independent variable is changed. For example, if $m = 2$, a given change in $x$ equal to $\Delta x$, produces a change in $y$ equal to $2\Delta x$.

Figure 3.2 shows two particular values of $x$, called $x_1$ and $x_2$, and their corresponding values of $y$, called $y_1$ and $y_2$. A line is drawn through the two points. If $m > 0$, then $y_2 > y_1$ and the line slopes upward to the right. If $m < 0$, then $y_2 < y_1$ and the line slopes downward to the right.

---

### Example 3.1

Show that the slope is given by

$$m = \frac{y_2 - y_1}{x_2 - x_1} = \frac{\Delta y}{\Delta x}. \tag{3.5}$$

---

### Solution

$$y_2 - y_1 = mx_2 + b - (mx_1 + b) = m(x_2 - x_1)$$

or

$$m = \frac{y_2 - y_1}{x_2 - x_1}.$$

---

Another important property of the slope is that

$$m = \tan(\alpha), \tag{3.6}$$

where $\alpha$ is the angle between the horizontal axis and the straight line of the function. The angle $\alpha$ is taken to be between $-90°$ and $90°$, and the slope can range from $-\infty$ to $\infty$.

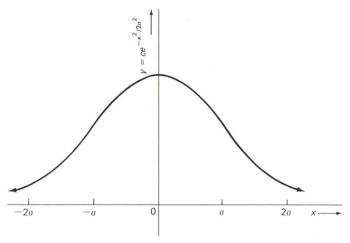

**FIGURE 3.3**  The graph of the Gaussian function.

Another important family of functions is the *quadratic function* or *second-degree polynomial*:

$$y = ax^2 + bx + c. \tag{3.7}$$

The graph of a function from this family is a parabola. You should also be familiar with the graphs of the trigonometric functions and the exponential function which are shown in Figs. 2.2 through 2.5.

Another important function is the *Gaussian function*:

$$y = ce^{-x^2/2\sigma^2}. \tag{3.8}$$

This is the function whose graph is the *bell-shaped curve* shown in Fig. 3.3. This function is proportional to the probability that a value of $x$ will occur in a number of statistical applications and is discussed in Chapter 10. The constant $\sigma$ is called the *standard deviation* and is a measure of the width of the "hump" in the curve.

The functions just listed form a repertoire of functions which you can use to generate approximate graphs.

---

**Example 3.2**

Sketch a rough graph of the function

$$y = x\cos(\pi x).$$

---

**Solution**

The given function is a product of the function $x$, which is shown in Fig. 3.4a, and $\cos(\pi x)$, which is shown in Fig. 3.4b. Where either factor vanishes, the product vanishes. Since the cosine oscillates between $-1$ and $+1$, the product oscillates between $-x$ and $+x$. A rough graph of the product is shown in Fig. 3.4c.

---

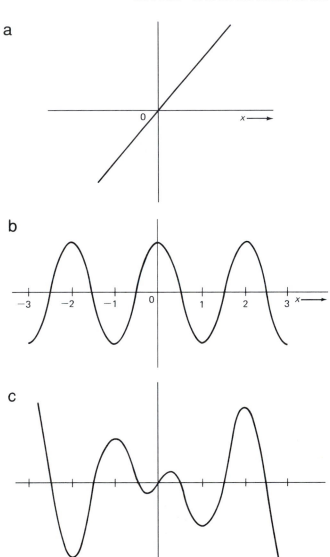

**FIGURE 3.4** (a) The first factor in the function of Example 3.2. (b) The second factor in the function of Example 3.2. (c) The function of Example 3.2.

## Problem 3.2

Sketch rough graphs of the following functions. Verify your graphs using graphing software if available.

a. $e^{-x} \sin(x)$

b. $\sin^2(x) = \sin(x)^2$

   c. $x^2 e^{-x^2/2}$

   d. $1/x^2$

   e. $(1-x)e^{-x}$.

---

## Graphical Solution of Equations

If an equation to be solved is written in the form

$$f(x) = 0 \qquad\qquad (3.9)$$

then a graph of $f(x)$ will cross the $x$ axis at any real values of $x$ which are solutions to the equation.

---

### Problem 3.3

Draw a graph and use it to find an approximate solution to the following equation. Use graphing software if it is available.

$$e^x - 3x = 0.$$

---

## SECTION 3.2. THE DERIVATIVE OF A FUNCTION

A function other than a linear function has a graph with a curve other than a straight line. Such a curve has a different direction, or different steepness, at different points on the curve.

### The Line Tangent to a Curve

At most points on the curve, the line tangent to the curve at that point is the line that has the point in common with the curve but does not cross it at that point, as shown in Fig. 3.5. We can speak of the direction of the curve at $x = x_1$ as the direction of the line which is tangent to the curve at that point. In addition to the tangent line, we have drawn a horizontal line intersecting the curve at $x = x_1$.

If the tangent line is represented by the formula

$$y = mx + b$$

then the slope of the tangent line is equal to m. In the figure, we have labeled another point at $x_2$, a finite distance away from $x_1$. At $x = x_2$, the vertical distance from the horizontal line to the tangent line is given by $m(x_2 - x_1)$. This is not necessarily equal to the distance from the horizontal line to the curve, which is given by

$$y(x_2) - y(x_1) = y_2 - y_1 = \Delta y.$$

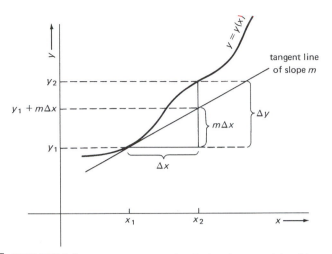

**FIGURE 3.5**   The curve representing the function $y = y(x)$ and its tangent line.

## Problem 3.4

Using graph paper or graphing software, plot the curve representing $y = \sin(x)$ for values of $x$ lying between 0 and $\pi/2$ radians. Using a ruler, draw the tangent line at $x = \pi/4$. By drawing a right triangle on your graph and measuring its sides, find the slope of the tangent line.

There is a case in which the definition of the tangent line to a curve at a point $x_1$ is more complicated than in the case shown in Fig. 3.5. In this case, the point $x_1$ lies between a region in which the curve is concave downward and a region in which the curve is concave upward. Such a point is called an *inflection point*. For such a point, we must consider tangent lines at points which are taken closer and closer to $x_1$. As we approach closer and closer to $x_1$ the tangent line will approach more and more closely to a line which is the tangent line at $x_1$. This line is the same whether we approach from the left or the right, and it does cross the curve at the point which it shares with the curve.

If $\Delta x$ is not too large, we can write as an approximation

$$\Delta y \approx m \, \Delta x. \tag{3.10}$$

We divide both sides of Eq. (3.10) by $\Delta x$ and write

$$m \approx \frac{\Delta y}{\Delta x} = \frac{y_2 - y_1}{x_2 - x_1} = \frac{y(x_2) - y(x_1)}{x_2 - x_1}. \tag{3.11a}$$

If the curve is smooth, this equation becomes a better and better approximation as $\Delta x$ becomes smaller, and if we take the mathematical limit as $\Delta x \to 0$, it becomes exact,

$$\boxed{m = \lim_{\Delta x \to 0} \frac{\Delta y}{\Delta x} = \frac{dy}{dx}}. \tag{3.11b}$$

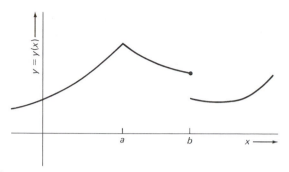

**FIGURE 3.6**   A function which is not differentiable at $x = a$ and at $x = b$.

If the limit in Eq. (3.11b) exists, it is called a *derivative* and is given the symbol $dy/dx$. The symbol resembles a fraction, but a derivative is not a fraction. It is a limit which a fraction approaches. It is not permissible to cancel what looks like a numerator or a denominator against another factor.

Another symbol is often used, especially by mathematicians:

$$\left(\frac{dy}{dx}\right)_{x_1} = y'(x_1)\,. \tag{3.12}$$

In this equation we have used the subscript to indicate that the derivative is to be evaluated at $x = x_1$. The notation $y'$ has the advantage that an argument rather than a subscript is used to show a value of $x$ at which the derivative is evaluated. We will use both notations.

If the function giving $y$ as a function of $x$ is discontinuous at $x = x_1$, the limit will not exist at that point. A function is therefore not differentiable at a discontinuity. There are also cases in which the limit exists but has a different value if $x_2 > x_1$ than it does if $x_2 < x_1$. In this case, the function is not differentiable at $x = x_1$. Figure 3.6 shows the graph of a function that is not differentiable at $x = b$, because the function is discontinuous at $x = b$ and is also not differentiable at $x = a$, even though the function is continuous at $x = a$, because the limit has different values for $x_2 > x_1$ and for $x_2 < x_1$ if $x_1 = a$. A "corner" or sharp bend in a curve such as at $x = a$ is called a *cusp*.

### Example 3.3

Decide where the following functions are differentiable.

a.  $y = |x|$
b.  $y = \sqrt{x}$.

### Solution

a.  Differentiable everywhere except at $x = 0$, where the limit has different values when $x = 0$ is approached from the two different directions.

b. This function has real values for $x \geq 0$. It is differentiable for all positive values of $x$, but at $x = 0$ the limit does not exist, so it is not differentiable at $x = 0$.

---

## Problem 3.5

Decide where the following functions are differentiable.

a. $y = \ln(x)$
b. $y = \tan(x)$
c. $y = \sec(x)$.

---

Now that we have defined the derivative, let us see how it is applied to a particular function.

---

## Example 3.4

Find the derivative of the function

$$y = y(x) = ax^2.$$

---

### Solution

$$\Delta y = y_2 - y_1 = ax_2^2 - ax_1^2 = a(x_1 + \Delta x)^2 - ax_1^2$$
$$= a\left[x_1^2 + 2x_1 \Delta x + (\Delta x)^2\right] - ax_1^2 = 2ax_1 \Delta x + (\Delta x)^2$$
$$\frac{\Delta y}{\Delta x} = 2ax_1 + \Delta x.$$

We now take the limit as $x_2 \to x_1$ or $\Delta x \to 0$. The first term, $2ax_1$ is not affected. The second term, $\Delta x$, vanishes. Thus, if we use the symbol $x$ instead of $x_1$

$$\frac{dy}{dx} = \frac{d(ax^2)}{dx} = \lim_{\Delta x \to 0} \frac{\Delta y}{\Delta x} = 2ax. \tag{3.13}$$

---

Figure 3.7 shows a graph of the function $y = ax^2$ and a graph of its derivative, $dy/dx = 2ax$. This graph exhibits some important characteristics:

1. Where the function has a horizontal tangent line, the derivative is equal to zero.
2. The derivative is positive where the function increases as $x$ increases.
3. A positive derivative is larger when the tangent line is steeper.
4. The derivative is negative where the function decreases as $x$ increases, as it does in this example for negative values of $x$.
5. A negative derivative is more negative (has a larger magnitude) when the tangent line is steeper.

These are general properties of the derivative of any function.

Table 3.1 gives the derivatives of other simple functions, derived in much the same way as Eq. (3.13). Additional derivatives are given in Appendix 2.

**TABLE 3.1   Some Elementary Functions and Their Derivatives.**[*]

| Function, $y = y(x)$ | Derivative, $dy/dx = y'(x)$ |
|---|---|
| $ax^n$ | $nax^{n-1}$ |
| $ae^{bx}$ | $abe^{bx}$ |
| $a$ | $0$ |
| $a\sin(bx)$ | $ab\cos(x)$ |
| $a\cos(bx)$ | $-ab\sin(bx)$ |
| $a\ln(x)$ | $a/x$ |

[*] In these formulas, $a$, $b$, and $n$ are constants, not necessarily integers.

---

### Problem 3.6

Make rough graphs of several functions from Table 3.1. Below each graph, on the same sheet of paper, make a rough graph of the derivative of the same function.

---

The derivative of a function is the rate of change of the dependent variable with respect to the independent variable. Because it is the slope of the tangent line, it has a large magnitude when the curve is steep and a small magnitude when the curve is nearly horizontal.

## SECTION 3.3.  DIFFERENTIALS

In Section 3.2, we talked about a change in a dependent variable produced by a change in an independent variable. If $y$ is a function of $x$, we wrote in Eq. (3.10)

$$\Delta y \approx m\,\Delta x, \tag{3.14}$$

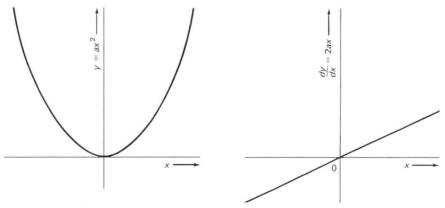

**FIGURE 3.7**   A graph of a function and its derivative.

where $\Delta y = y(x_2) - y(x_1)$, $\Delta x = x_2 - x_1$, and $m$ was the slope of the tangent line at $x = x_1$. Using Eq. (3.12), this becomes

$$\Delta y \approx \left(\frac{dy}{dx}\right)\Delta x. \tag{3.15}$$

This approximate equality will generally be more nearly correct when $\Delta x$ is made smaller, but may be quite badly incorrect if $\Delta x$ is fairly large.

---

### Example 3.5

Using Eq. (3.15), estimate the change in the pressure of 1.000 mol of an ideal gas at 0°C when its volume is changed from 22.414 liters to 21.414 liters.

---

### Solution

An ideal gas obeys

$$P = \frac{nRT}{V} \tag{3.16}$$

so that if $n$ and $T$ are kept fixed, we can differentiate with respect to $V$:

$$\frac{dP}{dV} = \frac{-nRT}{V^2} \tag{3.17}$$

$$\frac{dP}{dV} = \frac{-(1.000\,\text{mol})(0.08206\,\text{L atm mol}^{-1}\,\text{K}^{-1})(273.15\,\text{K})}{(22.414\,\text{L})^2}$$

$$= -0.0446\,\text{atm L}^{-1}.$$

We can approximate an increment in $P$:

$$\Delta P \approx \left(\frac{dP}{dV}\right)\Delta V = (-0.0446\,\text{atm L}^{-1})(-1.000\,\text{L})$$
$$\approx 0.0446\,\text{atm}.$$

### Example 3.6

Determine the accuracy of the result of Example 3.5.

---

### Solution

$$\Delta P = P(21.414\,\text{L}) - P(22.414\,\text{L}) = 1.0468\,\text{atm} - 1.0000\,\text{atm}$$
$$= 0.0468\,\text{atm}.$$

Our estimate in Example 3.5 was wrong by about 5%. If the change in volume bad been 0.1 L, the error would have been about 0.5%.

---

Since Eq. (3.15) becomes more nearly exact as $\Delta x$ is made smaller, we make it into an exact equation by making $\Delta x$ become smaller than any finite quantity that anyone can name. We do not make $\Delta x$ strictly vanish, but we say that we make it become *infinitesimal*. That is, we make it smaller in magnitude than any nonzero quantity one might specify. In this limit, $\Delta x$ becomes the *differential $dx$*,

$$dy = \left(\frac{dy}{dx}\right)dx. \tag{3.18}$$

The infinitesimal quantity $dy$ is the change in $y$ that results from the infinitesimal increment $dx$. It is proportional to $dx$ and to the slope of the tangent line, which is equal to $dy/dx$. Since $x$ is an independent variable, $dx$ is arbitrary, or subject to our control. Since $y$ is a dependent variable, its differential, $dy$, is determined by $dx$, as specified by Eq. (3.18).

Equation (3.18) has the appearance of an equation in which the $dx$ in the denominator is canceled by the $dx$ in the numerator. This is not the case. The symbol $dy/dx$ is not a fraction or a ratio. It is the limit which a ratio approaches, and that is not the same thing.

In numerical calculations, differentials are not of direct use, since they are smaller than any finite quantities that you can specify. Their use lies in the construction of formulas, especially through the process of integration, in which infinitely many infinitesimal quantities are added up to produce something finite. We discuss integration in Chapter 4.

---

### Problem 3.7

The number of atoms of a radioactive substance at time $t$ is given by

$$N(t) = N_o e^{-t/\tau},$$

where $N_o$ is the initial number of atoms and $\tau$ is the relaxation time. For $^{14}C$, $\tau = 8320$ years. Calculate the fraction of an initial sample of $^{14}C$ that remains after 10.0 years, using Eq. (3.15). Calculate the correct fraction and compare it with your first answer.

---

## SECTION 3.4. SOME USEFUL FACTS ABOUT DERIVATIVES

In this section we present some useful theorems about derivatives, which, together with the formulas for the derivatives of simple functions presented in Table 3.1, will enable you to obtain the derivative of almost any function that you will encounter in physical chemistry.

### The Derivative of a Product of Two Functions

If $y$ and $z$ are both functions of $x$,

$$\boxed{\frac{d(yz)}{dx} = y\frac{dz}{dx} + z\frac{dy}{dx}.} \tag{3.19}$$

## The Derivative of the Sum of Two Functions

If $y$ and $z$ are both functions of $x$,

$$\frac{d(y+z)}{dx} = \frac{dy}{dx} + \frac{dz}{dx}.$$

(3.20)

## The Derivative of the Difference of Two Functions

If $y$ and $z$ are both functions of $x$,

$$\frac{d(y-z)}{dx} = \frac{dy}{dx} - \frac{dz}{dx}.$$

(3.21)

## The Derivative of the Quotient of Two Functions

If $y$ and $z$ are both functions of $x$, and $z$ is not zero,

$$\frac{d(y/z)}{dx} = \frac{x(dy/dx) - y(dz/dx)}{z^2}.$$

(3.22a)

An equivalent result can be obtained by considering $y/z$ to be a product of $1/z$ and $y$ and using Eq. (3.19),

$$\frac{d}{dx}\left(y\frac{1}{z}\right) = \frac{1}{z}\frac{dy}{dx} - y\frac{1}{z^2}\frac{dz}{dx}.$$

(3.22b)

Many people think that Eq. (3.22b) is more convenient to use than Eq. (3.22a).

## The Derivative of a Constant

If $c$ is a constant,

$$\frac{dc}{dx} = 0.$$

(3.23)

A constant is a linear function of any variable, with zero slope. From this follows the important but simple fact

$$\frac{d(y+c)}{dx} = \frac{dy}{dx}.$$

(3.24)

If we add any constant to a function, we do not change its derivative.

## The Derivative of a Function Times a Constant

If $y$ is a function of $x$ and $c$ is a constant,

$$\boxed{\frac{d(cy)}{dx} = c\frac{dy}{dx}}. \tag{3.25}$$

This can be deduced by substituting into the definition of the derivative, or by using Eqs. (3.19) and (3.23).

## The Derivative of a Function of a Function (the Chain Rule)

If $u$ is a differentiable function of $x$, and $f$ is a differentiable function of $u$,

$$\boxed{\frac{df}{dx} = \frac{df}{du}\frac{du}{dx}}. \tag{3.26}$$

The function $f$ is sometimes referred to as a *composite function*. It is a function of $x$, because if a value of $x$ produces a value of $u$, this value of $u$ produces a value of $f$. This is communicated by the notation

$$f(x) = f[u(x)]. \tag{3.27}$$

Here we have used the same letter for the function $f$ whether it is expressed as a function of $u$ or of $x$. In the two cases, it would represent two different mathematical formulas which would give the same numerical value.

We now illustrate how these facts about derivatives can be used to obtain formulas for the derivatives of various functions.

---

### Example 3.7

Find the derivative of $\tan(ax)$ by using the formulas for the derivatives of the sine and cosine.

---

**Solution**

$$\frac{d}{dx}\tan(ax) = \frac{d}{dx}\left[\frac{\sin(ax)}{\cos(ax)}\right] = \frac{\cos(ax)a\cos(ax) + \sin(ax)a\sin(ax)}{\cos^2(ax)}$$

$$= a\left[\frac{\cos^2(ax) + \sin^2(ax)}{\cos^2(ax)}\right] = \frac{a}{\cos^2(ax)}$$

$$= a\sec^2(ax).$$

We have used several trigonometric identities in the solution.

### Example 3.8

Find $dP/dT$ if $P(T) = ke^{-Q/T}$.

**Solution**

Let $u = -Q/T$. From the chain rule,

$$\frac{dP}{dT} = \frac{dP}{du}\frac{du}{dT}$$

$$= ke^u \frac{Q}{T^2} = ke^{-Q/T}\left(\frac{Q}{T}\right).$$

**Problem 3.8**

Find the following derivatives. All letters stand for constants except for the dependent and independent variables indicated.

a.  $dy/dx$, where $y = (ax^2 + bx + c)^{-3/2}$
b.  $d\ln(P)/dT$, where $P = ke^{-Q/T}$
c.  $dy/dx$, where $y = a\cos(bx^3)$
d.  $d(yz)/dx$, where $y = ax^2$, $z = \sin(bx)$
e.  $dP/dV$, where $P = nRT/(V - nb) - an^2/V^2$
f.  $d\eta/d\lambda$, where $\eta = 2\pi hc^2/\lambda^5(e^{hc/\lambda kT} - 1)$.

## SECTION 3.5. HIGHER-ORDER DERIVATIVES

Since the derivative of a function is itself a function, a derivative is usually differentiable. The derivative of a derivative is called a *second derivative*, and the derivative of a second derivative is called a *third derivative*, etc. We use the notation

$$\frac{d^2y}{dx^2} = \frac{d}{dx}\left(\frac{dy}{dx}\right) \tag{3.28}$$

and

$$\frac{d^3y}{dx^3} = \frac{d}{dx}\left(\frac{d^2y}{dx^2}\right), \tag{3.29}$$

etc. The $n$th-order derivative is

$$\frac{d^ny}{dx^n} = \frac{d}{dx}\left(\frac{d^{n-1}y}{dx^{n-1}}\right). \tag{3.30}$$

The notation $y''(x)$ is sometimes used for the second derivative. The third and higher order derivatives are sometimes denoted by a lower-case Roman numeral superscript in parentheses, as $y^{(iii)}$, $y^{(iv)}$, etc.

**Example 3.9**

Find $d^2y/dx^2$ if $y = a\sin(bx)$.

**Solution**

$$\frac{d^2y}{dx^2} = \frac{d}{dx}[ab\cos(bx)] = -ab^2\sin(bx).$$

The result of Example 3.9 is sometimes useful: the sine is proportional to the negative of its second derivative. The cosine has the same behavior. The exponential function is proportional to all of its derivatives.

**Problem 3.9**

Find the second and third derivatives of the following functions. All letters stand for constants except for the indicated dependent and independent variables.

a. $y = y(x) = ax^n$
b. $y = y(x) = ae^{bx}$
c. $v_{rms} = v_{rms}(T) = \sqrt{3RT/M}$
d. $P = P(V) = nRT/(V - nb) - an^2/V^2$
e. $\eta = \eta(\lambda) = 2\pi hc^2/\lambda^5(e^{hc/\lambda kT} - 1).$

## The Curvature of a Function

Figure 3.8 shows a rough graph of a function, a rough graph of the first derivative of the function, and a rough graph of the second derivative of the function in the

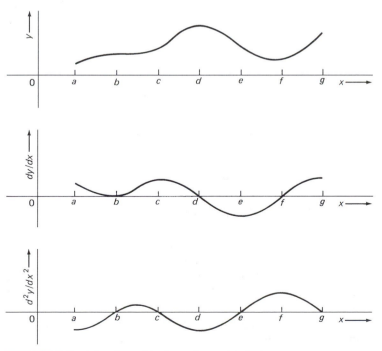

**FIGURE 3.8**   A function and its first and second derivative.

interval $a < x < g$. Where the function is concave downward, the second derivative is negative, and where the function is concave upward, the second derivative is positive. Where the graph of the function is more sharply curved, the magnitude of the second derivative is larger. The second derivative provides a measure of the curvature of the function curve.

For any function that possesses a second derivative, the *curvature K* is defined by

$$K = \frac{d^2y/dx^2}{[1 + (dy/dx)^2]^{3/2}}. \tag{3.31}$$

The magnitude of the curvature is equal to the reciprocal of the radius of the circle which fits the curve at that point.

Since the denominator in Eq. (3.31) is always positive, the curvature has the same sign as the second derivative and is therefore positive when the curve is concave upward and negative when it is concave downward.

---

**Example 3.10**

Find the curvature of the function $y = x^2$ at $x = 0$ and at $x = 2$.

---

**Solution**

$$\frac{d^2y}{dx^2} = \frac{d}{dy}(2x) = 2$$

$$K = \frac{2}{(1 + 4x^2)^{3/2}} = \begin{cases} 2 & \text{at } x = 0 \\ 0.0285 & \text{at } x = 2. \end{cases}$$

---

**Problem 3.10**

a. Find the curvature of the function $y = \sin(x)$ at $x = 0$ and at $x = \pi/2$.
b. Find a formula for the curvature of the function

$$P(V) = \frac{nRT}{V - nb} - \frac{an^2}{V^2},$$

where $n$, $R$, $a$, $b$, and $T$ are treated as constants.

---

## SECTION 3.6. MAXIMUM–MINIMUM PROBLEMS

Sometimes it is necessary to find the largest or smallest value that a function attains or approaches in a certain interval, or to find the value of the independent variable at which this happens. For example, in Fig. 3.8 the maximum value of the function in the interval shown is at $x = d$ and the minimum value of the function in the interval is at the left end of the interval.

We now state the fact that enables us to find maximum and minimum values of a function: *The maximum or minimum value of a differentiable function in an interval will either occur at an end of the interval or at a place where the first derivative of the function vanishes.*

The minimum value of a function means the most negative value, not necessarily the smallest magnitude. The maximum value means the most positive value, not necessarily the largest magnitude. If a function happens to be negative in an entire interval, the maximum will correspond to the smallest magnitude and the minimum to the largest magnitude.

In Fig. 3.8, there are three points of horizontal tangent (where the first derivative vanishes). At $x = f$ we have a *relative minimum* or *local minimum*. At such a point the function has a smaller value than at any other point in the immediate vicinity. At a local maximum, the function has a larger value than at any other point in the immediate vicinity. However, the maximum value occurs at $x = d$.

There is a third point at $x = b$ at which the first derivative vanishes. This is an inflection point with a horizontal tangent line. Relative minima, relative maxima, and such inflection points can be distinguished from each other by finding the sign of the curvature from the second derivative. At a relative minimum, the second derivative is positive. At a relative maximum, the second derivative is negative. At an inflection point with horizontal tangent line, the second derivative vanishes.

These facts suggest the following procedure for finding the maximum and minimum value of a function in an interval:

1. Find all the points in the interval at which the first derivative vanishes.
2. Evaluate the function at these points and at the ends of the interval. The largest value in the list is the maximum value and the smallest value is the minimum.
3. If you want to find whether a point at which the first derivative vanishes is a relative maximum, a relative minimum, or an inflection point, evaluate the second derivative and use the fact that the second derivative is positive at a minimum, negative at a maximum, and zero at an inflection point with horizontal tangent line.

---

### Example 3.11

Find the maximum and minimum values of the function

$$y = x^2 - 4x + 6$$

in the interval $0 < x < 5$.

---

### Solution

The derivative is

$$\frac{dy}{dx} = 2x - 4.$$

Let the value of $x$ which satisfies the equation $dy/dx = 0$ be called $x_m$.

$$2x_m - 4 = 0 \qquad \text{or} \qquad x_m = 2.$$

We evaluate the function at the ends of the interval and at $x = x_m$:

$$y(0) = 6$$
$$y(x_m) = y(2) = 2$$
$$y(5) = 11.$$

The maximum value of the function is at $x = 5$, the end of the interval. The minimum value is at $x = 2$.

---

### Problem 3.11

a. For the interval $-10 < x < 10$, find the maximum and minimum values of

$$y = -x^3 + 3x^2 - 3x + 8.$$

b. The probability that a molecule in a gas will have a speed $v$ is proportional to the function

$$f_v(v) = 4\pi \left( \frac{m}{2\pi k_B T} \right)^{3/2} v^2 \exp\left( \frac{-v^2}{2mk_B T} \right),$$

where $m$ is the mass of the molecule, $k_B$ is Boltzmann's constant, and $T$ is the temperature on the Kelvin scale. The most probable speed is the speed for which this function is at a maximum. Find the expression for the most probable speed and find its value for $N_2$ molecules at $T = 298$ K.

c. According to the Planck theory of black-body radiation, the radiant spectral emittance is given by the formula

$$\eta = \eta(\lambda) = \frac{2\pi hc^2}{\lambda^5 (e^{hc/\lambda kT} - 1)},$$

where $h$ is Planck's constant, $k_B$ is Boltzmann's constant, $c$ is the speed of light, and $T$ is the tempearture on the Kelvin scale. Treat $T$ as a constant and find an equation which will give the wavelength of maximum emittance. Do not try to solve the equation. We will discuss the solution of such an equation in Chapter 9.

---

## SECTION 3.7. LIMITING VALUES OF FUNCTIONS: L'HÔPITAL'S RULE

We have already mentioned mathematical limits. The limit of $y$ as $x$ approaches $a$ is denoted by

$$\lim_{x \to a} [y(x)] \qquad (3.32)$$

and is defined as the number that $y$ approaches ever more closely as $x$ approaches ever more closely to $a$, if such a number exists. The number $a$ is not required to be finite, and a limit such as

$$\lim_{x \to \infty} [y(x)] \qquad (3.33)$$

exists if $y(x)$ approaches more closely to some number as $x$ is made larger and larger without bound.

An example of a limit that does not exist is

$$\lim_{x \to a} \left( \frac{1}{x-a} \right). \tag{3.34}$$

As $x$ approaches closer to $a$, $1/(x-a)$ becomes larger without bound if $x$ approaches $a$ from the right (from values larger than $a$) and $1/(x-a)$ becomes more negative without bound if $x$ approaches $a$ from the left.

Another limit that does not exist is

$$\lim_{x \to \infty} [\sin(x)]. \tag{3.35}$$

The sine function continues to oscillate between $-1$ and $1$ as $x$ becomes larger and larger. However, a limit that does exist as $x$ approaches infinity is

$$\lim_{x \to \infty} (1 - e^{-x}) = 1. \tag{3.36}$$

---

**Problem 3.12**

Decide which of the following limits exist and find the values of those that do exist.

   a. $\lim\limits_{x \to 0} (1 - e^{x})$

   b. $\lim\limits_{x \to \infty} (e^{-x^2})$

   c. $\lim\limits_{x \to \pi/2} [x \tan(x)]$

   d. $\lim\limits_{x \to 0} [\ln(x)]$.

---

Sometimes a limit exists but cannot be evaluated in a straightforward way by substituting into the expression the limiting value of the independent variable. For example, if we try to determine the limit

$$\lim_{x \to 0} \left[ \frac{\sin(x)}{x} \right] \tag{3.37}$$

we find that for $x = 0$ both the numerator and demoninator of the expression vanish. If an expression appears to approach $0/0$, it might approach 0, it might approach a finite constant of either sign, or it might diverge in either direction (approach $-\infty$ or $+\infty$). The same is true if it appears to approach $\infty/\infty$ or $0 \times \infty$.

The *rule of l'Hôpital* provides a way to determine the limit in such cases. This rule can be stated: *If the numerator and denominator of a quotient both approach zero or both approach infinity in some limit, the limit of the quotient is equal to the limit of the quotient of the derivatives of the numerator and denominator, if this limit exists.*

That is, if the limits exist, then

$$\lim_{x \to a} \left[ \frac{f(x)}{g(x)} \right] = \lim_{x \to a} \left[ \frac{df/dx}{dg/dx} \right]. \tag{3.38}$$

---

**Example 3.12**

Find the value of the limit in Eq. (3.37) by use of l'Hôpital's rule.

**Solution**

$$\lim_{x \to 0} \left[ \frac{\sin(x)}{x} \right] = \lim_{x \to 0} \left[ \frac{d \sin(x)/(dx)}{dx/dx} \right] = \lim_{x \to 0} \left[ \frac{\cos(x)}{1} \right] = 1. \qquad (3.39)$$

l'Hôpital's rule does not necessarily give the correct limit if it is applied to a case in which the limit does not appear to approach $0/0$ or $\infty/\infty$ or $0 \times \infty$. One author put it, "As a rule of thumb, l'Hôpital's rule applies when you need it, and not when you do not need it."[1]

If the expression appears to approach $0 \times \infty$, it can be put into a form that appears to approach $0/0$ or $\infty/\infty$ by using the expression for the reciprocal of one factor. In the following example, we use this technique, as well as illustrating the fact that sometimes the rule must be applied more than once in order to find the value of the limit.

**Example 3.13**

Find the limit

$$\lim_{x \to \infty} (x^3 e^{-x}).$$

**Solution**

$$\lim_{x \to \infty} (x^3 e^{-x}) = \lim_{x \to \infty} \left( \frac{x^3}{e^x} \right) = \lim_{x \to \infty} \left( \frac{3x^2}{e^x} \right) = \lim_{x \to \infty} \left( \frac{6x}{e^x} \right) = \lim_{x \to \infty} \left( \frac{6}{e^x} \right) = 0. \qquad (3.40)$$

By applying l'Hôpital's rule $n$ times, we can show

$$\lim_{x \to \infty} (x^n e^{-x}) = 0 \qquad (3.41)$$

for any finite value of $n$. This is a result that is sometimes useful. The exponential function overwhelms any finite power of $x$ in the limit that $x$ becomes large.

**Problem 3.13**

Investigate the limit

$$\lim_{x \to \infty} (x^{-n} e^x)$$

for any finite value of $n$.

Another interesting limit is that of the next example.

[1] Sherman K. Stein, *Calculus and Analytic Geometry*, 2nd ed., p. 239, McGraw–Hill, New York, 1977.

**Example 3.14**

Find the limit

$$\lim_{x \to \infty} \left( \frac{\ln(x)}{x} \right).$$

**Solution**

$$\lim_{x \to \infty} \left( \frac{\ln(x)}{x} \right) = \lim_{x \to \infty} \left( \frac{1/x}{1} \right) = 0. \tag{3.42}$$

**Problem 3.14**

Find the limit

$$\lim_{x \to \infty} \left[ \frac{\ln(x)}{\sqrt{x}} \right].$$

**Problem 3.15**

A collection of $N$ harmonic oscillators at thermal equilibrium at absolute temperature $T$ is shown by statistical mechanics to have the thermodynamic energy

$$U = \frac{Nh\nu}{e^{h\nu/k_B T} - 1}, \tag{3.43}$$

where $k_B$ is Boltzmann's constant, $h$ is Planck's constant, and $\nu$ is the vibrational frequency.

a. Find the limit of $U$ as $\nu \to 0$.
b. Find the limit of $U$ as $T \to 0$.

There are a number of applications of limits in physical chemistry, and l'Hôpital's rule is useful in some of them. There is an article by Missen which lists some of these applications.[2]

**Problem 3.16**

Draw a rough graph of the function

$$y = \frac{\sin(x)}{x}$$

in the interval $-\pi < x < \pi$. Locate the zeros of the function accurately, but do not attempt to locate the relative extrema exactly. Use l'Hôpital's rule to evaluate the function at $x = 0$.

---

[2]Ronald W. Missen, "Applications of the l'Hôpital–Bernoulli Rule in Chemical Systems," *J. Chem. Educ.* **54**, 448 (1977).

## SUMMARY OF THE CHAPTER

A function is a rule which provides a value for a dependent variable for any given value of a dependent variable. The derivative of a function is another function of the same independent variable, which specifies the rate of change of the first function with respect to the independent variable.

We have discussed the definition of the derivative and the geometrical interpretation of the derivative as the slope of the tangent line to a curve representing the function. We have also presented the use of the derivative in calculating approximate changes in a dependent variable and its use in locating maxima and minima in a function.

## ADDITIONAL READING

James Newton Butler and Daniel Gureasko Bobrow, *The Calculus of Chemistry*, Benjamin, New York, 1965. This book is an elementary introduction to calculus aimed particularly at students of chemistry.

Daniel A. Greenberg, *Mathematics for Introductory Science Courses*, Benjamin, New York, 1965. This book is also an elementary introduction to calculus.

Clifford E. Swartz, *Used Math for the First Two Years of College Science*, Prentice–Hall, Englewood Cliffs, NJ, 1973. This book is a survey of various mathematical topics at the beginning college level.

Sherman K. Stein, *Calculus and Analytic Geometry*, 2nd ed., McGraw–Hill, New York, 1977. This is one of a number of standard introductory calculus textbooks. You can read about mathematical functions and differential calculus in this or whatever introductory calculus textbook you have access to.

## ADDITIONAL PROBLEMS

### 3.17

Using the definition of the derivative and Eq. (20) of Appendix 6, show that

$$\frac{d\ln(x)}{dx} = \frac{1}{x}.$$

### 3.18

Find the first and second derivatives of the following functions.

a. $P = P(V) = RT(1/V + B/V^2 + C/V^3)$
b. $G = G(x) = G^0 + RTx\ln(x) + RT(1-x)\ln(1-x)$, where $G^0$, $R$, and $T$ are constants
c. $y = a\ln(x^{1/3})$, where $a$ is a constant
d. $y = 3x^3\ln(x)$
e. $y = 1/(c - x^2)$, where $c$ is a constant
f. $y = \ln[\tan(2x)]$
g. $y = (1/x)(1/(1+x))$

h. $f = f(v) = ce^{-mv^2/(2kt)}$ where $m, c, k$, and $T$ are constants

i. $y = 3 \sin^2(2x) = 3 \sin(2x)^2$

j. $y = a_0 + a_1 x + a_2 x^2 + a_3 x^3 + a_4 x^4 + a_5 x^5$, where $a_0, a_1$, etc., are constants.

## 3.19

Find the following derivatives and evaluate them at the points indicated.

a. $(dy/dx)_{x=0}$ if $y = \sin(bx)$, where $b$ is a constant

b. $(df/dt)_{t=0}$ if $f = Ae^{-kt}$, where $A$ and $k$ are constants

c. $(dy/dx)_{x=1}$, if $y = (ax + bx^2)^{-1/2}$ , where $a$ and $b$ are constants

d. $(d^2 y^2/dx)_{x=0}$, if $y = ae^{-bx}$, where $a$ and $b$ are constants.

## 3.20

The volume of a cube is given by

$$V = V(a) = a^3,$$

where $a$ is the length of a side. Estimate the error in the volume if a 1% error is made in measuring the length, using the formula

$$\Delta V \approx \left(\frac{dV}{da}\right) \Delta a.$$

Check the accuracy of this estimate by computing $V(a_0)$ and $V(a_0 + 0.01a_0)$.

## 3.21

a. Draw a rough graph of the function $y = y(x) = e^{-|x|}$.

b. Is the function differentiable at $x = 0$?

c. Draw a rough graph of the derivative of the function.

## 3.22

Show that the function $\psi = \psi(x) = A \sin(kx)$ satisfies the equation

$$\frac{d^2 \psi}{dx^2} = -k^2 \psi$$

if $A$ and $k$ are constants. This is an example of a differential equation.

## 3.23

Draw rough graphs of the third and fourth derivatives of the function whose graph is given in Fig. 3.8.

## 3.24

The Gibbs energy of a mixture of two enantiomorphs (optical isomers of the same substance) is given by

$$G = G(x) = G^0 + RTx\ln(x) + RT(x_0 - x)\ln(x_0 - x),$$

where $x_0$ is the sum of the concentrations of the enantiomorphs and $x$ is the concentration of one of them. $G^0$ is a constant, $R$ is the gas constant, and $T$ is the temperature.

If the temperature is maintained constant, what is the concentration of each enantiomorph when $G$ has its minimum value? What is the maximum value of $G$ in the interval $0 \leq x \leq x_0$?

**3.25**

a. A rancher wants to enclose a rectangular part of a large pasture so that 1 km² is enclosed with the minimum amount of fence. Find the dimensions of the rectangle that he should choose. The area is

$$A = xy$$

but $A$ is fixed at 1 km², so that $y = A/x$.

b. The rancher now decides that the fenced area must lie along a road and finds that the fence costs \$10 per meter along the road and \$6 per meter along the other edges. Find the dimensions of the rectangle that would minimize the cost of the fence.

**3.26**

Using

$$\Delta y \approx \left( \frac{dy}{dx} \right) \Delta x$$

show that

$$e^{\Delta x} - 1 \approx \Delta x \qquad \text{if } \Delta x \ll 1.$$

**3.27**

The sum of two nonnegative numbers is 100. Find their values if their product plus twice the square of the first is to be a maximum.

**3.28**

A cylindrical tank in a chemical factory is to contain $1.000$ m³ of a corrosive liquid. Because of the cost of the material, it is desirable to minimize the area of the tank. Find the optimum radius and height and find the resulting area.

**3.29**

Find the following limits.

a. $\lim_{x \to \infty} [ln(x)/x^2]$

b. $\lim_{x \to 3}((x^3 - 27)/(x^2 - 9))$

c. $\lim_{x \to 0^+} [sin(x)ln(x)]$

d. $\lim_{x \to \infty} (e^{-x^2}/e^{-x})$

e. $\lim_{x \to 0}[x^2/(1 - \cos(2x))]$

f. $\lim_{x \to \pi} [sin(x)/sin(3x/2)]$.

### 3.30

If a hydrogen atom is in a 2s state, the probability of finding the electron at a distance r from the nucleus is proportional to $4\pi r^2 \psi_{2s}^2$ where

$$\psi_{2s} = \frac{1}{4\sqrt{2\pi}} \left(\frac{1}{a_0}\right)^{3/2} \left(2 - \frac{r}{a_0}\right) e^{-r/a_0},$$

where $a_0$ is a constant, the Bohr radius, equal to $0.529 \times 10^{-10}$ m.

    a. Draw a rough graph of $\psi_{2s}$ locating maxima and minima correctly.
    b. Draw a rough graph of $4\pi r^2 \psi_{2s}^2$, locating maxima and minima correctly.

### 3.31

The thermodynamic energy of a collection of $N$ harmonic oscillators (approximate representations of molecular vibrations) is given in Problem 3.15.

    a. Draw a rough sketch of the thermodynamic energy as a function of $T$.
    b. The heat capacity of this system is given by

$$C = \frac{dU}{dT}.$$

Show that the heat capacity is given by

$$C = Nk_B \left(\frac{h\nu}{k_B T}\right)^2 \frac{e^{h\nu/k_B T}}{\left(e^{h\nu/k_B T} - 1\right)^2}.$$

    c. Find the limit of the heat capacity as $T \to 0$ and as $T \to \infty$. Note that the limit as $T \to \infty$ is the same as the limit $\nu \to 0$.
    d. Draw a graph of $C$ as a function of $T$.

### 3.32

Find the relative maximum and minimum of the function $f(x) = x^3 + 3x^2 - 10x$.

### 3.33

The van der Waals equation of state is given in Problem 2.2. When the temperature of a given gas is equal to its critical temperature, the gas has a state at which the pressure as a function of $V$ at constant $T$ and $n$ exhibits an inflection point at which $dP/dV = 0$ and $d^2P/dV^2 = 0$. This inflection point is called the *critical point* of the gas. Write $P$ as a function of $T$, $V$, and $n$ and write expressions for $dP/dV$ and $d^2P/dV^2$, treating $T$ and $n$ as constants. Set these two expressions equal to zero and solve the simultaneous equations to find an expression for the pressure at the critical point.

# 4

# INTEGRAL CALCULUS

## Preview

In this chapter we survey the parts of integral calculus that are most useful in physical chemistry. We first discuss the antiderivative, a function that possesses a given derivative. We then define integration as the limit of a summation process and discuss the process of constructing a finite increment in a function from knowledge of its derivative but without necessarily knowing in advance what the function is. We discuss the role of the antiderivative as an indefinite integral and the use of tables of indefinite and definite integrals. We discuss several methods of working out integrals without the use of a table. Finally, we discuss the use of integration to find mean values with a probability distribution.

## Principal Facts and Ideas

1. The antiderivative $F(x)$ of a function $f(x)$ is the function such that $dF/dx = f(x)$.
2. A definite integral is the limit of a sum of terms $f(x)\Delta x$ in the limit that $\Delta x$ approaches zero, where $f(x)$ is the integrand function.
3. An indefinite integral is the same thing as the antiderivative function.
4. A definite integral equals the indefinite integral evaluated at the upper limit minus the indefinite integral evaluated at the lower limit:

$$\int_a^b f(x)\,dx = F(b) - F(a).$$

5. An improper integral has at least one infinite limit or has an integrand function that is infinite somewhere in the interval of integration. If an improper integral does not approach a finite limit as the limit of integration approaches infinity or a point at which the integrand becomes infinite, it is said to diverge.

6. Analytical methods exist to transform an integral into a more easily computed form.

7. Numerical methods exist to compute accurate approximations to integrals that cannot be analytically performed.

8. A mean value over a continuously distributed variable can be computed as an integral using a probability distribution.

## Objectives

After studying this chapter, you should be able to:

1. obtain the indefinite integral of an integrand function using a table;
2. calculate a definite integral using the indefinite integral and understand its role as an increment in the antiderivative function;
3. understand the relationship of a definite integral to an area in a graph of the integrand function;
4. obtain an approximate value for a definite integral using numerical analysis;
5. manipulate integrals into tractable forms by use of partial integration, the method of substitution, and the method of partial fractions;
6. calculate mean values of quantities using a probability distribution with one random variable.

## SECTION 4.1. THE ANTIDERIVATIVE OF A FUNCTION

In Chapter 3, we discussed the derivative of a function. We now review a little of that discussion, using a particular example. If a particle moves vertically, we can express its position as a function of time by

$$z = z(t), \tag{4.1}$$

where we use the coordinate $z$ to measure the height above a chosen origin. The velocity is the derivative of the position vector with respect to time, so that the $z$ component of the velocity is

$$v_z = v_z(t) = \frac{dz}{dt}. \tag{4.2}$$

The acceleration is the derivative of the velocity with respect to time, or the second derivative of the position vector, so that the $z$ component of the acceleration is

$$a_z = \frac{dv_z}{dt} = \frac{d^2z}{dt^2}. \tag{4.3}$$

**Example 4.1**

A particle falling in a vacuum near the surface of the earth has a position given by

$$z = z(t) = z(0) + v_z(0)t - \frac{gt^2}{2}, \qquad (4.4)$$

where $z(0)$ is the position at $t = 0$, $v_z(0)$ is the velocity at $t = 0$, and $g$ is the *acceleration due to gravity*,[1] equal to $9.80 \, \text{m s}^{-2}$. Find the velocity and the acceleration.

**Solution**

$$v_z(t) = \frac{dz}{dt} = v_z(0) - gt \qquad (4.5)$$

$$a_z(t) = \frac{d^2z}{dt^2} = -g. \qquad (4.6)$$

Now consider the reverse problem from that of Example 4.1. If we are given the acceleration as a function of time, how do we find the velocity? If we are given the velocity as a function of time, how do we find the position? In the following example, we see that the answer to these questions involves the *antiderivative function*, which is a function that possesses a particular derivative.

**Example 4.2**

Given that a particle moves in the $z$ direction and that its acceleration is $a_z = -g$, find its velocity and position.

**Solution**

We know that $-gt$ is a function that possesses $-g$ as its derivative, so that one possibility for the velocity is

$$v_z(t) = -gt.$$

However, the derivative of any constant is zero, so that the most general possibility is

$$v_z(t) = v_z(0) - gt, \qquad (4.7)$$

where $v_z(0)$ is a constant. Equation (4.7) represents a family of functions, one for each value of this constant. Every function in the family is an antiderivative of the given acceleration.

To find the position, we seek a function that has $v_z(0) - gt$ as its derivative. We know that

$$\frac{d}{dx}(ax^2) = 2ax \qquad \text{and} \qquad \frac{d}{dx}(ax) = a \qquad (4.8)$$

---

[1] The acceleration due to gravity varies slightly with latitude. This value holds for the latitude of Washington, DC, or San Francisco.

so that the second family of antiderivative functions that we need is

$$z = z(t) = z(0) + v_z(0)t - \frac{gt^2}{2}, \tag{4.9}$$

where $z(0)$ is again a constant. Equation (4.9) represents a family of functions that includes all of the functions that give the position $z$ as a function of time such that $-g$ is the acceleration in the $z$ direction. Since we went through two stages of antiderivatives, there are two arbitrary constants, $z(0)$ and $v_z(0)$, that would be assigned values to correspond to any particular case, as in the following example:

---

### Example 4.3

Find the expression for the velocity of a particle falling near the surface of the earth in a vacuum, given that the velocity at $t = 1.000$ s is $10.00\,\mathrm{m\,s^{-1}}$.

---

### Solution

The necessary family of functions is given by Eq. (4.7), with $v_z(0) = 19.80\,\mathrm{m\,s^{-1}}$:

$$v_z(t) = 19.80\,\mathrm{m\,s^{-1}} - (9.80\,\mathrm{m\,s^{-2}})t. \tag{4.10}$$

If the initial position had been given, $z(0)$ would be determined.

---

### Example 4.4

Find the antiderivative of

$$f(x) = a\sin(bx),$$

where $a$ and $b$ are constants.

---

### Solution

The antiderivative function is, from Table 3.1,

$$F(x) = -\frac{a}{b}\cos(bx) + c, \tag{4.11}$$

where $c$ is an arbitrary constant. You can differentiate to verify that

$$dF/dx = f(x). \tag{4.12}$$

---

### Problem 4.1

Find the family of functions whose derivative is $ae^{bx}$.

---

### Problem 4.2

Find the function whose derivative is $-10e^{-5x}$ and whose value at $x = 0$ is 20.

---

## SECTION 4.2. THE PROCESS OF INTEGRATION

Consider the general version of the problem of finding a function that possesses a certain derivative. Say that we have a function $f = f(x)$, and we want to find its antiderivative function, which we call $F(x)$. That is,

$$\frac{dF}{dx} = f(x). \tag{4.13}$$

We first describe a process of finding an increment in the function, that is, the value of $F(x_1) - F(x_0)$, where $x_0$ and $x_1$ are two values of $x$. Assume that we know the value of $F$ at $x = x_0$. The increment in the function will be called a *definite integral*.

In Section 3.4, we discussed the approximate calculation of an increment in a function, using the slope of the tangent line, which is equal to the derivative of the function. Equation (3.15) is

$$\Delta F = F(x_1) - F(x_0) \approx \left(\frac{dF}{dx}\right)_{x=x_0} \Delta x, \tag{4.14}$$

where $\Delta x = x_1 - x_0$. This approximation becomes more and more nearly exact as $\Delta x$ is made smaller.

If we want the value of $F$ at some point that is not close to $x = x_0$, we can get a better approximation by repeating the use of Eq. (4.14) several times. Say that we want the value of $F$ at $x = x'$. We divide the interval $(x_0, x')$ into $n$ equal subintervals. This is shown in Fig. 4.1, with $n$ equal to 3. We now write

$$F(x') - F(x_0) \approx \left(\frac{dF}{dx}\right)_{x=x_0} \Delta x + \left(\frac{dF}{dx}\right)_{x=x_1} \Delta x$$
$$+ \left(\frac{dF}{dx}\right)_{x=x_2} \Delta x + \cdots + \left(\frac{dF}{dx}\right)_{x=x_{n-1}} \Delta x, \tag{4.15}$$

where $\Delta x$ is the length of each subinterval:

$$\Delta x = x_1 - x_0 = x_2 - x_1 = x_3 - x_2 = \cdots = x' - x_{n-1}. \tag{4.16}$$

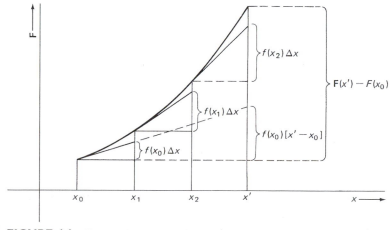

**FIGURE 4.1**    Figure to illustrate Eq. (4.15).

This approximation to $F(x') - F(x_0)$ is generally a better approximation than the one obtained by multiplying the slope at the beginning of the interval by the length of the whole interval, as you can see in Fig. 4.1. In fact, if we make $n$ fairly large, we can make the approximation nearly exact.

Let us now rewrite Eq. (4.15), using the symbol $f = f(x)$ instead of $dF/dx$,

$$F(x') - F(x_0) \approx f(x_0)\Delta x + f(x_1)\Delta x + f(x_2)\Delta x + \cdots + f(x_{n-1})\Delta x \quad (4.17)$$

$$\approx \sum_{k=0}^{n-1} f(x_k)\Delta x. \quad (4.18)$$

In Eq. (4.18), we have introduced the standard notation for a sum, a capital Greek sigma ($\Sigma$). The letter $k$ is called the *summation index*. Its initial value is given under the capital sigma and its final value is given above it. Unless otherwise stated, $k$ is incremented by unity for each term. There are $n$ terms indicated in this sum.

We can use Eq. (4.16) to write

$$x_k = x_0 + k\Delta x \quad (4.19)$$

and use this in Eq. (4.18):

$$F(x') - F(x_0) \approx \sum_{k=0}^{n-1} f(x_0 + k\Delta x)\Delta x. \quad (4.20)$$

We now make Eq. (4.20) into an exact equation by taking the limit as $n$ becomes larger and larger without bound, meanwhile making $\Delta x$ smaller and smaller so that $n\,\Delta x$ remains fixed and equal to $x' - x_0$.

$$F(x') - F(x_0) = \lim_{\substack{\Delta x \to 0 \\ n \to \infty \\ (n\Delta x = x' - x_0)}} \left[ \sum_{k=0}^{n-1} f(x_0 + k\Delta x)\Delta x \right]. \quad (4.21)$$

The limiting value of the right-hand side of Eq. (4.21) is called a *definite integral*. The function $f(x)$ is called the *integrand function*. The notation in this equation is cumbersome, so another symbol is used:

$$\lim_{\substack{\Delta x \to 0 \\ n \to \infty \\ (n\Delta x = x' - x_0)}} \left[ \sum_{k=0}^{n-1} f(x_0 + k\Delta x)\Delta x \right] = \int_{x_0}^{x'} f(x)\,dx. \quad (4.22)$$

The *integral sign* on the right-hand side of Eq. (4.22) is a stretched-out letter S, standing for a sum. However, an integral is not an ordinary sum. It is the limit that a sum approaches as the number of terms in the sum becomes infinite in a particular way. The value $x_0$ at the bottom of the integral sign is called the *lower limit of integration*, and the value $x'$ at the top is called the *upper limit of integration*. Equation (4.21) is now

$$\boxed{F(x') - F(x_0) = \int_{x_0}^{x'} f(x)\,dx}. \quad (4.23)$$

The finite increment $F(x) - F(x_0)$ is equal to the sum of infinitely many infinitesimal increments, each given by the differential $dF = (dF/dx)\,dx = f(x)\,dx$ evaluated at the appropriate value of $x$. The integral on the right-hand side of Eq. (4.23) is called a *definite integral*, because the limits of integration are definite values. Equation (4.23) is often called the *fundamental theorem of integral calculus*. It is equivalent to saying that a finite increment in $F$ is constructed by adding up infinitely many infinitesimal increments.

Equation (4.23) is a very important equation. In many applications of calculus to physical chemistry, we will be faced with an integral equivalent to the right-hand side of this equation. If we can by inspection or by use of a table find the function $F$ which possesses the function $f$ as its derivative, we can evaluate the function $F$ at the two limits of integration and take the difference to obtain the value of the integral. Because of this, the antiderivative function $F$ is called the *indefinite integral* of the integrand function $f$.

---

**Example 4.5**

Find the value of the definite integral

$$\int_0^\pi \sin(x)\, dx.$$

---

**Solution**

The indefinite integral of $\sin(x)$ is

$$-\cos(x) + C = F(x),$$

where $C$ is an arbitrary constant. The integral is the difference between the value of the antiderivative function at the upper limit and at the lower limit:

$$I = F(\pi) - F(0) = -\cos(\pi) + C - [-\cos(0) + C]$$
$$= -\cos(\pi) + \cos(0) = -(-1) + 1 = 2.$$

The constant $C$ cancels because it occurs with the same value in both occurrences of the antiderivative function.

---

This example illustrates an important fact: A definite integral is not a function of its integration variable, called $x$ in this case. Its value depends only on what values are chosen for the limits and on what function occurs under the integral sign. It is sometimes called a *functional*, or a function of a function, because its value depends on what function is chosen for the integrand function.

---

**Problem 4.3**

Find the value of the definite integral

$$\int_0^1 e^x\, dx.$$

---

## The Definite Integral as an Area

We will now show that a definite integral is equal in value to an area between the $x$ axis and the curve representing the integrand function in a graph. We return to Eq. (4.18), which gives an approximation to a definite integral. Figure 4.2 shows the situation with $n = 3$.

The curve in the figure is the curve representing the integrand function $f(x)$. Each term in the sum is equal to the area of a rectangle with height $f(x_k)$ and width $\Delta x$, so that the sum is equal to the shaded area under the bar graph in the figure.

As the limit of Eq. (4.21) is taken, the number of bars between $x = x_0$ and $x = x'$ becomes larger and larger, while the width $\Delta x$ becomes smaller and smaller. The roughly triangular areas between the bar graph and the curve become smaller and smaller, and although there are more and more of them their total area shrinks to zero as the limit is taken. The integral thus becomes equal to the area bounded by the $x$ axis, the curve of the integrand function, and the vertical lines at the limits $x = x_0$ and $x = x'$.

Figure 4.2 shows a case in which all values of the integrand function are positive. If $x_0 < x'$, the increment $\Delta x$ is always taken as positive. If the integrand function is negative in some region, we must take the area in that region as negative.

---

### Example 4.6

Find the area bounded by the $x$ axis and the curve representing

$$f(x) = \sin(x),$$

a. between $x = 0$ and $x = \pi$
b. between $x = 0$ and $x = 2\pi$.

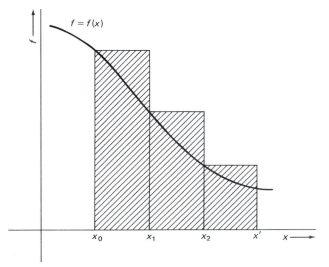

**FIGURE 4.2**   The area under a bar graph and the area under a curve.

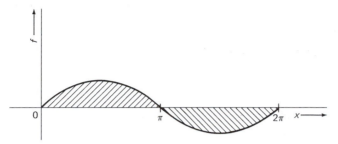

**FIGURE 4.3**  A graph of $f = \sin(x)$ for Example 4.6.

**Solution**

The graph of the function is shown in Fig. 4.3.

a. area $= \displaystyle\int_0^\pi \sin(x)\,dx = -\cos(\pi) - [-\cos(0)]$

$= 2.$

This integral was obtained in Example 4.5.

b. area $= \displaystyle\int_0^{2\pi} \sin(x)\,dx = -\cos(2\pi) - [-\cos(0)]$

$= -(1) - [-(1)] = 0.$

When we found the value of the integral by finding the antiderivative function, we omitted the constant term that generally must be present in the antiderivative function. This constant canceled out in Example 4.5 and would have canceled out in Example 4.6, so we left it out from the beginning.

**Problem 4.4**

Find the following areas by computing the values of definite integrals:

a. The area bounded by the curve representing $y = x^3$, the $x$ axis, and the line $x = 3$.
b. The area bounded by the straight line $y = 2x + 3$, the $x$ axis, the line $x = 1$, and the line $x = 4$.
c. The area bounded by the parabola $y = 4 - x^2$ and the $x$ axis. You will have to find the limits of integration.

Numerical approximations to integrals can be made by drawing an accurate graph of the integrand function and directly measuring the appropriate area in the graph. There are three practical ways to do this. One is by counting squares on your graph paper. Another is by cutting out the area to be determined and weighing this piece of graph paper and another piece of known area from the same sheet. A third is by using a mechanical device called a planimeter, which registers an area on a dial after a stylus is moved around the boundary of the area. Such procedures are now seldom used, since numerical approximations can be done more easily with a computer.

**Problem 4.5**

Find the approximate value of the integral

$$\int_0^1 e^{-x^2}\, dx$$

by making a graph of the integrand function and measuring an area.

## Facts about Integrals

The following facts can be understood by considering the relation between integrals and areas in graphs of integrand functions:

1. A definite integral over the interval $(a, c)$ is the sum of the definite integrals over the intervals $(a, b)$ and $(b, c)$:

$$\int_a^c f(x)\, dx = \int_a^b f(x)\, dx + \int_b^c f(x)\, dx. \qquad (4.24)$$

   This fact is illustrated in Fig. 4.4. The integral on the left-hand side of Eq. (4.24) is equal to the entire area shown, and each of the two terms on the right-hand side is equal to one of the two differently shaded areas that combine to make the entire area. For this equation to be correct, it is not necessary for $b$ to lie between $a$ and $c$.

2. The presence of a finite step discontinuity in an integrand function does not prevent us from carrying out the process of integration. In this regard, integration differs from differentiation. Figure 4.5 illustrates the situation. If the discontinuity is at $x = b$, we simply apply Eq. (4.24) and find that the integral is given by the integral up to $x = b$ plus the integral from $x = b$ to the end of the interval.

3. If the two limits of integration are interchanged, the resulting integral is the negative of the original integral. In Eq. (4.23), we assumed that $x_0 < x'$, so

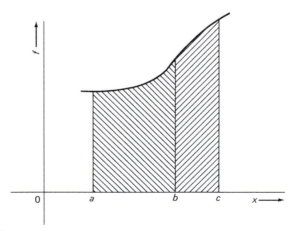

**FIGURE 4.4**   Figure to illustrate Eq. 4.24.

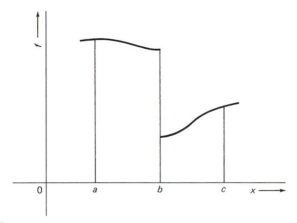

**FIGURE 4.5**   An integrand function that is discontinuous at $x = b$.

that $\Delta x$ would be positive. If the lower limit is larger than the upper limit of integration, $\Delta x$ must be taken as negative, reversing the sign of the area in the graph. Therefore,

$$\int_a^b f(x)\,dx = -\int_b^a f(x)\,dx. \tag{4.25}$$

Use of this fact makes Eq. (4.24) usable for any real values of a, b, and c.

4. If an integrand function consists of a constant times some function, the constant can be factored out of the integral:

$$\int_a^b cf(x)\,dx = c\int_a^b f(x)\,dx. \tag{4.26}$$

5. The integral of an odd function from $-c$ to $c$ vanishes. An *odd function* is one that obeys the relation

$$f(-x) = -f(x). \tag{4.27}$$

In this case, the area above the axis exactly cancels the area below the axis, so that

$$\int_{-c}^c f(x)\,dx = 0 \qquad (f(x)\ \text{odd}), \tag{4.28}$$

where $c$ is any real constant. The sine function and the tangent function are examples of odd functions, as stated in Eqs. (2.22) and (2.24).

---

**Problem 4.6**

Draw a rough graph of $f(x) = xe^{-x^2}$ and satisfy yourself that this is an odd function. Identify the area in this graph that is equal to the following integral and satisfy yourself that the integral vanishes:

$$\int_{-2}^2 xe^{-x^2}\,dx = 0.$$

---

6. The integral of an even function from $-c$ to $c$ is twice the integral from $0$ to $c$. An even function is one that obeys the relation

$$f(-x) = f(x). \tag{4.29}$$

For such an integrand function, the area between $-c$ and $0$ is equal to the area between $0$ and $c$, so that

$$\int_{-c}^{c} f(x)\,dx = 2 \int_{0}^{c} f(x)\,dx \qquad (f(x) \text{ even}), \tag{4.30}$$

where $c$ is any real constant. The cosine function is an example of an even function, as stated in Eq. (2.23).

## Problem 4.7

Draw a rough graph of $f(x) = e^{-x^2}$. Satisfy yourself that this is an even function. Identify the area in the graph that is equal to the definite integral

$$I_1 = \int_{-1}^{1} e^{-x^2}\,dx$$

and satisfy yourself that this integral is equal to twice the integral

$$I_2 = \int_{0}^{1} e^{-x^2}\,dx.$$

If you have an integrand that is a product of several factors, you can use the following facts:

a. The product of two even functions is an even function.
b. The product of two odd functions is an even function.
c. The product of an odd function and an even function is an odd function.

The facts about odd and even functions are valid if the function is either even or odd about the center of the integration interval, even if the center of the interval is not at the origin.

## Problem 4.8

The quantum-mechanical wave functions of a particle in a box of length $a$ are either even or odd functions. For example, the two lowest-energy wave functions are

$$\psi_1 = \sqrt{\frac{2}{a}} \sin\left(\frac{\pi x}{a}\right)$$

$$\psi_2 = \sqrt{\frac{2}{a}} \sin\left(\frac{2\pi x}{a}\right).$$

a. By drawing rough graphs, satisfy yourself that $\psi_1$ is even about the center of the interval from $x=0$ to $x=a$. That is, $\psi_1(x) = \psi_1(a-x)$. Satisfy yourself that $\psi_2$ is odd about the center of the interval.

b. Draw a rough graph of the product $\psi_1\psi_2$ and satisfy yourself that the integral of this product from $x=0$ to $x=a$ vanishes.

---

7. If the limits of a definite integral are considered to be variables, we can write

$$\frac{d}{db}\left[\int_a^b f(x)\,dx\right] = f(b),\tag{4.31a}$$

where $a$ is considered to be a constant, and

$$\frac{d}{da}\left[\int_a^b f(x)\,dx\right] = -f(a),\tag{4.31b}$$

where $b$ is considered to be a constant. If $a$ and $b$ are both functions of some variable $c$ (not the variable of integration, $x$), then

$$\boxed{\frac{d}{dc}\left[\int_a^b f(x)\,dx\right] = f(b)\frac{db}{dc} - f(a)\frac{da}{dc}}.\tag{4.31c}$$

These equations follow from Eq. (4.23), and Eq. (4.31c) also comes from the chain rule, Eq. (3.26).

## SECTION 4.3. INDEFINITE INTEGRALS: TABLES OF INTEGRALS

Let us consider the upper limit in Eq. (4.23) to be variable and the lower limit to be fixed and equal to $a$:

$$\int_a^{x'} f(x)\,dx = F(x') - C.$$

The quantity $C$ is equal to $f(a)$ and is arbitrary if $a$ is arbitrary. We omit mention of $a$ and write

$$\int^{x'} f(x)\,dx = F(x') - C.\tag{4.32}$$

This is called an *indefinite integral*, since the lower limit is unspecified and the upper limit is variable. The indefinite integral is the same as the antiderivative function. Large tables of indefinite integrals have been compiled. Appendix 5 is a brief version of such a table. In most such tables, the notation of Eq. (4.32) is not maintained. The entries are written in the form

$$\int f(x)\,dx = F(x).\tag{4.33}$$

This equation is an abbreviation for Eq. (4.32). The upper limit and the constant $C$, called the *constant of integration*, are usually omitted from table entries, and the same symbol is usually used for the variable of integration and for the argument of the integral function $F$. However, you should remember that an arbitrary constant C can be added to the right-hand side of Eq. (4.33).

The same information is contained in a table of indefinite integrals as is contained in a table of derivatives. However, we can get by with a fairly short table of derivatives, since we have the chain rule and other facts listed in Section 3.5. Antiderivatives are harder to find, so it is good to have a separate table of indefinite integrals, arranged so that similar integrand functions occur together.

---

### Example 4.7

Using a table, find the indefinite integrals:

a. $\displaystyle\int (dx/(a^2 + x^2))$

b. $\displaystyle\int x \sin^2(x)\, dx$

c. $\displaystyle\int x e^{ax}\, dx.$

---

### Solution

From Appendix 5, or any table of indefinite integrals,

a. $\displaystyle\int^{x'} \frac{dx}{a^2 + x^2} = \frac{1}{a} \arctan\left(\frac{x'}{a}\right) + C$

b. $\displaystyle\int^{x'} x \sin^2(x)\, dx = \frac{x'^2}{4} - \frac{x' \sin(2x')}{4} - \frac{\cos(2x')}{8} + C$

c. $\displaystyle\int^{x'} x e^{ax}\, dx = \frac{e^{ax'}}{a^2}(ax' - 1) + C.$

Notice that we have written $+C$ instead of $-C$ as in Eqs. (4.31) and (4.32). Since $C$ is an arbitrary constant, this makes no difference. We have also maintained the distinction between $x$ and $x'$.

---

### Problem 4.9

Show by differentiation that the functions on the right-hand sides of the equations in Example 4.7 yield the integrand functions when differentiated.

---

Since the indefinite integral is the antiderivative function, it is used to find a definite integral in the same way as in Section 4.2. If $x_1$ and $x_2$ are the limits of the definite integral,

$$\int_{x_1}^{x_2} f(x)\, dx = \int_{a}^{x_2} f(x)\, dx - \int_{a}^{x_1} f(x)\, dx$$

$$= F(x_2) - C - [F(x_1) - C] = F(x_2) - F(x_1). \tag{4.34}$$

Equation (4.34) is the same as Eq. (4.23).

### Example 4.8

Using a table of indefinite integrals, find the definite integral

$$\int_0^{\pi/2} \sin(x)\cos(x)\,dx.$$

### Solution

From Eq. (36) of Appendix 3, we find that the indefinite integral is $\sin^2(x)/2$,

$$\int_0^{\pi/2} \sin(x)\cos(x)\,dx = \frac{\sin^2(x)}{2}\Big|_0^{\pi/2} = \frac{1}{2}\left[\sin^2\left(\frac{\pi}{2}\right) - \sin^2(0)\right]$$

$$= \frac{1}{2}(1 - 0) = \frac{1}{2}.$$

We have used the common notation

$$F(x)|_a^b = F(b) - F(a). \tag{4.35}$$

### Problem 4.10

Using a table of indefinite integrals, find the definite integrals.

a. $\displaystyle\int_0^3 \cosh(2x)\,dx$

b. $\displaystyle\int_1^2 (\ln(3x)/x)\,dx$

c. $\displaystyle\int_0^5 4^x\,dx.$

In addition to tables of indefinite integrals, there exist tables of definite integrals. Some are listed at the end of the chapter, and Appendix 6 is a short version of such a table. Some of the entries in these tables are integrals that could be worked out by using a table of indefinite integrals, but others are integrals that cannot be obtained as indefinite integrals, but by some particular method can be worked out for one set of limits. An example of such an integral is worked out in Appendix 7.

Tables of definite integrals usually include only sets of limits such as $(0, 1)$, $(0, \pi)$, $(0, \pi/2)$, and $(0, \infty)$. The last set of limits corresponds to an improper integral, which is discussed in the next section.

## SECTION 4.4. IMPROPER INTEGRALS

Our discussion of integrals has thus far been based on the assumptions that both limits of integration are finite and that the integrand function does not become infinite inside the interval of integration. If either of these conditions is not met, an integral is said to be an *improper integral*. For example,

$$I = \int_0^\infty f(x)\,dx \tag{4.36}$$

is an improper integral because its upper limit is infinite.

We must decide what is meant by the infinite upper limit in Eq. (4.36), because Eq. (4.21) cannot always be modified by inserting $\infty$ into the equation instead of $x'$. We define

$$\int_0^\infty f(x)\,dx = \lim_{b \to \infty} \int_0^b f(x)\,dx. \tag{4.37}$$

In this mathematical limit, the upper limit of integration becomes larger and larger without bound. Notice that the word "limit" has several definitions, and two of them unfortunately occur here in the same sentence.

If the integral approaches more and more closely to some finite value as the upper limit is made larger and larger, we say that the limit exists and that the improper integral is equal to this finite value. The improper integral is said to *converge* to the value that is approached. Some improper integrals do not converge. The magnitude of the integral can become larger and larger without bound as the limit of integration is made larger and larger. In other cases, the integral oscillates repeatedly in value as the limit of integration is made larger and larger. We say in both of these cases that the integral *diverges*.

In some improper integrals, the lower limit of integration is made to approach $-\infty$ and in some improper integrals, the lower limit is made to approach $-\infty$ while the upper limit is made to approach $+\infty$. Just as in the case of Eq. (4.37), if the integral approaches a finite value more and more closely as the limit or limits approach infinite magnitude, the improper integral is said to converge to that value.

Another kind of improper integral has an integrand function that becomes infinite somewhere in the interval of integration. For example,

$$I = \int_0^1 \frac{1}{x}\,dx \tag{4.38}$$

is an improper integral because the integrand function becomes infinite at $x = 0$. In this case, the improper integral is defined by

$$\int_0^1 \frac{1}{x}\,dx = \lim_{a \to 0^+} \int_a^1 \frac{1}{x}\,dx. \tag{4.39}$$

Just as in the other cases, if the integral grows larger and larger in magnitude as the limit is taken, or if it oscillates in value without approaching closer and closer to some value, we say that it diverges. If the limit exists, we say that the improper integral converges to that limit. The situation is similar if the point at which the integrand becomes infinite is at the upper limit of integration. If it is within the interval of integration, break the interval into two subintervals so that the infinite value of the integrand is at the lower limit of one subinterval and at the upper end of the other subinterval.

The two principal questions that we need to ask about an improper integral are:

1. Does it converge?
2. If so, what is its value?

---

### Example 4.9

Determine whether the following improper integral converges, and if so, find its value:

$$\int_0^\infty e^{-x}\,dx.$$

---

### Solution

$$\int_0^\infty e^{-x}\,dx = \lim_{b\to\infty}\int_0^b e^{-x}\,dx = \lim_{b\to\infty}[-e^{-x}]_0^b$$
$$= \lim_{b\to\infty} -(e^{-b}-1) = 0 + 1 = 1.$$

The integral converges to the value 1.

---

### Problem 4.11

Determine whether each of the following improper integrals converges, and if so, determine its value:

a. $\displaystyle\int_0^1 (1/x)\,dx$

b. $\displaystyle\int_0^\infty \sin(x)\,dx$

c. $\displaystyle\int_0^\infty (1/(1+x))\,dx$

d. $\displaystyle\int_0^\infty (1/x^3)\,dx$

e. $\displaystyle\int_{-\infty}^0 e^x\,dx.$

---

## SECTION 4.5. METHODS OF INTEGRATION

In this section, we discuss three methods that can be used to transform an integral that is not exactly like any integral you can find in a table into one that you can handle.

### The Method of Substitution

In this method a *change of variables* is performed. The integrand function is expressed in terms of a different independent variable, which becomes the variable of integration.

**Example 4.10**

Find the integral

$$\int_0^\infty x e^{-x^2} \, dx$$

without using a table of integrals.

---

**Solution**

We have $x^2$ in the exponent, which suggests using $y = x^2$ as a new variable. If $y = x^2$, then $dy = 2x \, dx$, or $x \, dx = \frac{1}{2} \, dy$,

$$\int_0^\infty x e^{-x^2} \, dx = \frac{1}{2} \int_{x=0}^{x=\infty} e^{-y} \, dy = \frac{1}{2} \int_0^\infty e^{-y} \, dy$$

$$= -\frac{1}{2} e^{-y} \Big|_0^\infty = -\frac{1}{2}(0 - 1) = \frac{1}{2}.$$

Example 4.10 was solved as follows: first a new variable was chosen that looked as though it would give a simpler integrand function. Next, the integrand function was expressed in terms of this variable. The differential of the integration variable must also be reexpressed. The limits of integration were then expressed in terms of the new variable, making the limits equal to the values of the new variable that correspond to the values of the old variable at the old limits. The final thing was to compute the new integral, which is equal to the old integral.

---

**Example 4.11**

Find the integral

$$\int_0^{1/2} \frac{dx}{2 - 2x}$$

without using a table of integrals.

---

**Solution**

We let $y = 2 - 2x$, in order to get a simple denominator. With this, $dy = -2 \, dx$, or $dx = -dy/2$. When $x = 0$, $y = 2$, and when $x = \frac{1}{2}$, $y = 1$,

$$\int_0^{1/2} \frac{1}{2 - 2x} \, dx = -\frac{1}{2} \int_2^1 \frac{1}{y} \, dy = \frac{1}{2} \int_1^2 \frac{1}{y} \, dy = \frac{1}{2} \ln(y) \Big|_1^2$$

$$= \frac{1}{2}[\ln(2) - \ln(1)] = \frac{1}{2} \ln(2).$$

---

**Problem 4.12**

Find the integral

$$\int_0^\pi e^{\cos(\theta)} \sin(\theta) \, d\theta$$

without using a table of integrals.

## Integration by Parts

This method, which is also called *partial integration*, consists of application of the formula

$$\int u \frac{dv}{dx} \, dx = uv - \int v \frac{du}{dx} + C \, . \tag{4.40}$$

or the corresponding formula for definite integrals

$$\int_a^b u \frac{dv}{dx} \, dx = u(x)v(x)|_a^b - \int_a^b v \frac{du}{dx} \, dx \, . \tag{4.41}$$

In these formulas, $u$ and $v$ must be functions of $x$ that are differentiable everywhere in the interval of integration.

We can derive Eq. (4.40) by use of Eq. (3.19), which gives the derivative of the product of two functions:

$$\frac{d}{dx}(uv) = u \frac{dv}{dx} + v \frac{du}{dx} \, .$$

The antiderivative of either side of this equation is just $uv + C$, where $C$ is an arbitrary constant, so we can write the indefinite integral

$$\int \frac{d(uv)}{dx} \, dx = \int u \frac{dv}{dx} \, dx + \int v \frac{du}{dx} \, dx = u(x)v(x) + C \, .$$

This is the same as Eq. (4.40).

___

### Example 4.12

Find the indefinite integral

$$\int x \sin(x) \, dx$$

without using a table.

___

### Solution

Let $u(x) = x$ and $\sin(x) = dv/dx$. We make this choice because the antiderivative of $x$ is $x^2/2$, which will lead to a worse integral than the one containing $x$. With this choice

$$\frac{dv}{dx} = 1 \qquad \text{and} \qquad v = -\cos(x)$$

$$\int x \sin(x) \, dx = -x \cos(x) + \int \cos(x) \, dx$$

$$= -x \cos(x) + \sin(x) + C \, .$$

**Problem 4.13**

Find the integral

$$\int_0^\pi x^2 \sin(x)\, dx$$

without using a table. You will have to apply partial integration twice.

The fundamental equation of partial integration, Eq. (4.40), is sometimes written with differentials:

$$\boxed{\int u\, dv = uv - \int v\, du}. \tag{4.42}$$

## The Method of Partial Fractions

This method consists of an algebraic procedure for turning a difficult integrand into a sum of two or more easier functions. It works with an integral of the type

$$I = \int \frac{P(x)}{Q(x)}\, dx, \tag{4.43}$$

where $P(x)$ and $Q(x)$ are polynomials in $x$. The highest power of $x$ in $P$ must be lower than the highest power of $x$ in $Q$. However, if this is not the case, you can proceed by performing a long division, obtaining a polynomial plus a remainder, which will be a quotient of polynomials that does obey the condition. The polynomial can be integrated easily, and the remainder quotient can be handled with the method of partial fractions.

The first step in the procedure is to factor the denominator, $Q(x)$, into a product of polynomials of degree 1 and 2. A polynomial of degree 1 is an expression of the form $ax + b$, and a polynomial of degree 2 is $ax^2 + bx + c$. Let us first assume that all of the factors are of degree 1, so that

$$Q(x) = (a_1x + b_1)(a_2x + b_2)(a_3x + b_3) \cdots (a_nx + b_n), \tag{4.44}$$

where all the $a$'s and $b$'s are constants.

The fundamental formula of the method of partial fractions is a theorem of algebra that says that if $Q(x)$ is given by Eq. (4.44) and $P(x)$ is of lower degree than $Q(x)$, then

$$\boxed{\frac{P(x)}{Q(x)} = \frac{A_1}{a_1x + b_1} + \frac{A_2}{a_2x + b_2} + \cdots + \frac{A_n}{a_nx + b_n}}, \tag{4.45}$$

where $A_1, A_2, \ldots, A_n$ are all constants.

Equation (4.45) is applicable only if all the factors in $Q(x)$ are distinct from each other. If the same factor occurs more than once, Eq. (4.45) must be modified. If the factor $a_1x + b_1$ occurs $m$ times, we must write

$$\frac{P(x)}{Q(x)} = \frac{A_1}{a_1x + b_1} + \frac{A_2}{(a_1x + b_1)^2} + \cdots + \frac{A_m}{(a_1x + b_1)^m}$$
$$+ \text{ other terms as in Eq. (4.45).} \tag{4.46}$$

Sometimes a factor of degree 2 occurs that cannot be factored. If $Q$ contains a factor $a_1x^2 + b_1x + c_1$, we must write

$$\frac{P(x)}{Q(x)} = \frac{A_1x + B_1}{a_1x^2 + b_1x + c_1} + \text{other terms as in Eqs. (4.45) and (4.46).} \qquad (4.47)$$

---

### Example 4.13

Apply Eq. (4.45) to

$$\int \frac{6x - 30}{x^2 + 3x + 2}\, dx.$$

------

### Solution

$$\frac{6x - 30}{x^2 + 3x + 2} = \frac{A_1}{x + 2} + \frac{A_2}{x + 1}.$$

We need to solve for $A_1$ and $A_2$ so that this equation will be satisfied for all values of $x$. We multiply both sides of the equation by $(x + 2)(x + 1)$:

$$6x - 30 = A_1(x + 1) + A_2(x + 2).$$

Since this equation must be valid for all values of $x$, we can get a different equation for each value of $x$. If we let $x = 0$, we get

$$-30 = A_1 + 2A_2. \qquad (4.48)$$

If we let $x$ become very large, so that the constant terms can be neglected, we obtain

$$6x = A_1x + A_2x$$

or

$$6 = A_1 + A_2. \qquad (4.49)$$

Equations (4.48) and (4.49) can be solved simultaneously to obtain

$$A_1 = 42, \qquad A_2 = -36. \qquad (4.50)$$

Our result is

$$\int \frac{6x - 30}{x^2 + 3x + 2}\, dx = \int \frac{42}{x + 2}\, dx - \int \frac{36}{x + 1}\, dx. \qquad (4.51)$$

---

### Problem 4.14

Solve Eqs. (4.48) and (4.49) simultaneously to obtain Eq. (4.50).

---

### Problem 4.15

Find the indefinite integrals on the right-hand side of Eq. (4.51).

**Example 4.14**

Consider a chemical reaction

$$aA + bB \rightarrow cC,$$

where the capital letters are abbreviations for some chemical formulas and the lower-case letters are abbreviations for the stoichiometric coefficients that balance the equation. If the reaction is said to be first order in A and first order in B, the rate of the reaction is given by

$$-\frac{1}{a}\frac{d[A]}{dt} = k_f[A][B],$$

where $k_f$ is a constant called the rate constant and where [A] represents the concentration of A and [B] represents the concentration of B. To solve this equation and obtain an expression for the concentrations as a function of $t$, we must perform an integration. In order to proceed, we need to express [A] and [B] in terms of a single variable. We let

$$[A] = [A]_0 - ax$$

and

$$[B] = [B]_0 - bx,$$

where the initial values of the concentrations are labeled with a subscript 0. In the case that the reactants are not mixed in the stoichiometric ratio, we manipulate the rate expression into the form, where we have multiplied by $dt$ and recognized that $(dx/dt)\,dt = dx$,

$$\frac{1}{([A]_0 - ax)([B]_0 - bx)}\,dx = k_f\,dt.$$

This equation can be integrated by the method of partial fractions. We write

$$\frac{1}{([A]_0 - ax)([B]_0 - bx)} = \frac{G}{[A]_0 - ax} + \frac{H}{[B]_0 - bx}.$$

The constants $G$ and $H$ are found to be

$$G = \frac{1}{[B]_0 - b[A]_0/a} \qquad \text{and} \qquad H = \frac{1}{[A]_0 - a[B]_0/b}.$$

When these expressions are substituted into the rate expression, a definite integration gives

$$\frac{1}{a[B]_0 - b[A]_0} \ln\left(\frac{[B]_t[A]_0}{[A]_t[B]_0}\right) = k_f t.$$

The methods presented thus far in this chapter provide an adequate set of tools for the calculation of most integrals that will be found in a physical chemistry course. In applying these methods, it is probably best to proceed as follows:

1. If the limits are 0 and $\infty$ or 0 and $\pi$, or something else quite simple, look first in a table of definite integrals.

2. If this does not work, or if the limits were not suitable, look in a table of indefinite integrals.
3. If you do not find the integral in a table, try the method of substitution.
4. If you still have not obtained the integral, see if the method of partial fractions is applicable and use it if you can.
5. If this did not work, manipulate the integrand into a product of two factors and try the method of partial integration.
6. If all these things have failed, or if they could not be attempted, do a numerical approximation to the integral. This is discussed in the next section.

## SECTION 4.6. NUMERICAL INTEGRATION

Numerical methods of obtaining accurate approximate values of integrals have become very widely used since the advent of programmable computers. You can write your own computer program to carry out a numerical approximation, but software packages, such as Mathematica and MathCad, will carry out the calculation automatically.

There are two cases for which a numerical approximation to an integral must be sought. In one case the integrand function does not possess an antiderivative function that you can find in a table or can work out. In fact, there are integrand functions for which no known antiderivative function exists. One such integrand function is $e^{-x^2}$ (see Appendix 7). In the other case the integrand function is represented approximately by a set of data points instead of by a formula.

We begin with Eq. (4.22):

$$\int_a^b f(x)\,dx = \lim_{\substack{\Delta x \to 0 \\ n \to \infty \\ n\Delta x = b-a}} \sum_{j=0}^{n-1} f(a + j\Delta x)\Delta x. \tag{4.52}$$

If we do not take the limit but allow $n$ to be some convenient finite number, the resulting equation will be approximately correct.

### The Bar-Graph Approximation

Let us consider an example, the integral of $e^{-x^2}$ from 1 to 2. We apply an approximate version of Eq. (4.52) with $n = 10$. The result is

$$\int_1^2 e^{-x^2}\,dx \approx \sum_{j=0}^{9} \exp[-(1 + 0.1j)^2](0.1) = 0.15329. \tag{4.53}$$

This result is represented by the area under a bar graph such as in Fig. 4.2, and we call this approximation the bar-graph approximation. It is in error by the area of the roughly triangular areas between the bar graph and the curve of the integrand function. The error in Eq. (4.53) is about 15%, since the correct value of this integral is 0.13525726 to eight significant digits.

Each bar in Fig. 4.2 is called a *panel*. The bar-graph approximation can be made more nearly accurate by increasing the number of panels, but the rate of improvement can be quite slow. For example, if we take $n = 20$ and $\Delta x = 0.05$, we get 0.14413

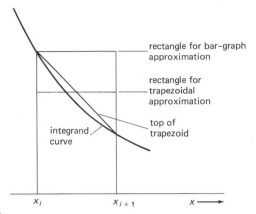

**FIGURE 4.6**    Figure to illustrate the trapezoidal approximation (enlarged view of one panel shown).

for the integral in Eq. (4.53), which is still in error by about 6%. If we take $n = 100$ and $\Delta x = 0.01$, we get 0.13701, which is still wrong by about 1%.

### The Trapezoidal Approximation

One way to improve on the bar-graph approximation is to take the height of the rectangles as equal to the value of the integrand function somewhere near the middle of the panel. In the trapezoidal approximation, the height of the bar is taken as the average of the values of the function at the two sides of the panel. This gives an area that is the same as that of a trapezoid whose upper corners match the integrand function at the sides of the panel, as shown in Fig. 4.6,

$$\int_a^b f(x)\,dx \approx \frac{f(a)\Delta x}{2} + \sum_{k=1}^{n-1} f(a + k\Delta x)\Delta x + \frac{f(b)\Delta x}{2}. \qquad (4.54)$$

As expected, the trapezoidal approximation gives more nearly correct values than does the bar-graph approximation, for the same number of panels. For 10 panels, the trapezoidal approximation gives a result of 0.135810 for the integral in Eq. (4.53). For 100 panels, the trapezoidal approximation is correct to five significant digits.

---

### Problem 4.16

Using the trapezoidal approximation with five panels, find the value of the integral

$$\int_{10}^{20} x^2\,dx.$$

Calculate the exact value of the integral for comparison.

---

### Simpson's Rule

In the bar-graph approximation, we used only one value of the integrand for each panel. In the trapezoidal approximation, we used two values for each panel, to determine a line segment fitting the integrand curve at both edges of the panel. If three points in a plane are given, there is one and only one parabola that can be drawn through all three. In *Simpson's rule*, we take the panels two at a time, construct a parabola through the three points, find the area under the parabola, and use this to approximate the integral. A parabolic curve is likely to fall closer to the integrand curve than a straight line, so we expect this to give a better approximation than the trapezoidal approximation, and it usually does. We must have an even number of panels to use this method.

We let $f_0 = f(a)$, $f_1 = f(a + \Delta x)$, $f_2 = f(a + 2\Delta x)$, etc., and use the formula for the area under a parabola to obtain as our final result

$$\int_a^b f(x)\,dx \approx \frac{(f_0 + 4f_1 + 2f_2 + 4f_3 + \cdots + 4f_{n-1} + f_n)\Delta x}{3}. \tag{4.55}$$

Notice the pattern, with alternating coefficients of 2 and 4, except for the first and last values of the integrand.

This version of Simpson's rule is sometimes called *Simpson's one-third rule* because of the 3 in the denominator. There is another version, called *Simpson's five-eighths rule*, which corresponds to fitting third-degree polynomials to four points at a time.

---

### Problem 4.17

Apply Simpson's rule to the integral of Problem 4.16, using two panels. Since the integrand curve is a parabola, your result should be exactly correct.

---

There is another widely used way to obtain a numerical approximation to a definite integral, known as *Gauss quadrature*. In this method, the integrand function must be evaluated at particular unequally spaced points on the interval of integration. We will not discuss this method, but you can read about it in books on numerical analysis.

So far, we have assumed that the integrand function was known so that it could be evaluated at the required points. Most of the applications of numerical integration in physical chemistry are to integrals where the integrand function is not known exactly, but is known only approximately from experimental measurements at a few points on the interval of integration. If there are an odd number of data points that are equally spaced, we can apply Simpson's rule.

---

### Problem 4.18

In thermodynamics, it is shown that the entropy change of a system that is heated at constant pressure from temperature $T_1$ to temperature $T_2$ is given by

$$\Delta S = S(T_2) - S(T_1) = \int_{T_1}^{T_2} \frac{C_p}{T}\,dT, \tag{4.56}$$

where $C_p$ is the constant-pressure heat capacity and $T$ is the temperature on the

Kelvin scale. Calculate $\Delta S$ for the heating of 1 mol of solid zinc from 20 K to 100 K, using the following data:

| T/K | $C_p$/J K$^{-1}$ mol$^{-1}$ |
|---|---|
| 20 | 1.70 |
| 30 | 4.966 |
| 40 | 8.171 |
| 50 | 11.18 |
| 60 | 13.60 |
| 70 | 15.43 |
| 80 | 16.87 |
| 90 | 18.11 |
| 100 | 19.15 |

## SECTION 4.7. A NUMERICAL INTEGRATION COMPUTER PROGRAM

This section describes the construction of a computer program in the BASIC language to carry out a numerical approximation to an integral, using Simpson's rule. A number of complete mathematical software packages, such as Mathematica and Mathcad, will carry out this kind of calculation automatically for you. The program is written in the BASIC language, which is discussed in Chapter 11. It is written so that almost any Basic processor will accept it without serious modification. For the TrueBasic processor on the Macintosh computer, you type in a new program as follows: open the folder containing TrueBasic and double-click on the TrueBasic icon. A blank window is opened, and you can type in your program. After you type in a few statements, you should save the program, using the "Save As" command in the "File" menu. We save the present program under the name NUMIT.

Interspersed between program statements are some explanations of the function of the next part of the program. It is possible to include such explanations in the program as "remarks" (see Chapter 11). You can read remarks when you read the program, but they will not be printed out when the program runs. You can use PRINT statements to have the computer print out explanations when the program runs, as follows:

```
100 PRINT "PROGRAM NUMIT APPROXIMATES AN INTEGRAL."
105 PRINT "IT USES SIMPSON'S ONE-THIRD RULE."
110 PRINT "IT PERMITS UP TO 101 DATA POINTS."
105 PRINT "DATA POINTS MUST BE EQUALLY SPACED."
120 PRINT "THE NUMBER OF DATA POINTS MUST BE ODD."
```

A DIMENSION statement is needed to reserve storage space for the arrays for the independent variable and the integrand function:

```
130   DIM X(101), Y(101)
```

The following part of the program will provide for input of some necessary values.

```
140 PRINT "TYPE IN THE NUMBER OF DATA POINTS."
145 INPUT N
150 PRINT "TYPE IN THE LOWER LIMIT."
155 INPUT X0
```

```
160 PRINT "TYPE IN THE UPPER LIMIT."
165 INPUT X9
```

The following loop allows the data points to be typed in when the program is run. Since the values of the independent variable must be equally spaced, it is not necessary to read them in.

```
170 FOR I = 1 TO N STEP 1
180 PRINT "TYPE IN THE VALUE OF THE INTEGRAND AT
POINT NUMBER"; I
190 INPUT Y(I)
200 NEXT I
```

We use the symbol W for the spacing between the values of the independent variable:

```
210 W = (X9 - X0)/(N - 1)
```

The number of panels is $N - 1$, not N. The next part of program allows the user to see if the data were entered correctly.

```
220 PRINT "DO YOU WANT TO SEE THE DATA POINTS (YES
OR NO)"
225 INPUT B$
230 IF B$ = "YES" THEN 240 ELSE 300
240 FOR I = 1 TO N STEP 1
250 J = I - 1
260 X(I) = X0 + J*W
270 PRINT "X("; I; ") = "; X(1),
Y("; I; ") = "; Y(I)
280 NEXT I
```

The next part of the program computes the integral. In Eq. (4.55), the points were numbered from 0 to n. However, some computers will not allow 0 to be used as an index for an array, so we number the points from 1 to N.

```
300 S = Y(1)
310 FOR I = 2 TO N - 1 STEP 2
320 S = S + 4*Y(I) + 2*Y(I + 1)
330 NEXT I
340 S = (S - Y(N))*W/3
```

The program ends with a statement to print out the answer and an END statement.

```
350 PRINT "THE VALUE OF THE INTEGRAL IS"; S
999 END
```

Notice statement number 340. After the loop, the last point will have been added in with a coefficient of 2. The formula requires that the last point be put in with a coefficient of 1. Statement 340 subtracts off Y(N) to restore the proper contribution.

 The program is now ready to run. In order to keep a copy of the program on file on the disk of the computer, you choose the "Save" command in the "File" menu. To run the program, choose "Run" from the "Run" menu, and the execution begins.

**Example 4.15**

Modify the program NUMIT so that it will perform the integration in Problem 4.18, including the division of $C_p$ by $T$.

**Solution**

The loop beginning with statement 240 can be used to carry out the divisions. We call up the program by double-clicking on the "Numit" icon.

We need to delete the statements 220 and 230 that make the loop optional. Unless we save the program again, this replacement will be temporary, affecting only the working version of the program, not the version stored in the disk file. Let us leave statement 270 in the program, so that we will see our data to make sure that it is entered correctly. We insert a statement to carry out the divisions by typing

```
275 Y(I) = Y(I)/X(I)
```

This will be inserted into the program between statement 270 and statement 280. The program is now modified. If we want to make the modifications permanent, we choose "Save" from the "File" menu.

---

**Problem 4.19**

Modify the program NUMIT so that it will evaluate the integral

$$\int_1^2 e^{-x^2}\, dx.$$

This will require replacement of the part of the program that reads in data points by statements that will evaluate the integrand function for each value of the independent variable.

---

A slightly more sophisticated numerical integration program, called NUMINT, is listed in Appendix 8. One of the options in this program provides for a modified Simpson's rule that allows for unequally spaced values of the independent variable and does not require the number of data points to be odd.

## SECTION 4.8. PROBABILITY DISTRIBUTIONS AND MEAN VALUES

There are several kinds of averages in common use. One type is the *median*, which is the value such that half of the set of values is greater than the median and half of the set is smaller than the median. The *mode* is the most frequently occurring value in the set. We discuss the most commonly encountered type of average, the *mean*. The mean of a set of $N$ values is defined as

$$\bar{x} = \frac{1}{N}(x_1 + x_2 + x_3 + x_4 + \cdots + x_N),$$
(4.57)

where $x_1$, $x_2$, $x_3$, etc., are the values to be averaged. The notation $\langle x \rangle$ is also used for the mean value.

---

### Problem 4.20

Calculate the mean of the integers beginning with 10 and ending with 20.

---

There is another way to write the mean of a set of values, which is convenient if several of the members of the list are equal to each other. Let us arrange the members of our list so that the first $M$ members of the list are all different from each other, and each of the other $N - M$ members is equal to some member of the first subset. Let $N_i$ be the total number of members of the entire list that are equal to $x_i$, where $x_i$ is one of the distinct values in the first subset. Equation (4.57) can be rewritten

$$\bar{x} = \frac{1}{N}(N_1 x_1 + N_2 x_2 + N_3 x_3 + \cdots + N_M x_M) \qquad (4.58)$$

$$\bar{x} = \frac{1}{N} \sum_{i=1}^{M} N_i x_i = \sum_{i=1}^{M} p_i x_i. \qquad (4.59)$$

We again use the standard notation for a sum, introduced in Eq. (4.18). The quantity $p_i$ is equal to $N_i/N$ and is the fraction of the members of the entire list that are equal to $x_i$. If we were to sample the entire list by choosing a member at random, the probability that this member would equal $x_i$ is given by $p_i$. The set of probabilities that we have defined adds up to unity:

$$\sum_{i=1}^{N} p_i = \sum_{i=1}^{N} \frac{N_i}{N} = \frac{1}{N} \sum_{i=1}^{N} N_i = 1. \qquad (4.60)$$

A set of probabilities that adds up to unity is said to be *normalized*.

---

### Example 4.16

A quiz was given in a class with 100 members. The scores were as follows:

| Score | Number of students |
|-------|--------------------|
| 100   | 8                  |
| 90    | 11                 |
| 80    | 35                 |
| 70    | 23                 |
| 60    | 14                 |
| 50    | 9                  |

Find the mean score.

---

### Solution

$$\bar{s} = (0.08)(100) + (0.11)(90) + (0.35)(80) + (0.23)(70)$$
$$+ (0.14)(60) + (0.09)(50) = 74.9.$$

---

If you have a set of values with considerable duplication, Eq. (4.59) is quicker and easier to use than Eq. (4.57).

It is also possible to take the mean of a function of the values in our set. For example, to form the mean of the squares of the values, we have

$$\overline{x^2} = \frac{1}{N}\left(x_1^2 + x_2^2 + x_3^2 + \cdots + x_N^2\right) \tag{4.61}$$

$$= \frac{1}{N}\sum_{i=1}^{M} N_i x_i^2 = \sum_{i=1}^{M} p_i x_i^2. \tag{4.62}$$

Similarly, if $g = g(x)$ is any function defined for all values of $x$ that occur in our list, the mean value of this function is given by

$$\boxed{\overline{g(x)} = \sum_{i=1}^{M} p_i g(x_i)}. \tag{4.63}$$

### Problem 4.21

Find the mean of the squares of the scores given in Example 4.16. Find also the square root of this mean, which is called the *root-mean-square* score.

### Probability Distributions

One of the important quantities in gas kinetic theory is the mean speed of molecules in a gas. The calculation of such a mean is a little more complicated than the case of Eq. (4.59). The reasons for this are: (1) the speed of a molecule can take on any real nonnegative value, and (2) there are many molecules in almost any sample of a gas.

Let us develop formulas to handle cases like this. Consider a variable $x$, which can take on any real values between $x = a$ and $x = b$. Let us divide the interval $(a, b)$ into $n$ subintervals. Look at one of the subintervals, say $(x_i, x_{i+1})$, which can also be written $(x_i, x_i + \Delta x_i)$, where

$$\Delta x = x_{i+1} - x_i. \tag{4.64}$$

Let the fraction of all members of our sample that have values of $x$ lying between $x_i$ and $x_{i+1}$ be called $p_i$. If $\Delta x$ is quite small, $p_i$ will be very nearly proportional to $\Delta x$. We write

$$p_i = f_i \Delta x. \tag{4.65}$$

The quantity $f_i$ will not depend strongly on $\Delta x$. The mean value of $x$ is given by Eq. (4.59):

$$\bar{x} \approx \sum_{i-0}^{n-1} p_i x_i = \sum_{i-0}^{n-1} x_i f_i \Delta x. \tag{4.66}$$

This equation is only approximately true, because we have multiplied the probability that $x$ is in the subinterval $(x_i, x_i + \Delta x)$ by $x_i$, which is only one of the values of $x$ in the subinterval. However, Eq. (4.66) can be made more and more nearly exact by making $n$ larger and larger and $\Delta x$ smaller and smaller in such a way that $n\Delta x$ is constant. In this limit, $f_i$ becomes independent of $\Delta x$. We replace the symbol $f_i$

by $f(x_i)$, and assume that $f(x_i)$ is an integrable function of $x_i$. It must be at least piecewise continuous, as discussed earlier in this chapter. Our formula is now becomes an integral as defined in Eq. (4.22):

$$\bar{x} = \lim_{\substack{n \to \infty \\ \Delta x \to 0 \\ n\Delta x = b-a}} \sum_{i=0}^{n-1} x_i f(x_i)\Delta x = \int_a^b x f(x)\, dx. \tag{4.67}$$

The function $f(x)$ is called the *probability density*, or *probability distribution*, or sometimes the *distribution function*.

If we desire the mean value of a function of $x$, say $g(x)$, the formula is analogous to Eq. (4.63),

$$\boxed{\overline{g(x)} = \int_a^b g(x) f(x)\, dx} \,. \tag{4.68}$$

For example, the mean of $x^2$ is given by

$$\overline{x^2} = \int_a^b x^2 f(x)\, dx. \tag{4.69}$$

As defined above, the probability density is *normalized*, which now means that

$$\int_a^b f(x)\, dx = 1. \tag{4.70}$$

It is possible to use a probability density that is not normalized, but if you do this, you must modify Eq. (4.68). For an unnormalized probability distribution

$$\boxed{\overline{g(x)} = \frac{\displaystyle\int_a^b g(x) f(x)\, dx}{\displaystyle\int_a^b f(x)\, dx} \quad \text{(unnormalized probability distribution)}} \,. \tag{4.71}$$

For a normalized probability distribution, the probability that $x$ lies in the infinitesimal interval $(x, x+dx)$ is $f(x)\, dx$, which is the probability per unit length times the length of the infinitesimal interval. The fact that $f(x)$ is a probability per unit length is the reason for using the name "probability density" for it.

Since all continuously variable values of $x$ in some range are possible, a continuous probability distribution must apply to a set of infinitely many members. Such a set is called the *population* to which the distribution applies. The probability $f(x')\, dx$ is the fraction of the population that has its value of $x$ lying in the region between $x'$ and $x' + dx$.

The most commonly used measure of the "spread" of a probability distribution is the *standard deviation*, $\sigma_x$,

$$\boxed{\sigma_x = [\overline{x^2} - (\bar{x})^2]^{1/2}} \,. \tag{4.72}$$

We label the standard deviation with a subscript to indicate what variable is being considered. Generally, about two-thirds of a population will have their values of $x$ within one standard deviation of the mean; that is, within the interval $(\bar{x} - \sigma_x, \bar{x} + \sigma_x)$.

### Example 4.17

If all values of $x$ between $a$ and $b$ are equally probable, find the mean value of $x$, the root-mean-square value of $x$, and the standard deviation of $x$.

### Solution

The probability density is a constant for values of $x$ between $a$ and $b$, and in order to be normalized, it is

$$f(x) = \frac{1}{b-a}$$

so that

$$\bar{x} = \int_a^b x \frac{1}{b-a} \, dx = \frac{1}{2}(b+a)$$

$$\overline{x^2} = \int_a^b x^2 \frac{1}{b-a} \, dx = \frac{1}{3}(b^2 + ab + a^2).$$

The root-mean-square value of $x$ is

$$(\overline{x^2})^{1/2} = \left[ \frac{1}{3}(b^2 + ab + a^2) \right]^{1/2}.$$

The standard deviation is found as

$$\sigma_x^2 = \overline{x^2} - \bar{x}^2 = \frac{1}{3}\left(b^2 + ab + a^2\right) - \frac{1}{4}(b^2 + 2ab + a^2)$$

$$= \frac{1}{12}(a - b)^2$$

$$\sigma_x = \sqrt{\frac{1}{12}(a - b)}.$$

### Problem 4.22

From the results of Example 4.17, find the numerical values of $\bar{x}$, $(\overline{x^2})^{1/2}$, and $\sigma_x$ for $a = 0$ and $b = 10$. Comment on your values. What fraction of the total probability is found between $\bar{x} - \sigma_x$ and $\bar{x} + \sigma_x$?

### Problem 4.23

If $x$ ranges from 0 to 10 and if $f(x) = cx^2$, find the value of $c$ so that $f(x)$ is normalized. Find the mean value of $x$ and the root-mean-square value of $x$.

## The Gaussian Distribution

The *Gaussian distribution* is represented by the formula

$$f(x) = \frac{1}{\sqrt{2\pi}\sigma} \exp\left[ -\frac{(x - \mu)^2}{2\sigma^2} \right], \tag{4.73}$$

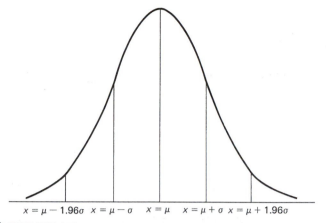

$x = \mu - 1.96\sigma \quad x = \mu - \sigma \quad x = \mu \quad x = \mu + \sigma \quad x = \mu + 1.96\sigma$

**FIGURE 4.7** The Gaussian probability distribution.

where $\mu$ is the mean value of $x$ and where $\sigma$ is the standard deviation. This distribution is also called the *normal distribution*. If $\sigma = 1$, then the distribution is called the *standard normal distribution*.

The Gaussian distribution is the most commonly encountered probability distribution. It is assumed to describe populations of various kinds, including the IQ scores of people and velocities of molecules in a gas.[2]

### Example 4.18

Show that the distribution in Eq. (4.73) satisfies the normalization condition of Eq. (4.70) with the limits of integration equal to $-\infty$ and $+\infty$.

### Solution

We have the following integral, which we modify by the method of substitution, and then look up in Appendix 7:

$$\int_{-\infty}^{\infty} \frac{1}{\sqrt{2\pi}\sigma} e^{-(x-\mu)^2/2\sigma^2} \, dx = \frac{1}{\sqrt{\pi}} \int_{-\infty}^{\infty} e^{-t^2} dt = 1.$$

### Problem 4.24

Calculate the mean and standard deviation of the Gaussian distribution, showing that $\mu$ is the mean and that $\sigma$ is the standard deviation.

Figure 4.7 shows a graph of the Gaussian distribution. Five values of $x$ have been marked on the $x$ axis: $x = \mu - 1.96\sigma$, $x = \mu - \sigma$, $x = \mu$, $x = \mu + \sigma$, and $x = \mu + 1.96\sigma$.

### Example 4.19

Assuming the Gaussian distribution, calculate the fraction of the population with $x$ lying between $x = \mu - \sigma$ and $x = \mu + \sigma$.

[2]The Gaussian distribution is named after Carl Friedrich Gauss, 1777–1855, a great German mathematician who made a number of important discoveries in addition to this one.

**Solution**

By integration of Eq. (4.73), we obtain

$$\text{(fraction between } \mu - \sigma \text{ and } \mu + \sigma) = \int_{\mu-\sigma}^{\mu+\sigma} \frac{1}{\sqrt{2\pi}\sigma} e^{-(x-\mu)^2/2\sigma^2} \, dx$$

$$= \int_{-\sigma}^{+\sigma} \frac{1}{\sqrt{2\pi}\sigma} e^{-y^2/2\sigma^2} \, dy$$

$$= \frac{2}{\sqrt{\pi}} \int_0^{1/\sqrt{2}} e^{-t^2} \, dt.$$

The last integral in this example is the *error function* with argument $1/\sqrt{2}$. The integrand function $e^{-t^2}$ does not possess an indefinite integral that can be written with a single formula, so the error function must be approximated numerically unless the upper limit is infinite. The error function is described in Appendix 7. From the table of values in that appendix,

$$\text{(fraction between } \mu - \sigma \text{ and } \mu + \sigma) = \text{erf}\left(\frac{1}{\sqrt{2}}\right) = \text{erf}(0.707\cdots) = 0.683\cdots$$

in agreement with our assertion that roughly two-thirds of the members of a population lie within one standard deviation of the mean.

---

**Problem 4.25**

Show that the fraction of a population lying between $\mu - 1.96\sigma$ and $\mu + 1.96\sigma$ is equal to 0.95 if the population is described by the Gaussian distribution.

---

You should remember the facts obtained in Example 4.19 and Problem 4.25. With a Gaussian distribution, 68% of the population lies within 1 standard deviation of the mean and 95% of the population lies within 1.96 standard deviations of the mean. Also, 99% of the population lies within 2.67 standard deviations of the mean. For other intervals, we can write

$$\text{fraction between } \sigma - x_1 \text{ and } \mu + x_1 = \text{erf}\left(\frac{x_1}{\sqrt{2}\sigma}\right). \tag{4.74}$$

There are a number of other common probability distributions in addition to the Gaussian distribution, including the binomial distribution, the Poisson distribution, and the Lorentzian distribution.[3] However, the Gaussian distribution is generally used in discussing experimental errors, and we return to this topic in Chapter 10.

## Probability Distributions in Gas Kinetic Theory

In gas kinetic theory, the probability density for the molecular velocity is a Gaussian distribution. The normalized probability distribution for $v_x$, the $x$ component of the

---

[3] See Philip R. Bevington, *Data Reduction and Error Analysis for the Physical Sciences*, Chap. 3, McGraw–Hill, New York, 1969, or Hugh D. Young, *Statistical Treatment of Experimental Data*, McGraw–Hill, New York, 1962, for discussions of various distributions.

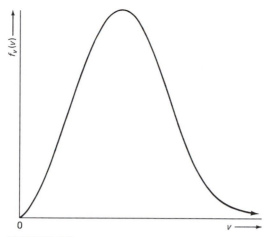

**FIGURE 4.8**    A graph of the probability density for speeds of molecules in a gas.

velocity, is given by

$$f(v_x) = \left(\frac{m}{2\pi k_B T}\right)^{3/2} \exp\left(-\frac{m v_x^2}{2k_B T}\right),$$  (4.75)

where $m$ is the molecular mass, $T$ the temperature on the Kelvin scale, and $k_B$ is Boltzmann's constant.

---

**Problem 4.26**

    a. Show that the mean value of $v_x$ is equal to zero. Explain this fact.
    b. Find the expression for the root-mean-square value of $v_x$.
    c. Find the expression for the standard deviation of $v_x$.

---

The speed $v$ is the magnitude of the velocity. Since velocities with the same magnitude but different directions are included in the same speed, the distribution of speeds is different from the velocity distribution of Eq. (4.75). We denote the speed distribution by $f_v(v)$. It is given by

$$f_v(v) = 4\pi \left(\frac{m}{2\pi k_B T}\right)^{3/2} v^2 \exp\left(-\frac{m v^2}{2k_B T}\right).$$  (4.76)

A graph of this function is shown in Fig. 4.8. The speed is never negative, so that the graph does not extend to the left of the origin. In ordinary gas kinetic theory, the requirements of special relativity are ignored, and speeds approaching infinity are included. The error due to this inclusion is insignificant at ordinary temperatures, because of the low probability ascribed to high speeds.

The mean speed is given by

$$\bar{v} = \int_0^\infty v f_v(v) \, dv.$$  (4.77)

---

**Example 4.20**

Obtain a formula for the mean speed of molecules in a gas.

**Solution**

$$\bar{v} = 4\pi \left( \frac{m}{2\pi k_B T} \right)^{3/2} \int_0^\infty v^3 \exp\left( -\frac{mv^2}{2k_B T} \right) dv. \tag{4.78}$$

This integral can be obtained from Eq. (7) of Appendix 7,

$$\bar{v} = 4\pi \left( \frac{m}{2\pi k_B T} \right)^{3/2} \frac{(2k_B T)^2}{2m^2} = \left( \frac{8k_B T}{\pi m} \right)^{1/2}. \tag{4.79}$$

---

**Problem 4.27**

Find the value of $\bar{v}$ for $N_2$ gas at 298 K.

---

**Example 4.21**

Find a formula for the mean of the square of the speed of molecules in a gas, and for the root-mean-square speed.

---

**Solution**

$$\overline{v^2} = 4\pi \left( \frac{m}{2\pi k_B T} \right)^{3/2} \int_0^\infty v^4 \exp\left( -\frac{mv^2}{2k_B T} \right) dv$$

$$= \frac{3k_B T}{m}$$

$$(\overline{v^2})^{1/2} = \left( \frac{3k_B T}{m} \right)^{1/2}.$$

---

**Problem 4.28**

For molecules in a gas, find the formula for $\sigma_v$ and find its value for $N_2$ gas at 298 K.

---

## Time Averages

If $g = g(t)$, the time average of $g$ is defined as

$$\overline{g(t)} = \int_{t_1}^{t_2} g(t) f(t) \, dt, \tag{4.80}$$

where we call $f(t)$ the *weighting function*. It plays the same role as a probability density. Most time averages are unweighted, which means that $f(t)$ is a constant,

$$\overline{g(t)} = \frac{1}{t_1 - t_2} \int_{t_1}^{t_2} g(t) \, dt. \tag{4.81}$$

Equation (4.81) is a version of the *mean value theorem* of integral calculus, which states that the mean value of a function is equal to the integral of the function divided by the length of the interval over which the mean is taken.

### Example 4.22

A particle falls in a vacuum near the surface of the earth. Find the average $z$ component of the velocity during the first 10.00 s of fall if the initial speed is zero.

### Solution

From Eq. (4.7),

$$v_z = -gt,$$

where $g$ is the acceleration due to gravity, $9.80\,\mathrm{m\,s^{-2}}$,

$$\overline{v_z} = -\frac{1}{10}\int_0^{10} gt\,dt = -\frac{g}{10}\frac{t^2}{2}\bigg|^{10} = -5g = -49.0\ \mathrm{m\,s^{-1}}.$$

### Problem 4.29

Find the time-average value of the $z$ coordinate of the particle in Example 4.22 for the first 10.00 seconds of fall if the initial position is $z = 0.00$ meters.

## SUMMARY OF THE CHAPTER

Integration is one of the two fundamental processes of calculus. It is essentially the reverse of differentiation, the other important process. The first kind of an integral is the indefinite integral, which is the antiderivative of the integrand function. The second kind of an integral is the definite integral, which is constructed as the sum of very many small increments of the antiderivative function $F$, constructed from the integrand function $f$ according to the formula from Chapter 3,

$$dF = f(x)\,dx = \frac{dF}{dx}\,dx.$$

The definite integral has a lower limit of integration, at which the summation process starts, and an upper limit, at which the process ends. The definite integral is therefore equal to the value of the antiderivative function (indefinite integral) at the upper limit minus its value at the lower limit

$$\int_a^b f(x)\,dx = F(b) - F(a).$$

Extensive tables of indefinite integrals exist, as well as tables of definite integrals. Some integrals have infinite limits or have integrands which attain infinite values, and such integrals are called improper integrals. Many such integrals diverge (have undefined values), but some improper integrals have finite values and are said to converge.

There are several techniques for manipulating integrals into a form which you can recognize or which you can look up in a table, and we presented a few of these. Some integrals cannot be worked out mathematically, but must be approximated numerically. We discussed some elementary techniques for carryhing out this approximation, including Simpson's rule, the most commonly used technique, and finally presented a simple computer program for implementing Simpson's rule.

Mean values of variables that take on all real values in a certain interval are calculated as integrals of the form

$$\bar{g} = \int_a^b g(x) f(x) \, dx,$$

where $g(x)$ is the function to be averaged and $f(x)$ is the probability distribution, or probability density, or distribution function.

## ADDITIONAL READING

Milton Abramowitz and Irene A. Stegun, Eds., *Handbook of Mathematical Functions with Formulas, Graphs and Mathematical Tables*, National Bureau of Standards Applied Mathematics Series No. 55, U.S. Government Printing Office, Washington, DC, 1964 (reprinted by Dover, New York). This large but inexpensive book contains a variety of different things, including integrals and useful formulas.

D. Bierens de Haan, *Nouvelles tables d'integrales definies* (New Tables of Definite Integrals), Hafner, New York, 1956. This book is a corrected reprint of an edition of 1867 and is an excellent large collection of definite integrals. However, some entries use an unfamiliar notation that is unfortunately not explained in the book.

Richard L. Burden and J. Douglas Faires, *Numerical Analysis*, 3rd ed., PWS–Kent, Boston, 1986. This is a standard numerical analysis textbook at the advanced undergraduate level. It contains explicit algorithms which can easily be converted into computer programs.

Herbert B. Dwight, *Tables of Integrals and Other Mathematical Data*, 4th ed., Macmillan Co., New York, 1962. This is a very good small compilation of definite and indefinite integrals, and also contains a lot of other useful formulas and tables.

I. S. Gradshteyn and I. M. Ryzhik, *Tables of Integrals, Series and Products*, 4th ed., prepared by Yu. V. Geronimus and M. Yu. Tseytlin, translated by Alan Jeffreys, Academic Press, New York, 1965. This is a large book with lots of definite and indefinite integrals in it.

Robert W. Hornbeck, *Numerical Methods*, Quantum Publishers, New York, 1975. This book discusses many topics of numerical analysis, including numerical integration, and appears as an inexpensive paperback.

Sherman K. Stein, *Calculus and Analytic Geometry*, 2nd ed., McGraw–Hill, New York, 1977. This is one of a number of standard calculus textbooks. All of them discuss integration in much more detail than the present book.

## ADDITIONAL PROBLEMS

### 4.30

Find the indefinite integrals without using a table:

a. $\displaystyle\int x \ln(x) \, dx$

b. $\int x \sin^2(x)\,dx$

c. $\int (1/x(x-a))\,dx$

d. $\int x^3 \ln(x^2)\,dx.$

### 4.31

Find the definite integrals without using a table:

a. $\int_0^{2\pi} \sin(x)\,dx$

b. $\int_1^2 x \ln(x)^2\,dx$

c. $\int_0^{\pi/2} \sin^2(x)\cos(x)\,dx$

d. $\int_0^{\pi/2} x \sin(x^2)\cos(x^2)\,dx.$

### 4.32

Evaluate the improper integrals:

a. $\int_1^\infty (1/x^2)\,dx$

b. $\int_1^{\pi/2} \tan(x)\,dx.$

### 4.33

At 298 K, what fraction of nitrogen molecules has speeds lying between 0 and the mean speed? Do a numerical approximation to the integral or use the identity

$$\int_0^x t^2 e^{-at^2}\,dt = \frac{\sqrt{\pi}}{4a^{3/2}} \mathrm{erf}(\sqrt{a}x) - \frac{x}{2a} e^{-ax^2}. \qquad (4.82)$$

### 4.34

Approximate the integral

$$\int_0^\infty e^{-x^2}\,dx$$

using Simpson's rule. You will have to take a finite upper limit, choosing a value large enough so that the error caused by using the wrong limit is negligible. The correct answer is $\sqrt{\pi}/2 = 0.886226926\ldots.$

### 4.35

Find the integrals

a. $\int \sin[x(x+1)](2x+1)\,dx$

b. $\int_0^\pi \sin[\cos(x)\sin(x)]\,dx.$

**4.36**

When a gas expands reversibly, the work that it does against its surroundings is given by the integral

$$w = \int_{v_3}^{v_2} P \, dV,$$

where $V_1$ is the initial volume, $V_2$ the final volume, and $P$ the pressure. Certain nonideal gases are described quite well by the van der Waals equation of state,

$$\left( P + \frac{a}{\bar{V}^2} \right)(\bar{V} - b) = RT$$

$$\left( P + \frac{a}{V^2} \right)(V - b) = RT,$$

where $\bar{V}$ is the volume of 1 mol of the gas (the molar volume), $T$ is the temperature on the Kelvin scale, and $a$, $b$, and $R$ are constants. $R$ is usually taken to be the ideal gas constant, $8.3145 \, \text{J K}^{-1} \, \text{mol}^{-1}$.

  a. Obtain a formula for the work done if 1 mol of such a gas expands reversibly at constant temperature from a molar volume $\bar{V}_1$ to a molar volume $\bar{V}_2$.
  b. If $T = 298$ K, $V_1 = 1.00$ liter $= 1.000 \times 10^{-3}$ m$^3$, and $V_2 = 100.$ liters $= 0.100$ m$^3$, find the value of the work done for 1.000 mol of $CO_2$, which has $a = 3.59 \, \text{liters}^2 \, \text{atm mol}^{-2}$, and $b = 42.7 \, \text{cm}^3 \, \text{mol}^{-1} = 0.0427 \, \text{liter mol}^{-1}$. Use the value of the ideal gas constant, $R = 8.3145 \, \text{J K}^{-1} \, \text{mol}^{-1} = 0.082057 \, \text{liter atm K}^{-1} \, \text{mol}^{-1}$.

**4.37**

The entropy change to bring a sample from 0 K (absolute zero) to a given state is called the *absolute entropy* of the sample in that state. Using Eq. (4.56) calculate the absolute entropy of 1.000 mol of solid silver at 270 K. For the region 0 to 30 K, use the approximate relation

$$C_p = aT^3,$$

where $a$ is a constant that you can evaluate from the value of $C_p$ at 30 K. For the region 30 to 270 K, use the following data:[4]

| $T(K)$ | $C_p(JK^{-1} \, mol^{-1})$ |
|---|---|
| 30 | 4.77 |
| 50 | 11.65 |
| 70 | 16.33 |
| 90 | 19.13 |
| 110 | 20.96 |
| 130 | 22.13 |
| 150 | 22.97 |
| 170 | 23.61 |
| 190 | 24.09 |
| 210 | 24.42 |
| 230 | 24.73 |
| 250 | 25.03 |
| 270 | 25.31 |

[4]Meads, Forsythe, and Giaque, *J. Am. Chem. Soc.* **63**, 1902 (1941).

# 5
# CALCULUS WITH SEVERAL INDEPENDENT VARIABLES

## Preview

In this chapter, we discuss functions of more than one independent variable. For example, if you have a function of three independent variables, the function will deliver a value of the dependent variable if a value of each independent variable is specified. Differential calculus of such functions begins with the differential of the function, which represents an infinitesimal change in the dependent variable resulting from infinitesimal changes in the independent variables and consists of a sum of terms. Each term consists of a partial derivative with respect to an independent variable multiplied by the differential of that variable (an infinitesimal change in that independent variable). Maximum and minimum values of functions can be found using partial derivatives. There are two principal kinds of integrals that have integrand functions depending on several independent variables: line integrals and multiple integrals.

## Principal Facts and Ideas

1. Functions of several independent variables occur frequently in physical chemistry, both in thermodynamics and in quantum mechanics.
2. A derivative of a function of several variables with respect to one independent variable is a partial derivative. The other variables are treated as constants during the differentiation.
3. There are some useful identities allowing manipulations of expressions containing partial derivatives.

4. The differential of a function of several variables (an exact differential) has one term for each variable, consisting of a partial derivative times the differential of the independent variable. This differential form delivers the value of an infinitesimal change in the function produced by infinitesimal changes in the independent variables.

5. Differential forms exist that are not the differentials of any function. Such a differential form delivers the value of an infinitesimal quantity, but it is not the differential of any function. Such a differential form is called an inexact differential.

6. An integral of a differential with several independent variables is a line integral, carried out on a specified path in the space of the independent variables.

7. The line integral of an exact differential depends only on the endpoints of the path, but the line integral of an inexact differential depends on the path.

8. A multiple integral has as its integrand function a function of several variables, all of which are integrated over.

9. The gradient operator is a vector derivative operator that produces a vector when applied to a scalar function.

10. The divergence operator is a vector derivative operator that produces a scalar when applied to a vector function.

11. Relative maxima and minima of a function of several variables are found by solving simultaneously the equations obtained by setting all partial derivatives equal to zero.

12. Constrained maxima and minima of a function of several variables can be found by the method of Lagrange multipliers.

## Objectives

After studying this chapter, you should be able to:

1. write formulas for the partial derivatives and for the differential of a function if given a formula for the function and use these in applications such as the calculation of small changes in a dependent variable;

2. perform a change of independent variables and obtain formulas relating different partial derivatives;

3. use identities involving partial derivatives to eliminate undesirable quantities from thermodynamic formulas;

4. identify an exact differential and an integrating factor;

5. perform a line integral with two independent variables;

6. perform a multiple integral;

7. change independent variables in a multiple integral;

8. use vector derivative operators;

9. find constrained and unconstrained maximum and minimum values of functions of several variables.

## SECTION 5.1. FUNCTIONS OF SEVERAL VARIABLES

A function of several independent variables is similar to a function of a single independent variable except that you must specify a value for each of the independent variables in order for the function to provide a value for the dependent variable. For example, the equilibrium thermodynamic properties of a fluid (gas or liquid) system of one component and one phase are functions of three independent variables. If we choose a set of values for the temperature, $T$, the volume, $V$, and the amount of the substance in moles, $n$, then the other thermodynamic properties, such as pressure, $P$, and thermodynamic energy, $U$, are functions of these variables. For example,

$$P = P(T, V, n). \tag{5.1}$$

We assume that the functions that represent the behavior of physical systems are piecewise continuous with respect to each variable. That is, if we temporarily keep all but one of the independent variables fixed, the function behaves as a piecewise continuous function of that variable. We also assume that the function is piecewise single-valued.

In physical chemistry, we sometimes work with mathematical formulas that represent specific functions, and we sometimes work with identities that allow us to obtain useful results without having a particular function at hand. For example, if the temperature of a gas is fairly high and its volume is large enough, the pressure of a gas is given to a good approximation by the ideal gas equation

$$P = \frac{nRT}{V}. \tag{5.2}$$

We will sometimes use this and other specific mathematical functional relations. However, in other cases, we will write general equations involving the pressure or some other variable without specifying whether the pressure is given by Eq. (5.2), by some other formula, or by some unknown function.

## SECTION 5.2. DIFFERENTIALS AND PARTIAL DERIVATIVES

Functions of several variables can be represented by mathematical formulas, as in Eq. (5.2). They can also be represented by graphs, by tables of values, or by infinite series. However, graphs, tables, and series become more complicated when used for a function of several variables than for functions of a single variable.

Figure 5.1 shows the dependence of the pressure of a nearly ideal gas as a function of the temperature $T$ and the molar volume, $\bar{V}$, defined as $V/n$. Equation (5.2) can be rewritten as a function of two variables, $T$ and $\bar{V}$:

$$P = \frac{RT}{\bar{V}}.$$

With only two axes on our graph, a curve can show the dependence of $P$ on $\bar{V}$ for a fixed value of $T$. The figure shows curves for several members of a family of functions of $\bar{V}$, each for a different value of $T$.

Figure 5.2 attempts to represent a three-dimensional graph. The value of $P$ is given by the height from the horizontal $\bar{V}$–$T$ plane to a surface, which plays the same role as the curve in a two-dimensional graph. It is fairly easy to read quantitative information from the two-dimensional graph in Fig. 5.1, but the perspective view in Fig. 5.2 is more difficult to get numbers from.

If you have more than two independent variables, a graph is even more difficult to use. Sometimes attempts are made to show roughly how functions of three variables

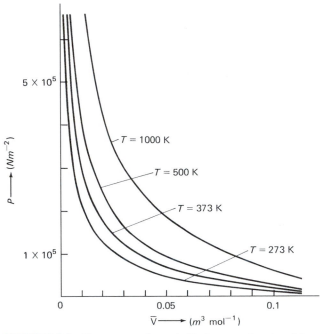

**FIGURE 5.1** The pressure of a nearly ideal gas as a function of the molar volume $\bar{V}$ at various fixed temperatures.

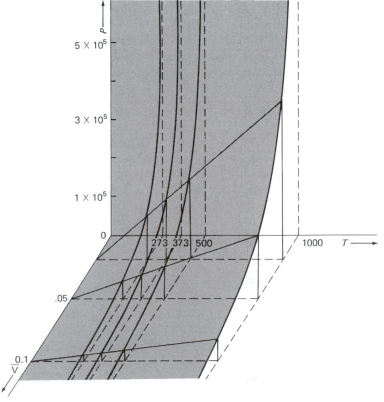

**FIGURE 5.2** The pressure of a nearly ideal gas a function of $\bar{V}$ and $T$.

depend on their independent variables by drawing a perspective view of three axes, one for each of the independent variables, and then trying to communicate the approximate value of the dependent variable by a density of shading or of dots placed in the diagram.

Tables of values are also cumbersome with two or more independent variables, since a function is now not a set of ordered pairs of numbers but a set of ordered sets of three numbers or four numbers, etc. For two independent variables, we need a rectangular array, with values of one independent variable along the top and values of the other along one side, and values of the dependent variable in the body of the array. For a third independent variable, we would need a different sheet of paper for each value of the third variable. The most common way to represent a function of several variables is with a mathematical formula.

## Changes in a Function of Several Variables

Consider a gas contained in a cylinder with a movable piston. Let the cylinder have a valve through which additional gas can be admitted or through which gas can be removed, and let the entire system be immersed in a constant-temperature bath with an adjustable thermoregulator. For the present, we keep the valve closed, so that $n$ is fixed (the system is now a *closed system*).

Let us now make an infinitesimal change $dV$ in the volume of the gas, keeping $T$ fixed. If $n$ and $T$ are both fixed, $P$ will behave just like a function of $V$ alone. The change in $P$ is given as in Chapter 3;

$$dP = \left(\frac{dP}{dV}\right) dV \qquad (n \text{ and } T \text{ fixed}), \tag{5.3}$$

where $dP/dV$ is the derivative of $P$ with respect to $V$.

We adopt a new notation, adding subscripts to remind us that $P$ is not just a function of $V$ but is also a function of $n$ and $T$, and replacing the $d$'s by special symbols that are slightly distorted Greek deltas.

$$dP = \left(\frac{\partial P}{\partial V}\right)_{n,T} dV \qquad (n \text{ and } T \text{ fixed}). \tag{5.4}$$

The quantity $(\partial P/\partial V)_{n,T}$ is called the *partial derivative of P with respect to V at constant n and T*. It is obtained by the differentiation techniques of Chapter 3, treating $n$ and $T$ like ordinary constants.

For an ideal gas

$$\left(\frac{\partial P}{\partial V}\right)_{n,T} = \left(\frac{\partial}{\partial V}\left[\frac{nRT}{V}\right]\right)_{n,T} = -\frac{nRT}{V^2}. \tag{5.5}$$

After making an infinitesimal change in the volume, we can now make an infinitesimal change in the temperature, $dT$, keeping $n$ and $V$ fixed. We can write

$$dP = \left(\frac{\partial P}{\partial T}\right)_{n,V} dT \qquad (n \text{ and } V \text{ fixed}). \tag{5.6}$$

Here $(\partial P/\partial T)_{n,V}$ is the partial derivative of $P$ with respect to $T$ at constant $n$ and $V$. It is obtained by the usual techniques of differentiation, treating $n$ and $V$ like constants.

For an ideal gas

$$\left(\frac{\partial P}{\partial T}\right)_{n,V} = \frac{nR}{V}. \tag{5.7}$$

If we now make a simultaneous change $dV$ in the volume and $dT$ in the temperature of the gas, the change in $P$ is the sum of the expressions in Eqs. (5.4) and (5.6).

$$dP = \left(\frac{\partial P}{\partial V}\right)_{n,T} dV + \left(\frac{\partial P}{\partial T}\right)_{n,V} dT \qquad (n \text{ fixed}). \tag{5.8}$$

Each term of this equation is the change due to the change in only one independent variable, and each partial derivative thus is taken with the other independent variables treated as constants.

If we make finite but small changes in $T$ and $V$, we can write as an approximation

$$\Delta P \approx \left(\frac{\partial P}{\partial V}\right)_{n,T} \Delta V + \left(\frac{\partial P}{\partial T}\right)_{n,V} \Delta T \qquad (n \text{ fixed}). \tag{5.9}$$

This equation becomes more nearly exact as $\Delta V$ and $\Delta T$ are made smaller and smaller, and likely is a better approximation for finite $\Delta V$ and $\Delta T$ if these increments are fairly small. We now consider simultaneous changes $dT$ in T, $dV$ in V, and $dn$ in $n$. For infinitesimal changes, these changes affect $P$ separately, and we can write for the total infinitesimal change in $P$,

$$dP = \left(\frac{\partial P}{\partial V}\right)_{n,T} dV + \left(\frac{\partial P}{\partial T}\right)_{n,V} dT + \left(\frac{\partial P}{\partial n}\right)_{V,T} dn. \tag{5.10}$$

The infinitesimal change $dP$ given by this expression is called the *differential* of $P$, or sometimes the *total differential* of $P$. It is a sum of terms, one for each independent variable. Each term gives the effect of one variable with the other treated as constants.

If we have a function $y$ that depends on $n$ independent variables, $x_1, x_2, x_3, \ldots, x_n$ its differential is

$$\boxed{dy = \sum_{i=1}^{n} \left(\frac{\partial y}{\partial x_i}\right)_{x'} dx_i}, \tag{5.11}$$

where the symbol $x'$ stands for keeping all of the variables except for $x_i$ fixed in the differentiation.

The expression for $dP$ for an ideal gas is

$$dP = -\frac{nRT}{V^2} dV + \frac{nR}{V} dT + \frac{RT}{V} dn. \tag{5.12}$$

For small but finite changes, an approximate version of this can be written

$$\Delta P \approx -\frac{nRT}{V^2} \Delta V + \frac{nR}{V} \Delta T + \frac{RT}{V} \Delta n. \tag{5.13}$$

**Example 5.1**

Use Eq. (5.13) to calculate approximately the change in pressure of an ideal gas if the volume is changed from 20.000 liters to 19.800 liters, the temperature is changed from 298.15 K to 299 K, and the number of moles is changed from 1.0000 to 1.0015. Compare with the correct value of the pressure change.

---

**Solution**

$$\Delta P \approx -\frac{(1.0000 \text{ mol})(8.3144 \text{ J K}^{-1} \text{ mol}^{-1})(298.15 \text{ K})}{(0.020000 \text{ m}^3)^2}(-0.200 \times 10^{-3} \text{ m}^3)$$

$$+ \frac{(1.0000 \text{ mol})(8.3144 \text{ J K}^{-1} \text{ mol}^{-1})}{0.020000 \text{ m}^3}(0.85 \text{ K})$$

$$+ \frac{(8.3144 \text{ J K}^{-1} \text{ mol}^{-1})(298.15 \text{ K})}{0.020000 \text{ m}^3}(0.0015 \text{ mol})$$

$$\approx 1.779 \times 10^3 \text{ N m}^{-2},$$

where we use the fact that $1 \text{ J} = 1 \text{ N m}$.

We can calculate the actual change as follows. Let the initial values of $n$, $V$, and $T$ be called $n_1$, $V_1$, and $T_1$, and the final values be called $n_2$, $V_2$, and $T_2$,

$$\Delta P = P(n_2, V_2, T_2) - P(n_1, V_1, T_1)$$

$$= \frac{n_2 R T_2}{V_2} - \frac{n_1 R T_1}{V_1}.$$

The result of this calculation is $1.797 \times 10^3 \text{ N m}^{-2} = 1.797 \times 10^3$ pascal (Pa), so that our approximate value is in error by about 1%.

---

In Example 5.1, the exact calculation could be made more easily than the approximation. However, in physical chemistry it is frequently the case that a formula for a function is not known, but values for the partial derivatives are available, so that approximation can be made while the exact calculation cannot. For example, there is usually no simple formula giving the thermodynamic energy as a function of its independent variables. However, the differential of the thermodynamic energy of a system containing only one substance as a function of $T$, $P$, and $n$ can be written as

$$dU = \left(\frac{\partial U}{\partial T}\right)_{P,n} dT + \left(\frac{\partial U}{\partial P}\right)_{T,n} dP + \left(\frac{\partial U}{\partial n}\right)_{P,T} dn. \tag{5.14}$$

Experimental values of these partial derivatives are usually available.

**Problem 5.1**

For a sample of 1.000 mole (0.15384 kg) of carbon tetrachloride, $CCl_4$ (with $n$ fixed so that $dn$ vanishes)

$$\left(\frac{\partial U}{\partial T}\right)_{P,n} = 129.4 \text{ J K}^{-1} \text{mol}^{-1}$$

$$\left(\frac{\partial U}{\partial P}\right)_{T,n} = 8.51 \times 10^{-4} \text{ J atm}^{-1} \text{mol}^{-1},$$

where these values are for a temperature of 20°C and a pressure of 1.000 atm. Estimate the change in the energy of 1.000 mole of $CCl_4$ if its temperature is changed from 20°C to 40°C and its pressure from 1 atm to 100 atm.

**Problem 5.2**

The volume of a right circular cylinder is given by

$$V = \pi r^2 h,$$

where $r$ is the radius and $h$ the height. Calculate the percentage error in the volume if the radius and the height are measured and a 1% error is made in each measurement in the same direction. Use the formula for the differential, and also direct substitution into the formula for the volume, and compare the two answers.

## SECTION 5.3. CHANGE OF VARIABLES

In thermodynamics there is usually the possibility of choosing between different sets of independent variables. For example, we can consider the thermodynamic energy $U$ of a one-component, one-phase system to be a function of $T$, $V$, and $n$,

$$U = U(T, V, n), \tag{5.15}$$

or a function of $T$, $P$, and $n$,

$$U = U(T, P, n). \tag{5.16}$$

The two choices lead to different expressions for $dU$,

$$dU = \left(\frac{\partial U}{\partial T}\right)_{V,n} dT + \left(\frac{\partial U}{\partial V}\right)_{T,n} dV + \left(\frac{\partial U}{\partial n}\right)_{T,V} dn \tag{5.17}$$

and

$$dU = \left(\frac{\partial U}{\partial T}\right)_{P,n} dT + \left(\frac{\partial U}{\partial P}\right)_{T,n} dP + \left(\frac{\partial U}{\partial n}\right)_{T,p} dn. \tag{5.18}$$

There are two different derivatives of $U$ with respect to $T$: $(\partial U/\partial T)_{V,n}$ and $(\partial U/\partial T)_{P,n}$. These derivatives have different values for most systems. If we did not use the subscripts, there would be no difference between the symbols for the two derivatives, which could lead to confusion.

**Example 5.2**

Express the function $z = x(x, y) = ax^2 + bxy + cy^2$ in terms of $x$ and $u$, where $u = xy$. Find the two partial derivatives $(\partial z/\partial x)_y$ and $(\partial z/\partial x)_u$.

**Solution**

$$z = z(x, u) = ax^2 + bu + \frac{cu^2}{x^2}$$

$$\left(\frac{\partial z}{\partial x}\right)_y = \left(\frac{\partial}{\partial x}(ax^2 + bxy + y^2)\right)_y = 2ax + by$$

$$\left(\frac{\partial z}{\partial x}\right)_u = 2ax - \frac{2cu^2}{x^3} = \left(\frac{\partial z}{\partial x}\right)_y - \frac{bu}{x} - \frac{2cu^2}{x^3}.$$

In Example 5.2, there was no difficulty in obtaining an expression for the difference between $(\partial z/\partial x)_y$ and $(\partial z/\partial x)_u$, because we had the formula to represent the mathematical function. In thermodynamics, it is unusual to have a functional form in front of us. More commonly we have measured values for partial derivatives and require a separate means for computing the difference between partial derivatives.

We will obtain a formula

$$\left(\frac{\partial U}{\partial T}\right)_{V,n} = \left(\frac{\partial U}{\partial T}\right)_{P,n} + ?, \tag{5.19}$$

where the question mark indicates an unknown term. The procedure that we use is not mathematically rigorous, but it does give the correct answer.

To construct the partial derivative on the left-hand side of our equation, we begin with an expression for $dU$ that contains the derivative on the right-hand side. This is Eq. (5.18). The first thing we do is to "divide" this differential expression by $dT$, because the derivative we want on the left-hand side is $(\partial U/\partial T)_{V,n}$. This cannot be done legitimately, because $dT$ is an infinitesimal quantity, but we do it anyway. We get

$$\frac{dU}{dT} = \left(\frac{\partial U}{\partial T}\right)_{P,n}\frac{dT}{dT} + \left(\frac{\partial U}{\partial P}\right)_{T,n}\frac{dP}{dT} + \left(\frac{\partial U}{\partial n}\right)_{P,T}\frac{dn}{dT}. \tag{5.20}$$

This equation contains several things that look like ordinary derivatives. However, we must interpret them as partial derivatives, since there are other independent variables besides $T$. We change the symbols to the symbols for partial derivatives and add the appropriate subscripts to indicate the other variables. We choose whatever we want the other variables to be, but they must be the same in all four of the derivatives that are now written as ordinary derivatives. We want $V$ and $n$ to be constant, so we write

$$\left(\frac{\partial U}{\partial T}\right)_{V,n} = \left(\frac{\partial U}{\partial T}\right)_{P,n}\left(\frac{\partial T}{\partial T}\right)_{V,n} + \left(\frac{\partial U}{\partial P}\right)_{T,n}\left(\frac{\partial P}{\partial T}\right)_{V,n} + \left(\frac{\partial U}{\partial n}\right)_{P,T}\left(\frac{\partial n}{\partial T}\right)_{V,n}.$$

$$\tag{5.21}$$

The partial derivative of $T$ with respect to $T$ is equal to unity, no matter what is held constant, and the partial derivative of $n$ with respect to anything is zero if $n$ is constant, so

$$\left(\frac{\partial U}{\partial T}\right)_{V,n} = \left(\frac{\partial U}{\partial T}\right)_{P,n} + \left(\frac{\partial U}{\partial P}\right)_{T,n}\left(\frac{\partial P}{\partial T}\right)_{V,n}. \tag{5.22}$$

### Example 5.3

Apply the foregoing method to the function in Example 5.2 and find the relation between $(\partial z/\partial x)_u$ and $(\partial z/\partial x)_y$.

### Solution

$$\left(\frac{\partial z}{\partial x}\right)_u = \left(\frac{\partial z}{\partial x}\right)_y + \left(\frac{\partial z}{\partial y}\right)_x\left(\frac{\partial y}{\partial x}\right)_u$$

$$\left(\frac{\partial z}{\partial y}\right)_x = bx + 2cy = bx + \frac{2cu}{x}$$

$$\left(\frac{\partial y}{\partial x}\right)_u = \left[\frac{\partial}{\partial x}\left(\frac{u}{x}\right)\right]_u = -\frac{u}{x^2}.$$

Thus

$$\left(\frac{\partial z}{\partial x}\right)_u = \left(\frac{\partial z}{\partial x}\right)_y - \frac{bu}{x} - \frac{2cu}{x^3}.$$

This agrees with Example 5.2, as it must.

### Problem 5.3

Complete the following equations.

a. $(\partial H/\partial T)_{P,n} = (\partial H/\partial T)_{V,n} +$ ?
b. $(\partial S/\partial T)_{U,n} = (\partial S/\partial T)_{U,V} +$ ?
c. $(\partial z/\partial u)_{x,y} = (\partial z/\partial u)_{x,w} +$ ?
d. Apply the equation of part c if $z = \cos(x/u) + e^{-y^2/u^2} + 4y/u$ and $w = y/u$.

## SECTION 5.4. SOME USEFUL RELATIONS BETWEEN PARTIAL DERIVATIVES

It is fairly common in thermodynamics to have measured values for some partial derivatives, such as $(\partial H/\partial T)_{P,n}$, the heat capacity at constant pressure. However, some other partial derivatives are difficult or impossible to measure. It is convenient to be able to express such partial derivatives in terms of quantities that are measurable. We now obtain some identities that can be used for this purpose. We will regard Eq. (5.22) and all analogous equations as the first kind of identity.

The next identity is the *reciprocal identity*, which states that a derivative is equal to the reciprocal of the derivative with the roles of dependent and independent variables reversed:

$$\left(\frac{\partial y}{\partial x}\right)_{z,u} = \frac{1}{(\partial x/\partial y)_{z,u}}. \tag{5.23}$$

The same variables must be held constant in the two derivatives.

---

**Example 5.4**

Show that

$$\left(\frac{\partial P}{\partial V}\right)_{n,T} = \frac{1}{(\partial V/\partial P)_{n,T}}$$

for an ideal gas.

---

**Solution**

$$\left(\frac{\partial P}{\partial V}\right)_{n,T} = \frac{nRT}{V^2}$$

$$\frac{1}{(\partial V/\partial P)_{n,T}} = \frac{1}{-nRT/P^2} = -\frac{P^2}{nRT} = -\frac{(nRT/V)^2}{nRT} = -\frac{nRT}{V^2}.$$

---

**Problem 5.4**

Show that the reciprocal identity is satisfied by $(\partial z/\partial x)$ and $(\partial x/\partial z)_y$ if

$$z = \sin\left(\frac{x}{y}\right) \qquad \text{and} \qquad x = y\sin^{-1}(z).$$

---

If $z = z(x, y)$ we can differentiate twice with respect to $x$:

$$\left(\frac{\partial^2 z}{\partial x^2}\right)_y = \left[\frac{\partial}{\partial x}\left(\frac{\partial z}{\partial x}\right)_y\right]_y. \tag{5.24}$$

In this case, $y$ is held fixed in both differentiations.

In addition, there are mixed second derivatives, such as the derivative with respect to $x$ and then with respect to $y$:

$$\frac{\partial^2 z}{\partial y \partial x} = \left[\frac{\partial}{\partial y}\left(\frac{\partial z}{\partial x}\right)_y\right]_x. \tag{5.25}$$

Since both variables are shown in the symbol, the subscripts are usually omitted, as in the symbol on the left. However, if there is a third independent variable, it must be held constant, and is listed as a subscript, as in

$$\left(\frac{\partial^2 U}{\partial V \partial T}\right)_n = \left[\frac{\partial}{\partial V}\left(\frac{\partial U}{\partial T}\right)_{V,n}\right]_{T,n}. \tag{5.26}$$

The next identity is a theorem known as the *Euler reciprocity relation*: If $z = z(x, y)$ is a differentiable function, then the two different mixed second partial derivatives must equal each other:

$$\boxed{\frac{\partial^2 z}{\partial y\, \partial x} = \frac{\partial^2 z}{\partial x\, \partial y}}. \tag{5.27}$$

An important set of identities obtained from the Euler reciprocity relation is the set of *Maxwell relations* of thermodynamics.

---

### Example 5.5

Show that $(\partial^2 P/\partial V\, \partial T)_n = (\partial^2 P/\partial T\, \partial V)_n$ for an ideal gas.

---

### Solution

$$\left(\frac{\partial^2 P}{\partial V \partial T}\right)_n = \left[\frac{\partial}{\partial V}\left(\frac{nR}{V}\right)\right]_{T,n} = -\frac{nR}{V^2}$$

$$\left(\frac{\partial^2 P}{\partial T \partial V}\right)_n = \left[\frac{\partial}{\partial T}\left(\frac{nRT}{V^2}\right)\right]_{V,n} = -\frac{nR}{V^2}.$$

---

### Problem 5.5

Show that $(\partial^2 z/\partial y\, \partial x) = (\partial^2 z/\partial x\, \partial y)$ if

$$z = e^{-xy^2}\sin(x)\cos(y).$$

---

Another identity is the *cycle rule*

$$\boxed{\left(\frac{\partial y}{\partial x}\right)_z \left(\frac{\partial x}{\partial z}\right)_y \left(\frac{\partial z}{\partial y}\right)_x = -1}. \tag{5.28}$$

We will "derive" this in the same nonrigorous way as was used to obtain Eq. (5.22). We write the differential of $y$ as a function of $x$ and $z$:

$$dy = \left(\frac{\partial y}{\partial x}\right)_z dx + \left(\frac{\partial y}{\partial z}\right)_x dz. \tag{5.29}$$

This equation delivers the value of $dy$ corresponding to arbitrary infinitesimal changes in $x$ and $z$, so it is still correct if we choose values of $dz$ and $dx$ such that $dy$ vanishes. We now "divide" by $dx$, and interpret the "quotients" of differentials as partial derivatives, remembering that $y$ is held fixed by our choice that $dy$ vanishes,

$$0 = \left(\frac{\partial y}{\partial x}\right)_z \left(\frac{\partial x}{\partial x}\right)_y + \left(\frac{\partial y}{\partial z}\right)_x \left(\frac{\partial z}{\partial x}\right)_y. \tag{5.30}$$

Since the partial derivative of $x$ with respect to $x$ is equal to unity, this equation becomes identical with Eq. (5.28) when the reciprocal identity is used.

---

**Problem 5.6**

For the particular function $y = x^2/z$, show that Eq. (5.28) is correct.

---

The final identity that we present in this section is the partial derivative version of the *chain rule*. If $z = z(u, x, y)$ but if $x$ can be expressed as a function of $u$, $v$, and $y$, then

$$\boxed{\left(\frac{\partial z}{\partial y}\right)_{u,v} = \left(\frac{\partial z}{\partial x}\right)_{u,v} \left(\frac{\partial x}{\partial y}\right)_{u,v}}. \tag{5.31}$$

This is very similar to Eq. (3.26). Notice that the same variables must be held fixed in all three derivatives.

---

**Example 5.6**

Show that if $z = ax^2 + bwx$ and $x = uy$ then Eq. (5.31) is correct.

---

**Solution**

$$\left(\frac{\partial z}{\partial x}\right)_{u,w} \left(\frac{\partial x}{\partial y}\right)_{u,w} = (2ax + bw)(u) = 2au^2 y + buw$$

$$\left(\frac{\partial z}{\partial y}\right)_{u,w} = \left[\frac{\partial}{\partial y}(au^2 y^2 + bwuy)\right]_{u,w} = 2au^2 y + buw.$$

---

The following are commonly measured quantities that are related to partial derivatives:

$$\text{heat capacity at constant pressure} = \left(\frac{\partial H}{\partial T}\right)_{P,n} = T\left(\frac{\partial S}{\partial T}\right)_{P,n} = C_P$$

$$\text{heat capacity at constant volume} = \left(\frac{\partial U}{\partial T}\right)_{V,n} = T\left(\frac{\partial S}{\partial T}\right)_{V,n} = C_V$$

$$\text{isothermal compressibility} = -\frac{1}{V}\left(\frac{\partial V}{\partial P}\right)_{T,n} = \kappa_T$$

$$\text{Adiabatic compressibility} = -\frac{1}{V}\left(\frac{\partial V}{\partial P}\right)_{S,n} = \kappa_S.$$

**Example 5.7**

Show that $C_P/C_V = \kappa_T/\kappa_S$.

**Solution**

$$\frac{C_P}{C_V} = \frac{(\partial S/\partial T)_{P,n}}{(\partial S/\partial T)_{V,n}} = \frac{-(\partial S/\partial P)_{T,n}(\partial P/\partial T)_{S,n}}{-(\partial S/\partial V)_{T,n}(\partial V/\partial T)_{S,n}},$$

where we have used the cycle rule twice. We use the reciprocal identity to write

$$\frac{C_P}{C_V} = \frac{(\partial V/\partial S)_{T,n}(\partial S/\partial P)_{T,n}}{(\partial V/\partial T)_{S,n}(\partial T/\partial P)_{S,n}}.$$

By the chain rule

$$\frac{C_P}{C_V} = \frac{(\partial V/\partial P)_{T,n}}{(\partial V/\partial P)_{S,n}} = \frac{-(1/V)(\partial V/\partial P)_{T,n}}{-(1/V)(\partial V/\partial P)_{S,n}} = \frac{\kappa_T}{\kappa_S}.$$

## SECTION 5.5. EXACT AND INEXACT DIFFERENTIALS

In the earlier sections of this chapter, we discussed the differential of a function. The differential of a function is called an *exact differential*. A general differential form or *Pfaffian form* can be written

$$du = M(x, y)\, dx + N(x, y)\, dy. \tag{5.32}$$

If this is the differential of a function, then $M$ and $N$ must be the appropriate derivatives of that function. However, Pfaffian forms exist in which $M$ and $N$ are not the appropriate partial derivatives of the same function. In this case $du$ is not the differential of any function and is called an *inexact differential*. It is an infinitesimal quantity that can be calculated from specified values of $dx$ and $dy$, but it is not the change in any function of $x$ and $y$ resulting from these changes.

In order to tell whether some differential form is an exact differential or not, we must have a way to tell whether the coefficients are the derivatives of the same function. The Euler reciprocity relation furnishes a means to do this. If there exists a function $u = u(x, y)$ such that

$$M(x, y) = \left(\frac{\partial u}{\partial x}\right)_y \qquad \text{and} \qquad N(x, y) = \left(\frac{\partial u}{\partial y}\right)_x$$

then from the Euler reciprocity relation,

$$\frac{\partial^2 u}{\partial x\, \partial y} = \frac{\partial^2 u}{\partial y\, \partial x}$$

which means that

$$\left(\frac{\partial N}{\partial x}\right)_y = \left(\frac{\partial M}{\partial y}\right)_x \qquad \text{(exact differential).} \tag{5.33}$$

Equation (5.33) represents a necessary and sufficient condition for the differential of Eq. (5.32) to be exact. That is, if the differential is exact, Eq. (5.33) will be obeyed, and if Eq. (5.33) is obeyed, the differential will be exact.

---

### Example 5.8

Show that the differential

$$du = \left(2xy + \frac{9x^2}{y}\right)dx + \left(x^2 - \frac{3x^2}{y^2}\right)dy$$

is exact.

---

### Solution

$$\left[\frac{\partial}{\partial y}\left(2xy + \frac{9x^2}{y}\right)\right]_x = 2x - \frac{9x^2}{y^2}$$

$$\left[\frac{\partial}{\partial x}\left(x^2 - \frac{3x^3}{y^2}\right)\right]_y = 2x - \frac{9x^2}{y^2}.$$

---

### Problem 5.7

Determine whether each of the following is an exact differential.

a. $du = (2ax + by^2)\,dx + (bxy)\,dy$
b. $du = (x + y)\,dx + (x + y)\,dy$
c. $du = (x^2 + 2x + 1)\,dx + (y^2 + 25y + 24)\,dy.$

---

Differential forms with three or more terms can also either be exact or inexact. The Euler reciprocity relation provides a test for such differentials. For example, if

$$du = M(x, y, z)\,dx + N(x, y, z)\,dy + P(x, y, z)\,dz \qquad (5.34)$$

then in order for this to be an exact differential, we must have all of the equations obeyed:

$$\left(\frac{\partial M}{\partial y}\right)_{x,z} = \left(\frac{\partial N}{\partial x}\right)_{y,z} \qquad (5.35a)$$

$$\left(\frac{\partial N}{\partial z}\right)_{x,y} = \left(\frac{\partial P}{\partial y}\right)_{x,z} \qquad (5.35b)$$

and

$$\left(\frac{\partial M}{\partial z}\right)_{x,y} = \left(\frac{\partial P}{\partial x}\right)_{y,z}. \qquad (5.35c)$$

### Problem 5.8

Show that the following is not an exact differential

$$du = (2y)\,dx + (x)\,dy + \cos(z)\,dz.$$

There are two important inexact differentials in thermodynamics. If a system undergoes an infinitesimal process (one in which the independent variables specifying the state of the system change infinitesimally), $dq$ denotes the amount of heat transferred to the system and $dw$ denotes the amount of work done on the system. Both of these quantities are inexact differentials. For a fluid system and reversible processes,

$$dw_{rev} = -P\,dV. \tag{5.36}$$

### Example 5.9

Show that for an ideal gas undergoing reversible processes with $n$ fixed, $dw_{rev}$ is inexact.

### Solution

We choose $T$ and $V$ as our independent variables,

$$dw = M\,dT + N\,dV \qquad (n \text{ fixed}). \tag{5.37}$$

Comparison with Eq. (5.36) shows that $M = 0$ and $N = P = nRT/V$. We apply the test for exactness, Eq. (5.33),

$$\left(\frac{\partial M}{\partial V}\right)_{T,n} = 0$$

$$\left(\frac{\partial N}{\partial T}\right)_{V,n} = \left[\frac{\partial}{\partial T}\left(\frac{nRT}{V}\right)\right]_{V,n} = \frac{nR}{V} \neq 0.$$

### Problem 5.9

The thermodynamic energy of a monatomic ideal gas is given approximately by

$$U = \frac{3nRT}{2}. \tag{5.38}$$

Find the partial derivatives and write the expression for $dU$ using $V$, $T$, and $n$ as independent variables. Show that the partial derivatives obey Eqs. (5.35).

In thermodynamics, quantities such as the thermodynamic energy, the volume, the pressure, the temperature, the amount of substances, etc., that are functions of the variables that can be used to specify the state of the system, are called *state functions*. The differentials of these quantities are exact differentials. Work and heat are not state functions. There is no such thing as an amount of work or an amount of heat in a system. We have already seen in Example 5.9 that $dw_{rev}$ is not an exact differential, and the same is true of $dw$ for an irreversible process. An infinitesimal amount of heat is also not an exact differential. However, the first law of thermodymics states that $dU$, which equals $dw + dq$, is an exact differential.

## Integrating Factors

Some inexact differentials are related to exact differentials in that an exact differential is produced if the inexact differential is multiplied by a function called an *integrating factor* for that differential.

---

### Example 5.10

Show that the differential

$$du = (2ax^2 + bxy)\,dx + (bx^2 + 2cxy)\,dy$$

is inexact, but that $1/x$ is an integrating factor, so that $du/x$ is exact.

---

### Solution

$$\left[\frac{\partial}{\partial y}(2ax^2 + bxy)\right]_x = bx$$

$$\left[\frac{\partial}{\partial x}(bx^2 + 2cxy)\right]_y = 2bx + 2cy \neq bx$$

so $du$ is inexact.

$$\left[\frac{\partial}{\partial y}(2ax + by)\right]_x = b$$

$$\left[\frac{\partial}{\partial x}(bx + 2cy)\right]_y = b$$

so $(1/x)\,du$ is exact.

---

### Problem 5.10

Show that the differential

$$(1 + x)\,dx + \left[\frac{x \ln(x)}{y} + \frac{x^2}{y}\right] dy$$

is inexact, and that $y/x$ is an integrating factor.

---

There is no general method for finding an integrating factor, although we will discuss a method that will work for a particular class of differential forms in Chapter 7. However, it is true that if a differential possesses one integrating factor, there are infinitely many integrating factors.

## SECTION 5.6. LINE INTEGRALS

In Chapter 4, we found that a finite increment in a function could be constructed by a definite integration. If $x_1$ and $x_0$ are particular values of the independent variable $x$, then

$$F(x_1) - F(x_0) = \int_{x_0}^{x_1} f(x)\,dx = \int_{x_0}^{x_1} dF, \tag{5.39}$$

where

$$f(x) = \frac{dF}{dx}. \tag{5.40}$$

We can think of the integral of Eq. (5.39) as being a sum of infinitesimal increments equal to $f(x)\,dx$.

We now consider the analogous process for a differential with two or more independent variables. For two independent variables, $x$ and $y$, we might try to define

$$\int_{x_0, y_0}^{x_1, y_1} dF = \int_{x_0, y_0}^{x_1, y_1} [M(x, y)\,dx + N(x, y)]\,dy. \tag{5.41}$$

However, this integral is not yet well defined. The situation is shown schematically in Fig. 5.3. For a pair of variables, $(x_0, y_0)$ represents one point in the $x$–$y$ plane, and $(x_1, y_1)$ represents another point in the plane, and many different paths can join the two points.

In order to complete the definition of the integral in Eq. (5.41), we must specify the path in the $x$–$y$ plane along which we integrate from the point $(x_0, y_0)$ to the point $(x_1, y_1)$. We introduce the notation

$$\int_C dF = \int_C [M(x, y)\,dx + N(x, y)\,dy], \tag{5.42}$$

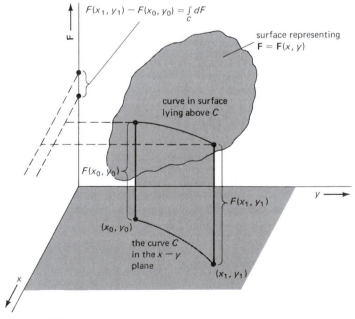

**FIGURE 5.3**   Diagram illustrating the line integral of an exact differential.

where the letter $C$ stands for the piece of a curve joining the two points. The integral is called a *line integral* or a *path integral*.

We can think of Eq. (5.42) as representing a sum of many infinitesimal contributions, each one given by the appropriate infinitesimal value of $dF$ resulting when $x$ is changed by $dx$ and $y$ is changed by $dy$. However, these changes $dx$ and $dy$ are not independent. They must be related so that we remain on the chosen curve during the integration process.

A curve in the $x$–$y$ plane specifies $y$ as a function of $x$, or $x$ as a function of $y$. For a given curve, we can write

$$y = y(x) \tag{5.43}$$
$$x = x(y). \tag{5.44}$$

In order to calculate a line integral such as that of Eq. (5.42), we replace $y$ in $M(x, y)$ by the function given in Eq. (5.43), and we replace $x$ in $N(x, y)$ by the function given in Eq. (5.44). With this replacement, $M$ is a function of $x$ only, and $N$ is a function of $y$ only, and each term becomes an ordinary one-variable integral:

$$\int_C dF = \int_{x_0}^{x_1} M[x, y(x)] \, dx + \int_{y_0}^{y_1} N[y, x(y)] \, dy. \tag{5.45}$$

In these integrals, specification of the curve $C$ determines not only what the beginning point $(x_0, y_0)$ and the final point $(x_1, y_1)$ are, but also what the functions are that replace $y$ in the $dx$ integral and $x$ in the $dy$ integral.

---

### Example 5.11

Find the value of the line integral

$$\int_C dF = \int_C [(2x + 3y) \, dx + (3x + 4y)] \, dy,$$

where $C$ is the straight-line segment given by $y = 2x + 3$ from $(0, 3)$ to $(2, 7)$.

---

### Solution

In the first term, $y$ must be replaced by $2x + 3$, and in the second term $x$ must be replaced by $(1/2)(y - 3)$,

$$\int_C dF = \int_0^2 [2x + 3(2x + 3)] dx + \int_3^7 \left[ \frac{3}{2}(y - 3) + 4y \right] dy$$

$$= \left( \frac{8x^2}{2} + 9x \right) \Big|_0^2 + \left( \frac{(11/2)y^2}{2} - \frac{9}{2}y \right) \Big|_3^7 = 126.$$

---

There is an important theorem, which we now state: *If $dF$ is an exact differential, then the line integral $\int_C dF$ depends only on the initial and final points and not on the choice of curve joining these points. Further, the line integral will have a value equal to the value of the function at the final point minus the value of the function at the beginning point.* For example, if $F = F(x, y)$ and if $(x_0, y_0)$ is the initial point of the line integration and $(x_1, y_1)$ is the final point of the line integration, then

$$\int_C dF = \int_C \left[ \left( \frac{\partial F}{\partial x} \right) dx + \left( \frac{\partial F}{\partial y} \right) dy \right] = F(x_1, y_1) - F(x_0, y_0) \qquad (dF \text{ exact}).$$

(5.46)

We have written $M$ and $N$ as the partial derivatives which they must be equal to in order for $dF$ to be exact (see Section 5.5). If $du$ is not an exact differential, there is no such things as a function $u$, and the line integral will depend not only on the beginning and ending points, but also on the curve of integration joining these points.

### Example 5.12

Show that the line integral of Example 5.11 has the same value as the line integral of the same differential on the rectangular path from $(0, 3)$ to $(2, 3)$ and then to $(2, 7)$.

### Solution

The path of this integration is not a single curve but two line segments, so we must carry out the integration separately for each segment. This is actually a simplification, because on the first line segment, $y$ is constant, so $dy = 0$ and the $dy$ integral vanishes. On the second line segment, $x$ is constant, so $dx = 0$ and the $dx$ integral vanishes. Therefore

$$\int_C dF = \int_0^2 (2x + 9) dx + \int_3^7 (6 + 4y) dy.$$

This follows from the fact that $y = 3$ on the first line segment, and from the fact that $x = 2$ on the second line segment. Performing the integrals yields

$$\int_C dF = \left( \frac{2x^2}{2} + 9x \right) \bigg|_0^2 + \left( 6y + \frac{4y^2}{2} \right) \bigg|_3^7 = 126.$$

### Problem 5.11

a. Show that the following differential is exact:

$$dz = (ye^{xy}) dx + (xe^{xy}) dy$$

b. Calculate the line integral $\int_C dz$ on the line segment from $(0,0)$ to $(2, 2)$. On this line segment, $y = x$.

c. Calculate the line integral $\int_C dz$ on the path going from $(0, 0)$ to $(0, 2)$ and then to $(2, 2)$ (a rectangular path).

### Example 5.13

Show that the differential

$$du = dx + x\, dy$$

is inexact and carry out the line integral from $(0, 0)$ to $(2, 2)$ by the two different paths: path 1, the straight-line segment from $(0, 0)$ to $(2, 2)$; and path 2, the rectangular path from $(0, 0)$ to $(2, 0)$ and then to $(2, 2)$.

---

**Solution**

Test for exactness:

$$\left[\frac{\partial}{\partial y}(1)\right]_x = 0$$

$$\left[\frac{\partial}{\partial x}(x)\right]_y = 1 \neq 0.$$

The differential is not exact.

Path 1,

$$\int_{C_1} du = \int_{C_1} dx + \int_{C_1} x\, dy = \int_0^2 dx + \int_0^2 y\, dy,$$

where we obtained the second integral by using the fact that $y = x$ on the straight-line segment of path 1,

$$\int_{C_1} du = x \Big|_0^2 + \frac{y^2}{2} \Big|_0^2 = 4.$$

Path 2,

$$\int_{C_2} du = \int_{C_2} dx + \int_{C_2} x\, dy = \int_0^2 dx + \int_0^2 2\, dy$$

$$x\Big|_0^2 + 2y\Big|_0^2 = 2 + 4 = 6.$$

The two line integrals have the same beginning point and the same ending point, but are not equal, because the differential is not an exact differential.

---

**Problem 5.12**

Carry out the two line integrals of $du$ from Example 5.13 from $(0, 0)$ to $(x_1, y_1)$

on path 1: rectangular path from $(0, 0)$ to $(0, y_1)$ and then to $(x_1, y_1)$;
on path 2: rectangular path from $(0, 0)$ to $(x_1, 0)$ and then to $(x_1, y_1)$.

---

There are also line integrals of functions of three independent variables. The line integral of the differential $du$ is

$$\int_C du = \int_C [M(x, y, z)\, dx + N(x, y, z)\, dy + P(x, y, z)\, dz], \qquad (5.47)$$

where $C$ specifies a curve that gives $y$ and $z$ as functions of $x$, or $x$ and $y$ as functions of $z$, or $x$ and $z$ as functions of $y$. If the beginning point of the curve $C$ is $(x_0, y_0, z_0)$

and the ending point is $(x_1, y_1, z_1)$, the line integral is

$$\int_C du = \int_{x_0}^{x_1} M[x, y(x), z(x)] \, dx + \int_{y_0}^{y_1} N[x, (y), y, z(y)] \, dy$$

$$+ \int_{z_0}^{z_1} P[x, (z), y(z)z] \, dz. \tag{5.48a}$$

If $du$ is an exact differential, then $u$ is a function, and

$$\int_C du = u(x_1, y_1, z_1) - u(x_0, y_0, z_0) \qquad \text{(if } du \text{ is an exact differential).} \tag{5.48b}$$

In thermodynamics, the equilibrium state of a system is represented by a point in a space whose axes represent the variables specifying the state of the system. A line integral in such a space represents a reversible process. A cyclic process is one that begins and ends at the same state of a system. A line integral that begins and ends at the same point is denoted by the symbol $\oint du$. Since the beginning and final points are the same, such an integral must vanish if $du$ is an exact differential:

$$\oint du = 0 \qquad \text{(if } du \text{ is exact).} \tag{5.49}$$

If $du$ is inexact, a line integral that begins and ends at the same point will not generally be equal to zero.


## SECTION 5.7. MULTIPLE INTEGRALS

While a line integral can be thought of as adding up infinitesimal contributions represented as a differential form, a multiple integral can be thought of as adding up contributions given by an integrand function times an infinitesimal element of area or of volume, etc. A *double integral*, which is the simplest kind of multiple integral, is written in the form

$$I = \int_{a_1}^{a_2} \int_{b_1}^{b_2} f(x, y) \, dy \, dx, \tag{5.50}$$

where $f(x, y)$ is the integrand function, $a_1$ and $a_2$ are the limits of the $x$ integration, and $b_1$ and $b_2$ are the limits of the $y$ integration. You should think of the product $dy \, dx$ as an infinitesimal element of area in the $x$–$y$ plane.

The double integral is carried out as follows: The "inside" integration is done first. This is the integration over the values of the variable whose differential and limits are written closest to the integrand function. During this integration, the other independent variable is treated as a constant if it occurs in the integrand. The result of the first integration is the integrand for the remaining integration. The limits $b_1$ and $b_2$ can depend on $x$, but the limits $a_1$ and $a_2$ must be constants.

### Example 5.14

Evaluate the double integral

$$I = \int_0^a \int_0^b (x^2 + 4xy)\,dy\,dx.$$

### Solution

The integration over $y$ is carried out, treating $x$ as a constant:

$$\int_0^b (x^2 + 4xy)\,dy = \left( x^2 y + \frac{4xy^2}{2} \right)\Big|_0^b = bx^2 + 2b^2 x.$$

This becomes the integrand for the second integration, so that

$$I = \int_0^b (bx^2 + 2b^2 x)\,dx = \left( \frac{bx^3}{3} + \frac{2b^2 x^2}{2} \right)\Big|_0^a$$

$$= \frac{ba^3}{3} + b^2 a^2.$$

### Example 5.15

Evaluate the double integral

$$\int_0^a \int_0^{3x} (x^2 + 2xy + y^2)\,dy\,dx.$$

### Solution

The result of the inside integration is

$$\int_0^{3x} (x^2 + 2xy + y^2)dy = \left( x^2 y + \frac{2xy^2}{2} + \frac{y^3}{3} \right)\Big|_0^{3x}.$$

$$= 3x^3 + 9x^3 + 9x^3 = 21x^3.$$

The $x$ integration gives

$$\int_0^a 21x^3 dx = \frac{21x^4}{4}\Big|_0^a = \frac{21a^4}{4}.$$

### Problem 5.13

Evaluate the double integral

$$\int_2^4 \int_0^\pi x \sin(y)\,dy\,dx.$$

## The Double Integral Represented as a Volume

In Section 4.3 we saw that a definite integral with one independent variable is equal to an area between the axis and the integrand curve. A double integral is equal to a volume in an analogous way. This is illustrated in Fig. 5.4, which is drawn to correspond to Example 5.16. In the $x$–$y$ plane, we have an infinitesimal element of area $dx\,dy$, drawn in the figure as though it were finite in size. The vertical distance from the $x$–$y$ plane to the surface representing the integrand function $f$ is the value of the integrand function, so that the volume of the small box lying between the element of area and the surface is equal to $f(x)dx\,dy$.

The double integral is the sum of the volume of all such infinitesimal boxes, and thus equals the volume of the solid bounded by the $x$–$y$ plane, the surface representing the integrand function, and surfaces representing the limits of integration. If the integrand function is negative in part of the region of integration, we must take the volume above the $x$–$y$ plane minus the volume below the plane as equal to the integral.

---

### Example 5.16

Calculate the volume of the solid shown in Fig. 5.4. The bottom of the solid is the $x$–$y$ plane. The flat surface corresponds to $y = 0$, the curved vertical surface corresponds to $y = x^2 - 4$, and the top of the solid corresponds to $f = 2 - y$.

### Solution

We carry out a double integral with $f = 2 - y$ as the integrand:

$$V = \int_{-2}^{2} \int_{x^2-4}^{0} (2 - y)\,dy\,dx.$$

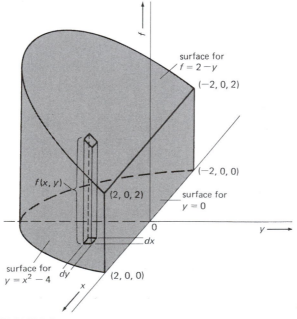

**FIGURE 5.4**   The diagram for Example 5.16.

The inside integral is

$$\int_{x^2-4}^{0} (2 - y)\,dy = \frac{x^4}{2} - 6x^2 + 16$$

so that

$$V = \int_{-2}^{2} \left( \frac{x^4}{2} - 6x^2 + 16 \right) dx = 38.4.$$

### Problem 5.14

Find the volume of the solid object shown in Fig. 5.5. The top of the object corresponds to $f = 5 - x - y$, the bottom of the object is the $x$–$y$ plane, the trapezoidal face is the $x$–$f$ plane, and the large triangular face is the $y$–$f$ plane. The small triangular face corresponds to $x = 3$.

Multiple integrals with three or more independent variables also occur. For example, if we have an integrand function $f$ depending on $x$, $y$, and $z$, we could have the triple integral

$$I = \int_{a_1}^{a_2} \int_{b_1}^{b_2} \int_{c_1}^{c_2} f(x, y, z)\,dz\,dy\,dx. \tag{5.51}$$

To evaluate the integral, we first integrate $z$ from $c_1$ to $c_2$ and take the result as the integrand for the double integral over $y$ and $x$. Then we integrate $y$ from $b_1$ to $b_2$, and take the result as the integrand for the integral over $x$ from $a_1$ to $a_2$.

The limits $c_1$ and $c_2$ can depend on $y$ and $x$, and the limits $b_1$ and $b_2$ can depend on $x$, but $a_1$ and $a_2$ must be constants. If the limits are all constants, and if the integrand function can be factored, the entire integral can be factored, as in the following example.

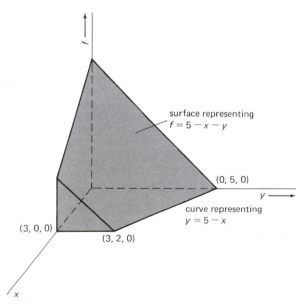

**FIGURE 5.5**   The diagram for Problem 5.13.

### Example 5.17

Find the triple integral

$$
I = A^2 \int_0^a \int_0^b \int_0^c \sin^2 \left( \frac{n\pi x}{a} \right) \sin^2 \left( \frac{m\pi y}{b} \right) \sin^2 \left( \frac{k\pi z}{c} \right) dz \, dy \, dx.
$$

This is a *normalization integral* from quantum mechanics. The integrand is the square of the wave function for one of the states of a particle in a three-dimensional box. The quantities $m$, $n$, and $k$ are integral quantum numbers specifying the state of the particle. The integral is equal to the total probability of finding a particle in the box. It is customary to choose the value of the constant $A$ so that the total probability equals unity, in which case the wave function is said to be *normalized*.

### Solution

The integrand function is a product of three factors, each of which depends on only one variable, and the limits of integration are constants. The entire integral can therefore be written in factored form:

$$
I = A^2 \left[ \int_0^a \sin^2 \left( \frac{n\pi x}{a} \right) dx \right] \left[ \int_0^b \sin^2 \left( \frac{m\pi x}{b} \right) dy \right] \left[ \int_0^c \sin^2 \left( \frac{k\pi x}{c} \right) dz \right].
$$

We first carry out the $z$ integration, using the substitution $u = k\pi z / c$,

$$
\int_0^c \sin^2 \left( \frac{k\pi z}{c} \right) dz = \frac{c}{k\pi} \int_0^{k\pi} \sin^2 (u) \, du.
$$

The integrand is a periodic function with period $\pi$, so that the integral from 0 to $k\pi$ is just $k$ times the integral from 0 to $\pi$, which is given as Eq. (8) of Appendix 6:

$$
\int_0^c \sin^2 \left( \frac{k\pi z}{c} \right) dz = \frac{c}{k\pi} k \frac{\pi}{2} = \frac{c}{2}.
$$

The other integrals are similar, except for having $a$ or $b$ instead of $c$, so that

$$
I = A^2 \frac{abc}{8}.
$$

Many triple integrals in quantum mechanics are factored in the same way as in this example.

### Problem 5.15

Find the value of the constant A so that the following integral equals unity.

$$
A \int_{-\infty}^{\infty} \int_{-\infty}^{\infty} e^{-x^2 - y^2} dy \, dx.
$$

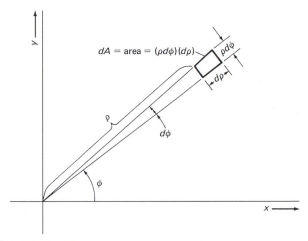

**FIGURE 5.6** An infinitesimal element of area in plane polar coordinates.

## Changing Variables in Multiple Integrals

Sometimes it is convenient to take a multiple integral over an area or over a volume using polar coordinates or spherical polar coordinates, etc., instead of cartesian coordinates. Figure 5.6 shows how this is done in polar coordinates. We require an infinitesimal element of area given in terms of the coordinates $\rho$ and $\phi$. One dimension of the element of area is $d\rho$ and the other dimension is $\rho\, d\phi$, from the fact that an arc length is the radius of the circle times the angle subtended by the arc, measured in radians. The element of area is $\rho\, d\phi\, d\rho$. If the element of area were finite, it would not quite be rectangular, and this formula would not be exact, but it is valid for an infinitesimal element of area.

We can think of the plane as being covered completely by infinitely many such elements of area, and a double integral over some region of the plane is just the sum of the value of the integrand function at each element of area in the region times the area of the element.

---

### Example 5.18

In cartesian coordinates, the wave function for the ground state of a two-dimensional harmonic oscillator is

$$\psi = B \exp[-a(x^2 + y^2)].$$

Transform this to plane polar coordinates and find the value of $B$ such that the integral of $\psi^2$ over the entire $x$–$y$ plane is equal to unity.

---

### Solution

$$\psi = Be^{-a\rho^2}.$$

The integral that is to equal unity is

$$B^2 \int_0^\infty \int_0^{2\pi} e^{-2a\rho^2} \rho \, d\phi \, d\rho.$$

The integral can be factored,

$$1 = B^2 \int_0^\infty e^{-2a\rho^2} \rho \, d\rho \int_0^{2\pi} d\phi = 2\pi B^2 \int_0^\infty e^{-2a\rho^2} \rho \, d\rho.$$

The $\rho$ integral is done by the method of substitution, letting $u = 2a\rho^2$ and $du = 4a\rho \, d\rho$. We obtain

$$1 = 2\pi B^2 \left( \frac{1}{4a} \right) \int_0^\infty e^{-u} du = \frac{B^2 \pi}{2a}$$

$$B = \sqrt{\frac{2a}{\pi}}.$$

---

### Problem 5.16

Use a double integral to find the volume of a cone of height $h$ and radius $a$ at the base. If the cone is standing with its point upward and with its base centered at the origin, the equation giving the surface of the cone is

$$f = h \left( 1 - \frac{\rho}{a} \right).$$

---

In transforming from cartesian to plane polar coordinates, the factor $\rho$ which is used with the product of the differentials $d\phi \, d\rho$ is called a *jacobian*. The symbol $\partial(x, y)/\partial(\rho, \phi)$ is used for this jacobian:

$$\iint f(x, y) \, dx \, dy = \iint f(\rho, \phi) \rho \, d\phi \, d\rho = \iint f(\rho, \phi) \frac{\partial(x, y)}{\partial(\rho, \phi)} d\rho \, d\phi. \quad (5.52)$$

We will not discuss the mathematical theory, but this jacobian is given by the following determinant (determinants are discussed in Chapter 8):

$$\frac{\partial(x, y)}{\partial(\rho, \phi)} = \begin{vmatrix} \partial x/\partial\rho & \partial x/\partial\phi \\ \partial y/\partial\rho & \partial y/\partial\phi \end{vmatrix}. \quad (5.53)$$

Equation (8.57a) gives the formula for a 2 by 2 determinant:

$$\begin{vmatrix} \partial x/\partial\rho & \partial x/\partial\phi \\ \partial y/\partial\rho & \partial y/\partial\phi \end{vmatrix} = \begin{vmatrix} \cos(\phi) & -\rho \sin(\phi) \\ \cos(\phi) & \rho \cos(\phi) \end{vmatrix}$$

$$= \rho \cos^2(\phi) + \rho \sin^2(\phi) = \rho, \quad (5.54)$$

where we have also used Eq. (7) of Appendix 5. Equation (5.54) gives us the same result as we had before,

$$dA = \text{element of area} = \rho \, d\phi \, d\rho. \quad (5.55)$$

The jacobian for transformation of coordinates in three dimensions is quite similar. If $u$, $v$, and $w$ are some set of coordinates such that

$$x = x(u, v, w)$$
$$y = y(u, v, w)$$
$$z = z(u, v, w)$$

then the jacobian for the transformation of a multiple integral from cartesian coordinates to the coordinates $u$, $v$, and $w$ is given by the determinant

$$\frac{\partial(x, y, z)}{\partial(u, v, w)} = \begin{vmatrix} \partial x/\partial u & \partial x/\partial v & \partial x/\partial w \\ \partial y/\partial u & \partial y/\partial v & \partial y/\partial w \\ \partial z/\partial u & \partial z/\partial v & \partial z/\partial w \end{vmatrix} . \qquad (5.56)$$

The rule for expanding a 3 by 3 determinant is discussed in Chapter 8.

### Example 5.19

Obtain the jacobian for the transformation from cartesian coordinates to spherical polar coordinates.

### Solution

The equations relating the coordinates are Eqs. (2.81), (2.82), and (2.83). From these equations

$$\frac{\partial(x, y, z)}{\partial(r, \theta, \phi)} = \begin{vmatrix} \sin(\theta)\cos(\phi) & r\cos(\theta)\cos(\phi) & -r\sin(\theta)\sin(\phi) \\ \sin(\theta)\sin(\phi) & r\cos(\theta)\sin(\phi) & r\sin(\theta)\cos(\phi) \\ \cos(\phi) & -r\sin(\theta) & 0 \end{vmatrix} .$$

From Example 8.12 for a 3 by 3 determinant,

$$\frac{\partial(x, y, z)}{\partial(r, \theta, \phi)} = \cos(\theta)[r^2 \cos(\theta)\sin(\theta)\cos^2(\phi) + r^2 \sin(\theta)\cos(\theta)\sin^2(\phi)]$$
$$+ r\sin(\theta)[r\sin^2(\theta)\cos^2(\phi) + r^2 \sin^2(\theta)(\phi)]$$
$$= r^2 \sin(\theta)\cos^2(\theta) + r^2 \sin^3(\theta) = r^2 \sin(\theta), \qquad (5.57)$$

where we have used Eq. (7) of Appendix 5 several times.

### Problem 5.17

Find the jacobian for the transformation from cartesian to cylindrical polar coordinates.

A triple integral in cartesian coordinates is transformed into a triple integral in spherical polar coordinates by

$$\iiint f(x, y, z)\, dx\, dy\, dz = \iiint f(r, \theta, \phi) r^2 \sin(\theta)\, d\phi\, d\theta\, dr \qquad (5.58)$$

or, equivalently, an element of volume is given by

$$dV = dx \, dy \, dz = r^2 \sin(\theta) \, d\phi \, d\theta \, dr \,. \tag{5.59}$$

To complete the transformation, the limits on $r$, $\theta$, and $\phi$ must be found so that they correspond to the limits on $x$, $y$, and $z$. Sometimes the purpose of transforming to spherical polar coordinates is to avoid the task of finding the limits in cartesian coordinates when they can be expressed easily in spherical polar coordinates. For example, if the integration is over the interior of a sphere of radius $a$ centered at the origin, $\phi$ ranges from 0 to $2\pi$, $\theta$ ranges from 0 to $\pi$, and $r$ ranges from 0 to $a$. If all of space is to be integrated over, $\phi$ ranges from 0 to $2\pi$, $\theta$ ranges from 0 to $\pi$, and $r$ ranges from 0 to $\infty$.

## SECTION 5.8. VECTOR DERIVATIVE OPERATORS

An operator is a symbol for carrying out a mathematical operation (see Chapter 8). There are several vector derivative operators. We first define them in cartesian coordinates.

### Vector Derivatives in Cartesian Coordinates

The *gradient operator* is defined by

$$\nabla = \mathbf{i}\frac{\partial}{\partial x} + \mathbf{j}\frac{\partial}{\partial y} + \mathbf{k}\frac{\partial}{\partial z} \,, \tag{5.60a}$$

where $\mathbf{i}$, $\mathbf{j}$, and $\mathbf{k}$ are the unit vectors in the directions of the $x$, $y$, and $z$ axes defined in Chapter 2. The gradient of a scalar function is a vector. The symbol $\nabla$, which is an upside-down capital Greek delta, is called "del." If $f$ is some scalar function of $x$, $y$, and $z$, the *gradient* of $f$ is given in cartesian coordinates by

$$\nabla f = \mathbf{i}\frac{\partial f}{\partial x} + \mathbf{j}\frac{\partial f}{\partial y} + \mathbf{k}\frac{\partial f}{\partial z} \,. \tag{5.60b}$$

The gradient of $f$ is sometimes denoted by grad $f$ instead of $\nabla f$. The direction of the gradient of a scalar function is the direction in which the function is increasing most rapidly, and its magnitude is the rate of change of the function in that direction.

---

**Example 5.20**

Find the gradient of the function

$$f = x^2 + 3xy + z^2 \sin\left(\frac{x}{y}\right) \,. \tag{5.61}$$

**Solution**

$$\nabla f = \mathbf{i} \left[ 2x + 3y + \frac{z^2}{y} \cos\left(\frac{x}{y}\right) \right] + \mathbf{j} \left[ 3x - \frac{xz^2}{y^2} \cos\left(\frac{x}{y}\right) \right] + \mathbf{k} 2z \sin\left(\frac{x}{y}\right).$$

**Problem 5.18**

Find the gradient of the function

$$g = ax^3 + ye^{bz},$$

where $a$ and $b$ are constants.

A common example of a gradient is found in mechanics. In a conservative system, the force on a particle is given by

$$\mathbf{F} = -\nabla \mathcal{V}, \tag{5.62}$$

where $\mathcal{V}$ is the potential energy of the entire system. The gradient is taken with respect to the coordinates of the particle being considered, and the coordinates of any other particles are treated as constants in the differentiations.

**Example 5.21**

The potential energy of an object of mass $m$ near the surface of the earth is

$$\mathcal{V} = mgz,$$

where $g$ is the acceleration due to gravity. Find the gravitational force on the object.

**Solution**

$$\mathbf{F} = -\mathbf{k}mg.$$

**Problem 5.19**

Neglecting the attractions of all other celestial bodies, the gravitational potential energy of the earth and the sun is given by

$$\mathcal{V} = -\frac{Gm_s m_e}{r},$$

where $G$ is the universal gravitational constant, equal to $6.673 \times 10^{-11}$ m$^3$ s$^{-2}$ kg$^{-1}$, $m_s$ the mass of the sun, $m_e$ the mass of the earth, and $r$ the distance from the sun to the earth,

$$r = (x^2 + y^2 + z^2)^{1/2}$$

with the center of the sun taken as the origin.

Find the force on the earth in cartesian coordinates. That is, find the force in terms of the unit vectors **i**, **j**, and **k** with the components expressed in terms of $x$, $y$, and $z$. Find the magnitude of the force.

---

The operator $\nabla$ can operate on vector functions as well as on scalar functions. An example of a vector function is the velocity of a compressible flowing fluid

$$\mathbf{v} = \mathbf{v}(x, y, z) \tag{5.63}$$

which is the same as the expression

$$\mathbf{v} = \mathbf{i}v_x(x, y, z) + \mathbf{j}v_y(x, y, z) + \mathbf{k}v_z(x, y, z). \tag{5.64}$$

There are two principal vector derivatives of vector functions. The *divergence* of **F** is defined in cartesian coordinates by

$$\boxed{\nabla \cdot \mathbf{F} = \left(\frac{\partial F_x}{\partial x}\right) + \left(\frac{\partial F_y}{\partial y}\right) + \left(\frac{\partial F_z}{\partial z}\right).} \tag{5.65}$$

where **F** is a vector function with cartesian components $F_x$, $F_y$, and $F_z$. The divergence of a vector function **F** is a scalar and is somewhat analogous to a scalar product (dot product) of two vectors. The divergence of **F** is sometimes denoted by div **F**.

One way to visualize the divergence of a function is to consider the divergence of the velocity of a compressible fluid. Curves that are followed by small portions of the fluid are called *stream lines*. In a region where the stream lines diverge (become farther from each other) as the flow is followed, the fluid will become less dense, and in such a region the divergence of the velocity is positive. The divergence thus provides a measure of the spreading of the stream lines. The *equation of continuity* of a compressible fluid expresses the effect this spreading has on the density of the fluid,

$$\nabla \cdot (\rho \mathbf{v}) = -\frac{\partial \rho}{\partial t}, \tag{5.66}$$

where $\rho$ is the density of the fluid and $t$ is the time.

---

## Example 5.22

Find the divergence of the function

$$\mathbf{F} = \mathbf{i}x^2 + \mathbf{j}yz + \mathbf{k}\frac{xz^2}{y}.$$

---

## Solution

$$\nabla \cdot \mathbf{F} = 2x + z + \frac{2xz}{y}.$$

---

## Problem 5.20

Find the divergence of

$$\mathbf{r} = \mathbf{i}x + \mathbf{j}y + \mathbf{k}z.$$

---

The *curl* of the vector function **F** is defined in cartesian coordinates by

$$\nabla \times \mathbf{F} = \mathbf{i}\left(\frac{\partial F_z}{\partial y} - \frac{\partial F_y}{\partial z}\right) + \mathbf{j}\left(\frac{\partial F_x}{\partial z} - \frac{\partial F_z}{\partial x}\right) + \mathbf{k}\left(\frac{\partial F_y}{\partial x} - \frac{\partial F_x}{\partial y}\right). \quad (5.67)$$

The curl is a vector and is somewhat analogous to the vector product (cross product) of two vectors. To remember which vector derivative is which, remember that "dot" and "divergence" both begin with the letter "d" and that "cross" and "curl" both begin with the letter "c." The symbol curl **F** is sometimes used for the curl.

The curl of a vector function is more difficult to visualize than is the divergence. In fluid flow, the curl of the velocity gives the *vorticity* of the flow, or the rate of turning of the velocity vector. Because of this, the symbol rot **F** is also sometimes used for the curl of **F**.

---

## Example 5.23

Find the curl of the vector function

$$\mathbf{F} = \mathbf{i}y + \mathbf{j}z + \mathbf{k}x.$$

---

### Solution

$$\nabla \times \mathbf{F} = \mathbf{i}(0 - 1) + \mathbf{j}(0 - 1) + \mathbf{k}(0 - 1) = -\mathbf{i} - \mathbf{j} - \mathbf{k}.$$

---

## Problem 5.21

Find the curl of

$$\mathbf{r} = \mathbf{i}x + \mathbf{j}y + \mathbf{k}z.$$

---

We can define derivatives corresponding to successive application of the del operator. The first such operator is the divergence of the gradient and is an operator that occurs in the Schrödinger equation of quantum mechanics and in electrostatics. If $f$ is a scalar function, the divergence of the gradient of $f$ is given in cartesian coordinates by

$$\nabla \cdot \nabla f = \nabla^2 f = \left(\frac{\partial^2 f}{dx^2}\right) + \left(\frac{\partial^2 f}{dy^2}\right) + \left(\frac{\partial^2 f}{dz^2}\right). \quad (5.68)$$

The operator $\nabla \cdot \nabla$ occurs so commonly that it has its own name, the *laplacian operator*,[1] and its own symbol, $\nabla^2$.

---

## Example 5.24

Find the laplacian of the function

$$f(x, y, z) = A \sin(ax)\sin(by)\sin(cz).$$

---

[1]After Pierre Simon, Marquis de Laplace, 1749–1827, a famous French mathematician and astronomer.

----------

## Solution

$$\nabla^2 f = -Aa^2 \sin(ax)\sin(by)\sin(cz) - Ab^2 \sin(ax)\sin(by)\sin(cz)$$
$$-Ac^2 \sin(ax)\sin(by)\sin(cz)$$
$$= -(a^2 + b^2 + c^2)f.$$

----------

**Problem 5.22**

Find $\nabla^2 f$ if $= \exp(x^2 + y^2 + z^2) = e^{x^2}e^{y^2}e^{z^2}$.

----------

Two other possibilities for successive operation of the del operator are the curl of the gradient and the gradient of the divergence. The curl of the gradient of any differentiable scalar function always vanishes.

----------

**Problem 5.23**

a. Show that the curl of the gradient of a differentiable scalar function equals zero.
b. Write the expression for the gradient of the divergence of a vector function **F**.

----------

## Vector Derivatives in Other Coordinate Systems

It is sometimes convenient to work in coordinate systems other than cartesian coordinates. For example, in the Schrödinger equation for the quantum mechanical motion of the electron in a hydrogen atom, the potential energy is a simple function of $r$, the distance from the nucleus, but a more complicated function of $x$, $y$, and $z$. This Schrödinger equation can be solved only if spherical polar coordinates are used. The complications produced by expressing the laplacian in the Schrödinger equation in spherical polar coordinates are more than outweighed by the simplifications produced by having a simple expression for the potential energy.

Coordinate systems such as spherical polar or cylindrical polar coordinates are called *orthogonal coordinates*, because an infinitesimal displacement produced by changing only one of the coordinates is perpendicular (orthogonal) to a displacement produced by an infinitesimal change in any one of the other coordinates.

Figure 5.7 shows displacements, drawn as though they were finite, produced by infinitesimal changes in $r$, $\theta$, and $\phi$. These displacements are lengths

$$ds_r = \text{displacement in } r \text{ direction} = dr$$
$$ds_\theta = \text{displacement in } \theta \text{ direction} = r\,d\theta$$
$$ds_\phi = \text{displacement in } \phi \text{ direction} = r\sin(\theta)\,d\phi.$$

We define three vectors of unit length, whose directions are those of the infinitesimal displacements in Fig. 5.7, called $\mathbf{e}_r$, $\mathbf{e}_\theta$, and $\mathbf{e}_\phi$.

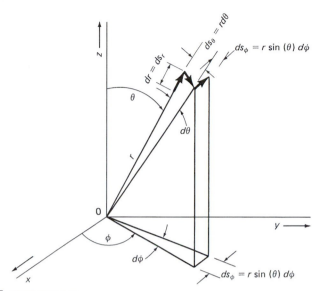

**FIGURE 5.7** Infinitesimal displacements $ds_r$, $ds_\theta$, and $ds_\phi$, produced by infinitesimal increments $dr$, $d\theta$, and $d\phi$.

An infinitesimal vector displacement is the sum of displacements in the three orthogonal directions. In cartesian coordinates,

$$d\mathbf{r} = \mathbf{i}\,dx + \mathbf{j}\,dy + \mathbf{k}\,dz. \tag{5.69}$$

In spherical polar coordinates,

$$d\mathbf{r} = \mathbf{e}_r\,dr + \mathbf{e}_\theta r\,d\theta + \mathbf{e}_\phi r\sin(\theta)\,d\phi. \tag{5.70}$$

We can write an expression for $d\mathbf{r}$ in a form that will hold for any set of orthogonal coordinates. Let the three coordinates of an orthogonal system in three dimensions be called $q_1$, $q_2$, and $q_3$. Let the displacements due to the infinitesimal increments be called $ds_1$, $ds_2$, and $ds_3$. Let the unit vectors in the directions of the displacements be called $\mathbf{e}_1$, $\mathbf{e}_2$, and $\mathbf{e}_3$. The equation analogous to Eq. (5.69) is

$$d\mathbf{r} = \mathbf{e}_1 ds_1 + \mathbf{e}_2 ds_2 + \mathbf{e}_3 ds_3 = \mathbf{e}_1 h_1 dq_1 + \mathbf{e}_2 h_2 dq_2 + \mathbf{e}_3 h_3 dq_3, \tag{5.71}$$

where the $h$'s are the factors needed to give the correct expression for each displacement. For cartesian coordinates, all three of the $h$'s are equal to unity. For spherical polar coordinates, $h_r = 1$, $h_\theta = r$, and $h_\phi = r\sin(\theta)$. For other systems, you can figure out what the $h$'s are geometrically so that $ds = h\,dq$ for each coordinate.

The gradient of a scalar function $f$ is written in terms of components in the direction of $ds_1$, $ds_2$, and $ds_3$ as

$$\nabla f = \mathbf{e}_1 \frac{\partial f}{\partial s_1} + \mathbf{e}_2 \frac{\partial f}{\partial s_2} + \mathbf{e}_3 \frac{\partial f}{\partial s_3}$$

or

$$\boxed{\nabla f = \mathbf{e}_1 \frac{1}{h_1}\frac{\partial f}{\partial q_1} + \mathbf{e}_2 \frac{1}{h_2}\frac{\partial f}{\partial q_2} + \mathbf{e}_3 \frac{1}{h_3}\frac{\partial f}{\partial q_3}}. \tag{5.72}$$

---

**Example 5.25**

Find the expression for the gradient of a function $f = f(r, \theta, \phi)$.

---

**Solution**

$$\nabla f = \mathbf{e}_r \frac{\partial f}{\partial r} + \mathbf{e}_\theta \frac{1}{r} \frac{\partial f}{\partial \theta} + \mathbf{e}_\phi \frac{1}{r \sin(\theta)} \frac{\partial f}{\partial \phi}. \tag{5.73}$$

---

**Problem 5.24**

a. Find the $h$ factors for cylindrical polar coordinates.
b. Find the expression for the gradient of a function of cylindrical polar coordinates, $f = f(\rho, \phi, z)$. Find the gradient of the function

$$f = e^{-(\rho^2 + z^2)/a^2} \sin(\phi).$$

---

The divergence of a vector function can similarly be expressed in orthogonal coordinates. If $\mathbf{F}$ is a vector function, it must be expressed in terms of the unit vectors of the coordinate system in which we are to differentiate,

$$\mathbf{F} = \mathbf{e}_1 F_1 + \mathbf{e}_2 F_2 + \mathbf{e}_3 F_3. \tag{5.74}$$

The components $F_1$, $F_2$, and $F_3$ are the components in the directions of $\mathbf{e}_1$, $\mathbf{e}_2$, and $\mathbf{e}_3$, not necessarily the cartesian components.

The divergence of the vector function $\mathbf{F}$ is given by

$$\nabla \cdot \mathbf{F} = \frac{1}{h_1 h_2 h_3} \left[ \frac{\partial}{\partial q_1}(F_1 h_2 h_3) + \frac{\partial}{\partial q_2}(F_2 h_1 h_3) + \frac{\partial}{\partial q_3}(F_3 h_1 h_2) \right]. \tag{5.75}$$

---

**Example 5.26**

a. Write the expression for the divergence of a function $\mathbf{F}$ expressed in terms of spherical polar coordinates.
b. Find the divergence of the position vector, which in spherical polar coordinates is

$$\mathbf{r} = \mathbf{e}_r r.$$

---

**Solution**

a.

$$\nabla \cdot \mathbf{F} = \frac{1}{r^2 \sin(\theta)} \left[ \frac{\partial}{\partial r}[F_r r^2 \sin(\theta)] + \frac{\partial}{\partial \theta}[F_\theta r \sin(\theta)] \frac{\partial}{\partial \phi}(F_\phi r) \right]$$

$$= \frac{1}{r^2} \frac{\partial}{\partial r}(r^2 F_r) + \frac{1}{r \sin(\theta)} \frac{\partial}{\partial \theta}[\sin(\theta) F_\theta] + \frac{1}{r \sin(\theta)} \frac{\partial F_\phi}{\partial \phi}. \tag{5.76}$$

b. The divergence of $\mathbf{r}$ is

$$\nabla \cdot \mathbf{r} = \frac{1}{r^2} 3r^2 + 0 + 0 = 3.$$

---

### Problem 5.25

Write the formula for the divergence of a function in cylindrical polar coordinates.

---

The curl of a vector function is

$$\nabla \times \mathbf{F} = \mathbf{e}_1 \frac{1}{h_2 h_3} \left[ \frac{\partial}{\partial q_2}(h_3 F_3) - \frac{\partial}{\partial q_3}(h_2 F_2) \right]$$

$$+ \mathbf{e}_2 \frac{1}{h_1 h_3} \left[ \frac{\partial}{\partial q_3}(h_1 F_1) - \frac{\partial}{\partial q_1}(h_3 F_3) \right]$$

$$+ \mathbf{e}_3 \frac{1}{h_1 h_2} \left[ \frac{\partial}{\partial q_1}(h_2 F_2) - \frac{\partial}{\partial q_2}(h_1 F_1) \right]. \qquad (5.77)$$

The expression for the laplacian of a scalar function, $f$, is

$$\nabla^2 f = \frac{1}{h_1 h_2 h_3} \left[ \frac{\partial}{\partial q_1} \left( \frac{h_2 h_3}{h_1} \frac{\partial f}{\partial q_1} \right) + \frac{\partial}{\partial q_2} \left( \frac{h_1 h_3}{h_2} \frac{\partial f}{\partial q_2} \right) + \frac{\partial}{\partial q_3} \left( \frac{h_1 h_2}{h_3} \frac{\partial f}{\partial q_3} \right) \right].$$

$$(5.78)$$

---

### Example 5.27

Write the expression for the laplacian in spherical polar coordinates.

---

### Solution

$$\nabla^2 f = \frac{1}{r^2} \frac{\partial}{\partial r} \left( r^2 \frac{\partial f}{\partial r} \right) + \frac{1}{r^2 \sin(\theta)} \frac{\partial}{\partial \theta} \left[ \sin(\theta) \frac{\partial f}{\partial \theta} \right] + \frac{1}{r^2 \sin^2(\theta)} \frac{\partial f}{\partial \phi^2}. \qquad (5.79)$$

---

## SECTION 5.9. MAXIMUM AND MINIMUM VALUES OF FUNCTIONS OF SEVERAL VARIABLES

It is frequently necessary to find the circumstance at which a given function has a maximum or a minimum value. A point at which either a maximum or a minimum occurs is sometimes called an *extremum*. For example, Fig. 5.8 shows a graph of the function $f = e^{-x^2 - y^2}$. The surface representing the function has a "peak" at the

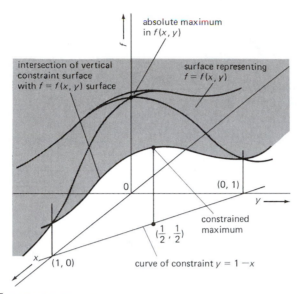

**FIGURE 5.8**   The surface representing a function of $x$ and $y$ with the absolute maximum and a constrained maximum shown.

origin, where the function attains its maximum value. Shown also in the figure is a curve at which the surface intersects with a plane representing the equation $y = 1 - x$. On this curve there is also a maximum, which has a smaller value than the maximum at the peak. We call this value the maximum subject to the constraint that $y = 1 - x$. We discuss the constrained maximum later.

A maximum at a peak such as the maximum at the origin in Fig. 5.8 is called a *local maximum* or a *relative maximum*. The value of the function at such a peak is larger than at any other point in the immediate vicinity. However, a complicated function can have more than one local maximum. Also, if we consider a finite region, the function might have a larger value somewhere on the boundary of the region that is larger than the value at a local maximum. To find the absolute maximum of the function for a given region, we must consider all local maxima and any points on the boundary of the region that might have greater values.

Points of minimum value are completely analogous to points of maximum value. Local minima are located at the bottom of depressions or valleys in the surface representing the function. To find an absolute minimum for a given region, you must consider all local minima and any points on the boundary of the region that might have smaller values.

To find a local maximum or minimum, we use the fact that the plane that is tangent to the surface will be horizontal at any local maximum or minimum. Therefore, the curve representing the intersection of any vertical plane with the surface will have a local maximum or a minimum at the same place. The partial derivative with respect to one independent variable gives the slope of the curve in the plane corresponding to a constant value of the other independent variable, so we can find a local maximum or minimum by finding the places where all the partial derivatives of the function vanish simultaneously.

Our method for two variables is therefore:

1. Solve the simultaneous equations

$$\left(\frac{\partial f}{\partial x}\right)_y = 0 \tag{5.80}$$

$$\left(\frac{\partial f}{\partial y}\right)_x = 0 \tag{5.81}$$

2. Calculate the value of the function at all points satisfying these equations, and also at the boundaries of the region being considered. The maximum or minimum value in the region being considered must be in this set of values.

For more than two independent variables, the method is similar, except that there is one equation for each independent variable.

---

### Example 5.28

Find the maximum value of the function shown in Fig. 5.8: $f = e^{-x^2 - y^2}$.

---

### Solution

At a local maximum or minimum

$$\left(\frac{\partial f}{\partial x}\right)_y = e^{-x^2 - y^2}(-2x) = 0$$

$$\left(\frac{\partial f}{\partial d}\right)_x = e^{-x^2 - y^2}(-2y) = 0.$$

The only solution is $x = 0$, $y = 0$. Since no restricted region was specified, we consider all values of $x$ and $y$. For very large magnitudes of $x$ and $y$, the function vanishes, so we have found the desired absolute maximum, at which $f = 1$.

---

In the case of one independent variable, a local maximum could be distinguished from a local minimum or an inflection point by determining the sign of the second derivative. With more than one variable, the situation is more complicated. In addition to inflection points, we can have points corresponding to a maximum with respect to one variable and a minimum with respect to another. Such a point is called a saddle point, and at such a point, the surface representing the function resembles a mountain pass or the surface of a saddle. Such points are important in the transition-state theory of chemical reaction rates.

For two independent variables, the following quantity is calculated:

$$D = \left(\frac{\partial^2 f}{\partial x^2}\right)\left(\frac{\partial^2 f}{\partial y^2}\right) - \left(\frac{\partial^2 f}{\partial x \partial y}\right)^2. \tag{5.82}$$

The different cases are as follows:

1. If $D > 0$ and $(\partial^2 f / \partial x^2) > 0$, then we have a local minimum.
2. If $D > 0$ and $(\partial^2 f / \partial x^2) < 0$, then we have a local maximum.
3. If $D < 0$, then we have neither a local maximum nor a local minimum.
4. If $D = 0$, the test fails, and we cannot tell what we have.

### Problem 5.26

Evaluate $D$ at the point $(0, 0)$ for the function of Example 5.28 and establish that the point is a local maximum.

## Constrained Maximum–Minimum Problems

Sometimes we must find a maximum or a minimum value of a function subject to some condition. Such an extremum is called a *constrained maximum* or a *constrained minimum*. Generally, a constrained maximum is smaller than the unconstrained maximum of the function, and a constrained minimum is larger than the unconstrained minimum of the function. Consider the following example:

### Example 5.29

Find the maximum value of the function in Example 5.28 subject to the constraint

$$x + y = 1.$$

### Solution

The situation is shown in Fig. 5.8. The constraint corresponds to the specification of $y$ as a function of $x$ by

$$y = 1 - x. \tag{5.83}$$

This function is given by the line in the $x$–$y$ plane of the figure. We are now looking for the place along this curve at which the function has a larger value than at any other place on the curve. Unless the curve happens to pass through the unconstrained maximum, the constrained maximum will be smaller than the unconstrained maximum.

Since $y$ is no longer an independent variable on the curve of the constraint, the direct way to proceed is to replace $y$ by use of Eq. (5.83):

$$f = (x, 1 - x) = f(x) = e^{-x^2 - (1-x)^2} = e^{-2x^2 + 2x - 1}. \tag{5.84}$$

The local maximum is now at the point where $df/dx$ vanishes:

$$\frac{df}{dx} = e^{-2x^2 + 2x - 1}(-4x + 2) = 0. \tag{5.85}$$

The solution to this is $x = \frac{1}{2}$, which corresponds to $y = \frac{1}{2}$. At this point

$$f\left(\frac{1}{2}, \frac{1}{2}\right) \exp\left[-\left(\frac{1}{2}\right)^2 - \left(\frac{1}{2}\right)^2\right] = e^{-1/2} = 0.6065 \cdots.$$

As expected, this value is smaller than the unconstrained maximum, at which $f = 1$.

### Problem 5.27

a. Find the minimum in the function

$$f(x, y) = x^2 + y^2 + 2x.$$

b. Find the constrained minimum subject to the constraint

$$x + y = 0.$$

---

If we have a constrained maximum–minimum problem with more than two variables, the direct method of substituting the constraint relation into the function is usually not practical.

## Lagrange's Method of Undetermined Multipliers[2]

This is a method for finding a constrained maximum or minimum without substituting the constraint relation into the function. If the constraint is written in the form $g(x, y) = 0$, the method for finding the constrained maximum or minimum in $f(x, y)$ is as follows:

1. Form the new function

$$u(x, y) = f(x, y) + \lambda g(x, y), \tag{5.86}$$

where $\lambda$ is a constant called an *undetermined multiplier*. Its value is unknown at this point of the analysis.

2. Form the partial derivatives of $u$, and set them equal to zero,

$$\left(\frac{\partial u}{\partial x}\right)_y = \left(\frac{\partial f}{\partial x}\right)_y + \lambda \left(\frac{\partial g}{\partial x}\right)_y = 0 \tag{5.87a}$$

$$\left(\frac{\partial u}{\partial y}\right)_x = \left(\frac{\partial f}{\partial y}\right)_x + \lambda \left(\frac{\partial g}{\partial y}\right)_x = 0. \tag{5.87b}$$

3. Solve the set of equations consisting of $g = 0$ and the two equations of Eq. (5.87) as a set of simultaneous equations for the value of $x$, the value of $y$, and the value of $\lambda$ that correspond to the local maximum or minimum. The method is equivalent to finding the ordinary maximum of $u$ as a function of $x$, $y$, and $\lambda$.

We will not present a proof of the validity of this method, but you can find such a proof in the book by Taylor and Mann listed at the end of the chapter.

---

### Example 5.30

Find the constrained maximum of Example 5.29 by the method of Lagrange.

---

### Solution

The constraining equation is written

$$g(x, y) = x + y - 1 = 0. \tag{5.88}$$

---

[2]Named for Joseph Louis Lagrange (born Guisepps Lodovico Lagrangia), 1736–1813, French–Italian physicist and mathematician.

The function $u$ is

$$u(x, y) = e^{-x^2-y^2} + \lambda(x + y - 1)$$

so that the equations to be solved are Eq. (5.88) and

$$\left(\frac{\partial u}{\partial x}\right)_y = (-2x)e^{-x^2-y^2} + \lambda = 0 \tag{5.89a}$$

$$\left(\frac{\partial u}{\partial y}\right)_x = (-2y)e^{-x^2-y^2} + \lambda = 0. \tag{5.89b}$$

Let us begin by solving for $\lambda$ in terms of $x$ and $y$. Multiply Eq. (5.89a) by $y$ and Eq. (5.89b) by $x$ and add the two equations. The result can be solved to give

$$\lambda = \frac{4xy}{x + y}e^{-x^2-y^2}. \tag{5.90}$$

Substitute this into Eq. (5.89a) to obtain

$$(-2x)e^{-x^2-y^2} + \frac{4xy}{x + y}e^{-x^2-y^2} = 0. \tag{5.91}$$

The exponential factor is not zero for any finite values of $x$ and $y$, so

$$-2x + \frac{4xy}{x + y} = 0. \tag{5.92a}$$

When Eq. (5.90) is substituted into Eq. (5.89b) in the same way, the result is

$$-2y + \frac{4xy}{x + y} = 0. \tag{5.92b}$$

When Eq. (5.92b) is subtracted from Eq. (5.92a), the result is

$$-2x + 2y = 0$$

which is solved for $y$ in terms of $x$ to obtain

$$y = x.$$

This is substituted into Eq. (5.88) to obtain

$$x + x - 1 = 0$$

which gives

$$x = \frac{1}{2}, \qquad y = \frac{1}{2}.$$

This is the same result as in Example 5.29. In this case, the method of Lagrange was more work than the direct method. In more complicated problems, the method of Lagrange will usually be easier.

---

The method of Lagrange also works if there is more than one constraint. If we desire the local maximum or minimum of the function

$$f = f(x, y, z) \tag{5.93}$$

subject to the two constraints

$$g_1(x, y, z) = 0 \tag{5.94}$$

and

$$g_2(x, y, z) = 0 \tag{5.95}$$

the procedure is similar, except that two undetermined multipliers are used.

One forms the function

$$u = u(x, y, z) = f(x, y, z) + \lambda_1 g_1(x, y, z) + \lambda_2 g_2(x, y, z)$$

and solves the set of simultaneous equations consisting of Eqs. (5.94), (5.95), and

$$\left(\frac{\partial u}{\partial x}\right)_{y,z} = 0 \tag{5.96a}$$

$$\left(\frac{\partial u}{\partial y}\right)_{x,z} = 0 \tag{5.96b}$$

$$\left(\frac{\partial u}{\partial z}\right)_{x,y} = 0. \tag{5.96c}$$

The result is a value for $\lambda_1$, a value for $\lambda_2$, and values for $x$, $y$, and $z$ which locate the constrained local maximum or minimum.

---

### Problem 5.28

Find the constrained minimum of Problem 5.27 using the method of Lagrange.

---

## SUMMARY OF THE CHAPTER

The calculus of functions of several independent variables is a natural extension of the calculus of functions of one independent variable. The partial derivative is the first important quantity. For example, a function of three independent variables has three partial derivatives. Each one is obtained by the same techniques as with ordinary derivatives, treating other independent variables temporarily as constants. The differential of a function of $x_1$, $x_2$, and $x_3$ is given by

$$df = \left(\frac{\partial y}{\partial x_1}\right)_{x_2,x_3} dx_1 + \left(\frac{\partial y}{\partial x_2}\right)_{x_1,x_3} dx_2 + \left(\frac{\partial y}{\partial x_3}\right)_{x_1,x_2} dx_3,$$

where $df$ is an infinitesimal change in the function $f$ produced by the changes $dx_1$, $dx_2$, and $dx_3$ imposed on the independent variables and where the coefficients are partial derivatives. This is called an exact differential, identifying $df$ as in increment in a function. A similar expression such as

$$dw = M dx_1 + N dx_2 + P dx_3$$

is an inexact differential if the coefficients $M$, $N$, and $P$ are not the appropriate derivatives of the same function. If not, they do not obey the Euler reciprocity relation, which is one of several relations that partial derivatives obey.

One application of partial derivatives is in the search for minimum and maximum values of a function. An extremum (minimum or maximum) of a function in a region is found either at a boundary of the region or at a point where all of the partial derivatives vanish. A constrained maximum or minimum is found by the method of Lagrange, in which a particular augmented function is maximized or minimized.

A line integral is denoted by

$$\int_c dw = \int_c (M\, dx_1 + N\, dx_2 + P\, dx_3).$$

In this integral, the variables $x_2$ and $x_3$ in $M$ must be replaced by functions of $x_1$ corresponding to the curve on which the integral is performed, with similar replacements in $N$ and $P$. The line integral of the differential of a function depends only on the end points of the curve, which the line integral of an inexact differential depends on the path of the curve as well as on the end points.

A multiple integral is an integral of the form

$$\iiint f(x, y, z)\, dz\, dy\, dx$$

in which the integrations are carried out one after another. As each integration is carried out, those integration variables not yet integrated are treated as constants.

## ADDITIONAL READING

Thomas H. Barr, *Vector Calculus*, Prentice Hall, Upper Saddle River, NJ, 1997. This is a textbook for a third-semester calculus course that emphasizes vector calculus.

Henry Margenau and George Mosely Murphy, *The Mathematics of Physics and Chemistry*, 2nd ed., Van Nostrand, Princeton, NJ, 1956. This reference book contains discussions of many of the mathematical topics needed for physical chemistry at the beginning graduate level.

Philip M. Morse and Herman Feshbach, *Methods of Theoretical Physics*, McGraw–Hill, New York, 1953. This is a large work which comes in two volumes and contains a lot of information about the topics in this chapter. It is at a somewhat higher level than Margenau and Murphy.

Louis A. Pipes, *Applied Mathematics for Engineers and Physicists*, McGraw–Hill, New York, 1946. This book is at about the same level as that of Margenau and Murphy.

Angus E. Taylor and Robert W. Mann, *Advanced Calculus*, 2nd ed., Ginn, and Lexington, MA, 1972. This book is a typical textbook for a calculus course to be taken after the elementary sequence and contains thorough discussions of most of the topics of this chapter. There are a number of similar books, including one by Kaplan listed at the end of Chapter 6.

## ADDITIONAL PROBLEMS

### 5.29

A certain nonideal gas has an equation of state

$$\frac{P\bar{V}}{RT} = 1 + \frac{B_2}{V},$$

where $B_2$ is given as a function of $T$ by

$$B_2 = [-1.00 \times 10^{-4} - (2.148 \times 10^{-6})e^{(1986\,\text{K})/T}]\ \text{m}^3\,\text{mol}^{-1},$$

where $T$ is the temperature on the Kelvin scale, $\bar{V}$ is the volume of 1 mol, $P$ is the pressure, and $R$ is the gas constant. Find $(\partial P/\partial \bar{V})_{n,T}$ and $(\partial P/\partial T)_{n,\bar{V}}$ and an expression for $dP$.

### 5.30

For a certain system, the thermodynamic energy $U$ is given as a function of $S$, $V$, and $n$ by

$$U = U(S, V, n) = Kn^{5/3}V^{-2/3}e^{2S/3nR},$$

where $S$ is the entropy, $V$ is the volume, $n$ is the number of moles, $K$ is a constant, and $R$ is the gas constant. Find $dU$ in terms of $dS$, $dV$, and $dn$.

### 5.31

Find $(\partial f/\partial x)_y$ and $(\partial f/\partial y)_x$ for each of the following functions. $a$, $b$, and $c$ are constants.

a. $f = axy\ln(y) + bx\cos(x + y)$
b. $f = ae^{-b(x^2+y^2)} + c\sin(x^2 y)$
c. $f = a(x + by)/(c + xy)$
d. $f = (x + y)^{-3}$.

### 5.32

Find $(\partial^2 f/\partial x^2)_y$, $(\partial^2 f/\partial x\,\partial y)$, and $(\partial^2 f/\partial y^2)$, for each of the following functions.

a. $f = (x + y)^{-1}$
b. $f = \cos(x/y)$
c. $f = e^{(ax^2+by^2)}$.

### 5.33

a. Find the area of the semicircle of radius a given by

$$y = +(a^2 - x^2)^{1/2}$$

by doing the double integral

$$\int_{-a}^{a} \int_{0}^{(a^2-x^2)^{1/2}} 1\,dy\,dx.$$

b. Change to polar coordinates and repeat the calculation.

### 5.34

Test each of the following differentials for exactness.

a. $du = by\cos(bx)\,dx + \sin(bx)\,dy$
b. $du = ay\sin(xy)\,dx + ax\sin(xy)\,dy$
c. $du = (y/(1 + x^2))\,dx - \tan^{-1}(x)\,dy$
d. $du = x\,dy + y\,dx$
e. $du = y\ln(x)\,dx + x\ln(y)\,dy$
f. $du = 2xe^{axy}\,dx + 2ye^{axy}\,dy$.

**5.35**

If $G = -RT \ln(aT^{3/2}V/n)$ find $dG$ in terms of $dT$, $dV$, and $dn$, where $R$ and $a$ are constants.

**5.36**

Perform the line integral

$$\int_C df = \int_C (x^2 y\, dx + xy^2\, dy),$$

where $C$ is the line segment from $(0, 0)$ to $(2, 2)$. Would another path with the same end points yield the same result?

**5.37**

Find the function $f(x, y)$ whose differential is

$$df = (x + y)^{-1}\, dx + (x + y)^{-1}\, dy$$

and which has the value $f(1, 1) = 0$. Do this by performing a line integral on a rectangular path from $(1, 1)$ to $(x_1, y_1)$ Assume $x_1 > 0$, $y_1 > 0$.

**5.38**

A wheel of radius $R$ has a mass per unit area given by

$$m(\rho) = a\rho^2 + b,$$

where $a$ and $b$ are constants, $m$ is the mass per unit area, and $\rho$ is the distance from the center. Assume that $m$ depends only on $\rho$. Find the moment of inertia, defined by

$$I = \int\int m(\rho)\rho^2\, dA,$$

where $dA$ represents the element of area and the integral is a double integral over the entire wheel. Carry out the integral twice, once in both cartesian and once in plane polar coordinates.

**5.39**

Complete the formula

$$\left(\frac{\partial S}{\partial V}\right)_{P,n} = \left(\frac{\partial S}{\partial V}\right)_{T,n} + ?$$

**5.40**

Find the location of the minimum in the function

$$f = f(x, y) = x^2 - 6x + 8y + y^2.$$

What is the value of the function at the minimum?

**5.41**

Find the minimum in the function of Problem 5.40 subject to the constraint

$$x + y = 2.$$

Do this by substitution and by the method of undetermined multipliers.

# 6 ■ MATHEMATICAL SERIES AND TRANSFORMS

**Preview**

A mathematical series is a sum of terms. A series can have a finite number of terms or can have an infinite number of terms. If a series has an infinite number of terms, an important question is whether it approaches a finite limit as more and more terms of the series are included (in which case we say that it converges) or whether it becomes infinite in magnitude or oscillates endlessly (in which case we say that it diverges).

A constant series has terms that are constants, so that it equals a constant. A functional series has terms that are functions of one or more independent variables, so that the series is a function of the same independent variables. Each term of a functional series contains a constant coefficient that multiplies a function from a set of basis functions. The process of constructing a functional series to represent a specific function is the process of determining the coefficients. We discuss two common types of functional series, power series and Fourier series.

An integral transform is similar to a functional series, except that it contains an integration instead of a summation. It therefore contains a variable over which the integration is carried out instead of a summation index. The integrand contains two factors, as does a term of a functional series. The first factor is the transform, which plays the same role as the coefficients of a power series. The second factor is the basis function, which plays the same role as the set of basis functions in a functional series. We discuss two types of transforms, Fourier transforms and Laplace transforms.

## Principal Facts and Ideas

1. A mathematical series is a sum of terms, either with a finite number of terms or an infinite number of terms.
2. A constant series has terms that are constants, and a functional series has terms that are functions.
3. An infinite series converges if the sum approaches a finite limit ever more closely as ever more terms are summed, or diverges if the series does not approach such a limit.
4. An infinite functional series can represent a function if it converges.
5. A Taylor series is a sum of terms that consist of coefficients times powers of $x - h$, where $x$ is a variable and $h$ is a constant. The coefficients can be determined to represent any analytic function within a region of convergence.
6. A Fourier series is an infinite series of terms that consist of coefficients times sine and cosine functions. It can represent almost any periodic function.
7. A Fourier transform is a representation of a function an integral instead of a sum. Many modern instruments use Fourier transforms to produce spectra from raw data in another form.
8. A Laplace transform is a representation of a function that is similar to a Fourier transform.

## Objectives

After studying this chapter, you should be able to:

1. determine whether an infinite constant series converges,
2. determine how large a partial sum must be taken to approximate a series,
3. compute the coefficients for a power series to represent a given function,
4. determine the region of convergence of a power series,
5. determine the coefficients of a Fourier series to represent some elementary functions,
6. determine the Fourier transform of some elementary functions,
7. determine the Laplace transform of some elementary functions,
8. manipulate Laplace transforms using various theorems.

## SECTION 6.1. CONSTANT SERIES

A *sequence* is a set of numerical quantities with a rule for generating one member of the set from the previous one. If the members of a sequence are added together, the result is a *series*. A *finite series* has a finite number of terms, and an *infinite series* has a infinite number of terms.

If a series has terms that are constants, it is a *constant series*. Such a series is written

$$s = a_0 + a_1 + a_2 + a_3 + a_4 + \cdots + a_n + \cdots. \tag{6.1}$$

For an infinite series, we define the $n$th *partial sum* as the sum of the first $n$ terms:

$$S_n = a_0 + a_1 + a_2 + a_3 + \cdots + a_{n-1}. \tag{6.2}$$

The entire infinite series is the limit

$$s = \lim_{n \to \infty} S_n. \tag{6.3}$$

If this limit exists and is finite, we say that the series *converges*. If the magnitude of $S_n$ becomes larger and larger without bound as $n$ becomes large, or if $S_n$ continues to oscillate as $n$ becomes large, we say that the series *diverges*.

The two questions that we generally ask about an infinite constant series are: (1) Does the series converge? (2) What is the value of the series if it does converge? Sometimes it is difficult to find the value of some convergent infinite series, and we then might ask how well we can approximate the series with a partial sum.

## Some Convergent Series

Let us consider a well-known convergent constant series:

$$s = 1 + \frac{1}{2} + \frac{1}{4} + \frac{1}{8} + \cdots + \frac{1}{2^n} + \cdots = \sum_{n=0}^{\infty} \frac{1}{2^n}. \tag{6.4}$$

The value of this series is calculated in the following example. However, there is no general method that is capable of finding the value of every series.[1]

---

**Example 6.1**

Find the value of the series in Eq. (6.4).

---

**Solution**

We write the sum as the first term plus the other terms, with a factor of $\frac{1}{2}$ factored out of all the other terms:

$$s = 1 + \frac{1}{2}\left(1 + \frac{1}{2} + \frac{1}{4} + \frac{1}{8} + \cdots\right).$$

The series in the parentheses is just the same as the original series. There is no problem due to the apparent difference that the series in the parentheses seems to have one less term than the original series, because both series have an infinite number of terms. We now write

$$s = 1 + \frac{1}{2}s$$

which can be solved to give

$$s = 2. \tag{6.5}$$

---

[1]See A. D. Wheelon, "On the Summation of Infinite Series," *J. Appl. Phys.*, **25**, 113 (1954), for a method that can be applied to a large number of series.

**Problem 6.1**

Show that in the series of Eq. (6.4) any term of the series is equal to the sum of all the terms following it. (*Hint:* Factor a factor out of all of the following terms so that they will equal this factor times the original series, whose value is now known.)

The result of Problem 6.1 is of interest in seeing how a series can be approximated by a partial sum. For the series of Eq. (6.4), we can use the result of Problem 6.1, that any term is equal to the sum of all following terms and write

$$s = S_n + a_{n-1} \qquad \text{(for the series of Eq. (6.4)).} \qquad (6.6)$$

In some cases it is necessary to approximate a series with a partial sum, and in such cases Eq. (6.6) may be applied to other series as a rough measure of the error in approximating $s$ by $S_n$.

**Example 6.2**

Determine which partial sums approximate the series of Eq. (6.4) to (a) 1% and (b) 0.001%.

**Solution**

    a.  Since 1% of 2 is 0.02, we find the first term of the series that is equal to or smaller than 0.02, and take the partial sum that ends with that term. We have

$$\frac{1}{2^6} = \frac{1}{64} = 0.015625$$

    so that the partial sum required is the one ending with $\frac{1}{64}$, or $S_7$. Its value is

$$S_7 = 1.984375.$$

    b.  Since 0.001% of 2 is $2 \times 10^{-5}$, and $1/2^n$ has the value $1.5259 \times 10^{-5}$ when $n = 16$, we need the partial sum $S_{17}$, which has the value $S_{17} = 1.999984741$.

**Problem 6.2**

Consider the series

$$s = 1 + \frac{1}{2^2} + \frac{1}{3^2} + \frac{1}{4^2} + \cdots + \frac{1}{n^2} + \cdots$$

which is known to be convergent. Using Eq. (6.6) as an approximation, determine which partial sum approximates the series to (a) 1% and (b) 0.001%.

## The Geometric Series

The series of Eq. (6.4) is an example of a *geometric series*, which is defined to be

$$s = a + ar + ar^2 + ar^3 + ar^4 + \cdots + ar^n + \cdots \qquad (6.7a)$$
$$= ar(1 + r + r^2 + r^3 + \cdots + r^n + \cdots), \qquad (6.7b)$$

where $a$ and $r$ are constants. In order for the infinite series of Eq. (6.7) to converge, the magnitude of $r$ must be less than unity. Otherwise, each term would be equal to or larger than the previous term, causing the sum to grow without bound (diverge) as more and more terms are added. If $r = 1$, all of the terms equal $a$, and the infinite series still diverges. However, $r$ can be positive or negative. If $r$ is negative, the sum has terms of alternating sign, but still converges if $|r| < 1$.

The value of a geometric series can be obtained in the same way as was the value of the series in Eq. (6.4),

$$s = a + r(a + ar + ar^2 + ar^3 + \cdots)$$
$$= a + rs$$

or

$$\boxed{s = \frac{a}{1-r}} \qquad (|r| < 1). \qquad (6.8)$$

The partial sums of the geometric series are given by

$$\boxed{S_n = a + ar + ar^2 + \cdots + ar^{n-1} = a\frac{1-r^n}{1-r}}. \qquad (6.9)$$

Equation (6.9) is valid for any value of $r$, since a finite series always converges.

## Example 6.3

The molecular partition function $z$ is defined in the statistical mechanics of non-interacting molecules as the sum over all the states of one molecule

$$z = \sum_{i=0}^{\infty} \exp\left(\frac{-E_i}{k_B T}\right), \qquad (6.10)$$

where $i$ is an index specifying the state, $E_i$ is the energy that the molecule has when in state number $i$, $k_B$ is Boltzmann's constant, and $T$ is the absolute temperature. If we consider only the vibration of a diatomic molecule, to a good approximation

$$E_i = E_v = h\nu\left(v + \frac{1}{2}\right), \qquad (6.11)$$

where $\nu$ is the vibrational frequency; $v$ is the vibrational quantum number, which can take on the integral values 0,1,2,3, etc.; and $h$ is Planck's constant. Use Eq. (6.8) to find the value of the partition function for vibration.

## Solution

$$z_{vib} = \sum_{v=0}^{\infty} \exp\left[\frac{-h\nu\left(v + \frac{1}{2}\right)}{k_B T}\right]$$
$$= \exp\left(\frac{-h\nu}{2k_B T}\right) \sum_{v=0}^{\infty} \left[\exp\left(\frac{h\nu}{k_B T}\right)\right]^v$$
$$= e^{-x/2} \sum_{v=0}^{\infty} (e^{-x})^v,$$

where $h\nu/k_B T = x$, a positive quantity.

The sum is a geometric series, so

$$z_{\text{vib}} = \frac{e^{-x/2}}{1 - e^{-x}}. \tag{6.12}$$

The series is convergent, because $e^{-x}$ is smaller than unity for all positive values of $x$ and is never negative.

## A Divergent Series

The *harmonic series* is

$$s = 1 + \frac{1}{2} + \frac{1}{3} + \frac{1}{4} + \cdots + \frac{1}{n} + \cdots. \tag{6.13}$$

Here are a few partial sums of this series:

$$S_1 = 1$$
$$S_2 = 1.5$$
$$S_{200} = 6.87803$$
$$S_{1000} = 8.48547$$
$$S_{100,000} = 13.0902.$$

However,

$$s = \lim_{n \to \infty} S_n = \infty.$$

Many people are surprised when they first learn that this series diverges, because the terms keep on getting smaller as you go further into the series. This is a necessary condition for a series to converge, but it is not sufficient. We will show that the harmonic series is divergent when we introduce tests for convergence.

### Problem 6.3

Write a computer program in the BASIC language to evaluate partial sums of the harmonic series and use it to verify the foregoing values.

There are a number of constant series listed in Appendix 3, and additional series can be found in the references listed at the end of the chapter.

## Tests for Convergence of a Series

There are several tests that will usually tell us whether an infinite series converges or not.

1. *The Comparison Test.* If a series has terms each of which is smaller in magnitude than the corresponding term of a series known to converge, it is convergent. If a series has terms each of which is larger in magnitude than the corresponding term of a series known to diverge, it is divergent.

2. *The Alternating Series Test*. If a series has terms that alternate in sign, it is convergent if the terms approach zero as you go further and further into the series and if each term is smaller in magnitude than the previous term.

3. *The nth-Term Test*. If the terms of a series approach some limit other than zero or do not approach any limit as you go further into the series, the series diverges.

4. *The Integral Test*. If a formula can be written to deliver the terms of a series

$$a_n = f(n) \tag{6.14}$$

then the series will converge if the improper integral

$$\int_1^\infty f(x)\,dx \tag{6.15}$$

converges and will diverge if the improper integral diverges.

5. *The Ratio Test*. For a series of positive terms, consider the limit

$$r = \lim_{n\to\infty} \frac{a_{n+1}}{a_n}. \tag{6.16}$$

If $r < 1$, the series converges. If $r > 1$, the series diverges. If $r = 1$, the test fails, and the series might either converge or diverge. If the ratio does not approach any limit but does not increase without bound, the test also fails.

## Example 6.4

Apply the ratio test and the integral test to the harmonic series, Eq. (6.13).

## Solution

Apply the ratio test:

$$r = \lim_{n\to\infty} \left[ \frac{1/n}{1/(n-1)} \right] = \lim_{n\to\infty} \frac{n-1}{n} = 1.$$

The ratio test fails. Apply the integral test:

$$\int_1^0 \frac{1}{x}\,dx = \ln(x)\Big|_1^\infty = \lim_{b\to\infty} [\ln(b) - \ln(1)] = \infty.$$

The series diverges by the integral test.

## Example 6.5

Determine whether the series converges:

$$s = 1 - \frac{1}{2} + \frac{1}{3} - \frac{1}{4} + \frac{1}{5} - \cdots = \sum_{n-1}^{\infty} \frac{(-1)^{n-1}}{n}.$$

## Solution

This is an alternating series, so the alternating series test applies. Since every term approaches more closely to zero than the previous term, the series is convergent.

**Example 6.6**

Determine whether the series converges:

$$s = 1 - \frac{1}{2} + \frac{2}{2} - \frac{1}{3} + \frac{2}{3} - \frac{1}{4} + \frac{2}{4} - \frac{1}{5} + \frac{2}{5} - \cdots .$$

**Solution**

This is a tricky series, because it is an alternating series, and the $n$th term approaches zero as $n$ becomes large. However, the series diverges. The alternating series test does not apply, because it requires that each term be closer to zero than the previous term. Half the time as you go from one term to the next in this series, the magnitude increases instead of decreasing. Let us manipulate the series by subtracting each negative term from the following positive term to obtain

$$s = 1 + \frac{1}{2} + \frac{1}{3} + \frac{1}{4} + \frac{1}{5} + \cdots .$$

This is the harmonic series, which we already found to diverge.

**Problem 6.4**

Show that the series in Eq. (6.7) converges.

**Problem 6.5**

Show that the geometric series converges if $r^2 < 1$.

**Problem 6.6**

Test the following series for convergence.

a. $\displaystyle\sum_{n=0}^{\infty} (1/n^2)$.

b. $\displaystyle\sum_{n=0}^{\infty} (1/n!)$.

c. $\displaystyle\sum_{n=1}^{\infty} ((-1)^n (n-1)/n^2)$.

d. $\displaystyle\sum_{n=0}^{\infty} ((-1)^n n/n!)$.

Note. $n!$ ($n$ factorial) is defined for positive integral values of $n$ to be $n(n-1)$ $(n-2)\cdots(2)(1)$, and is defined to equal 1 if $n = 0$.

## SECTION 6.2. POWER SERIES

A *functional series* has terms that are constants times functions. If a single independent variable is called $x$,

$$s(x) = a_0 g_0(x) + a_1 g_1(x) + a_2 g_2(x) + a_3 g_3(x) + \cdots. \qquad (6.17)$$

We call the set of constant quantities $a_0$, $a_1$, $a_2$, etc., the *coefficients* of the series and the set of functions $g_0$, $g_1$, $g_2$, $g_3$, ... the *basis functions*. Just as with constant series, a functional series such as that of Eq. (6.17) might converge or it might diverge. However, it might converge for some values of $x$ and diverge for others. If there is an interval of values of $x$ such that the series converges for all values of $x$ in that interval, we say that the series is convergent in that interval.

There is an important mathematical concept called *uniform convergence*. If a functional series converges in some interval, it is uniformly convergent in that interval if it converges with at least a certain fixed rate of convergence in the entire interval. We do not discuss the details of this concept.[2] However, if a functional series is uniformly convergent in some interval, it has been shown to have some useful mathematical properties, which we discuss later.

Since a functional series is a representation of a function, it encodes the same information as any other representation of the function, such as a mathematical formula. However, the information is encoded in a different way. In the series, the information about the nature of the function is encoded in the coefficients. You should think of the series as changing the independent variable. If the initial representation of the function is a formula containing the independent variable $x$, the independent variable in the series representation is the summation index. That is, the way in which the coefficients depend on the value of the index describes the nature of the function. The dependence on $x$ is contained in the basis functions, which are a fixed set of functions.

A common type of functional series is the *power series*, in which the basis functions are powers of $x - h$, where $x$ is the independent variable and $h$ is a constant (it can equal zero).

$$s(x) = a_0 + a_1(x - h) + a_2(x - h)^2 + a_3(x - h)^3 + \cdots, \qquad (6.18)$$

where $a_0$, $a_1$, etc., are constant coefficients. An infinite series of this form is called a *Taylor series*, and if $h = 0$ it is called a *Maclaurin series*. If we represent a function by a power series, we say that we *expand* the function in terms of the power series.

Several methods of theoretical chemistry, such as the perturbation method in quantum mechanics, use power series. In such theories it is usually assumed without proof that the series is uniformly convergent in some region, since it is not possible to write a formula that represents every term in such a way that convergence can be tested. Some important series that were once assumed to be convergent are now known to be divergent.

Power series with finite numbers of terms also occur and are called *polynomials*. Several important polynomial functions occur in physical chemistry, as, for example,

---

[2]David Bressoud, *A Radical Approach to Real Analysis*, pp. 191–194, Math. Assoc. of America, Washington, DC, 1994.

in the solution of the Schrödinger equation for the hydrogen atom and the harmonic oscillator.

There are two problems to be faced in constructing a series to represent a given function. The first is finding the values of the coefficients so that the function will be correctly represented. The second is finding the interval in which the series is convergent and in which it represents the given function.

## Maclaurin Series

Let us consider first a Maclaurin series to represent an arbitrary function $f(x)$. We want to require that

$$f(x) = s(x) = a_0 + a_1 x + a_2 x^2 + \cdots = s(x), \tag{6.19}$$

where $f(x)$ is the function and $s(x)$ is the series. In order for the function and the series to be equal at all values of $x$, they must be equal at $x = 0$,

$$\boxed{f(0) = s(0) = a_0} \tag{6.20}$$

which gives us a formula to determine $a_0$.

We also require that all derivatives of the function and the series be equal at $x = 0$. This is sufficient for the series to represent the function in some interval around $x = 0$, although we do not prove this fact. This means that only a function that possesses derivatives of all orders at $x = 0$ can be represented by a Maclaurin series. Such a function is said to be *analytic* at $x = 0$. The $n$th derivative of the series is, at $x = 0$,

$$\left( \frac{d^n s}{dx^n} \right)_0 = n! a_n, \tag{6.21}$$

where $n!$ ($n$ factorial) is, for $n \geq 1$

$$n! = n(n - 1)(n - 2) \cdots (2)(1).$$

This gives us a general formula for the coefficients in a Maclaurin series to represent the function $f(x)$:

$$\boxed{a_n = \frac{1}{n!} \left( \frac{d^n f}{dx^n} \right)_0 \qquad (n = 1, 2, 3 \ldots.)}. \tag{6.22}$$

A power series that is obtained by using Eq. (6.22) to obtain the coefficients faithfully represents the appropriate function in the vicinity of $x = 0$ if it converges and if infinitely many terms are taken. However, we must discuss later how far from $x = 0$ we can go and still represent the function by the series.

---

### Problem 6.7

Show that Eq. (6.22) is correct.

---

We now apply our formulas to obtain an example Maclaurin series.

**Example 6.7**

Find all the coefficients for the Maclaurin series representing $\sin(x)$.

---

**Solution**

Since $\sin(0) = 0$,

$$a_0 = 0$$

$$a_1 = \left[ \frac{d}{dx} \sin(x) \right]_0 = \cos(0) = 1, \qquad (6.23)$$

where the subscripts indicate that the derivative is evaluated at $x = 0$. The second partial sum of the series is therefore

$$S_2 = 0 + x = x$$

giving the same approximation as in Eq. (2.28). Figure 6.1 shows the function $\sin(x)$ and the approximation, $S_2 = x$. The derivatives of $\sin(x)$ follow a repeating pattern:

$$f(x) = \sin(x)$$

$$\frac{df}{dx} = \cos(x)$$

$$\frac{d^2 f}{dx^2} = -\sin(x)$$

$$\frac{d^3 f}{dx^3} = -\cos(x)$$

$$\frac{d^4 f}{dx^4} = \sin(x),$$

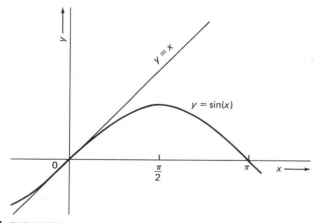

**FIGURE 6.1**    The function $\sin(x)$ and the approximation $S_2 = x$.

etc. When these are evaluated at $x = 0$, all of the even-numbered derivatives vanish, and the odd-numbered derivatives are alternately equal to 1 and $-1$:

$$\sin(x) = x - \frac{1}{3!}x^3 + \frac{1}{5!}x^5 - \frac{1}{7!}x^7 + \cdots .$$  (6.24)

## Problem 6.8

a. Show that the Maclaurin series for $e^x$ is

$$e^x = 1 + \frac{1}{1!}x + \frac{1}{2!}x^2 - \frac{1}{3!}x^3 + \frac{1}{4!}x^4 + \cdots .$$  (6.25)

b. Find the Maclaurin series for $\cos(x)$.

## Taylor Series

There are a number of important functions that are not analytic at $x = 0$, and these cannot be represented by a Maclaurin series. One such function is $\ln(x)$. The first derivative of this function is $1/x$, which becomes infinite as $x \rightarrow 0$, as do the other derivatives. Although there is no Maclaurin series for $\ln(x)$, you can find a Taylor series for a positive value of $h$.

In order to find the coefficients for a Taylor series, we require the function and the series to be equal at $x = h$ and to have the same derivatives of all orders at $x = h$. This gives

$$a_0 = f(h); \qquad a_n = \frac{1}{n!}\left(\frac{d^n f}{dx^n}\right)_h ,$$  (6.26)

where $f(x)$ is the function to be represented and the subscript $h$ indicates that the derivative is to be evaluated at $x = h$. We say that $h$ is the value of $x$ about which the function is expanded.

## Example 6.8

Find the Taylor series for $\ln(x)$, expanding about $x = 1$.

## Solution

The first derivative of $\ln(x)$ is $1/x$, which equals 1 at $x = 1$. The second derivative is $-1/x^2$, which equals $-1$ at $x = 1$. The derivatives follow a regular pattern,

$$\left(\frac{d^n f}{dx^n}\right)_1 = (-1)^{n-1}(n-1)!$$

so that

$$\ln(x) = (x-1) - \frac{1}{2}(x-1)^2 + \frac{1}{3}(x-1)^3 - \frac{1}{4}(x-1)^4 + \cdots. \qquad (6.27)$$

### Problem 6.9

Find the Maclaurin series for $\ln(1+x)$. You can save some work by using the result of Example 6.8.

### Problem 6.10

Find the Taylor series for $\cos(x)$, expanding about $x = \pi/2$.

## The Convergence of Power Series

If a series is to represent a function $f(x)$ in some interval, it must be convergent in the entire interval and must converge to the value of the function for every value of $x$ in the interval. For a fixed value of $x$, the series $s(x)$ is no different from a constant series, and all the tests for convergence of Section 6.1 can be applied. We can then consider different fixed values of $x$ and determine the *interval of convergence*, that interval in which the series is convergent.

### Example 6.9

Investigate the convergence of the series in Eq. (6.27).

### Solution

Let us consider three cases separately: $x = 1$, $x > 1$, and $x < 1$. If $x = 1$, the entire series vanishes, as does the function, so the series converges to the value of the function for $x = 1$. For $x > 1$, the series is an alternating series, and we can apply the alternating series test. The $n$th term of the series is

$$t_n = a_n(x-1)^n = \frac{(x-1)^n(-1)^{n-1}}{n}.$$

Look at the limit of this as $n$ becomes large:

$$\lim_{n \to \infty} t_n = \begin{cases} 0 & \text{if } |x-1| \le 1 \quad \text{or} \quad 2 < x \le 2 \\ \infty & \text{if } |x-1| > 1 \quad \text{or} \quad x > 2. \end{cases}$$

The interval of convergence for $x > 1$ extends up to and including $x = 2$.

For $x < 1$, the series is not alternating. We apply the ratio test.

$$r = \lim_{n \to \infty} \frac{t_n}{t_{n-1}} = \lim_{n \to \infty} \left[ -\frac{(x-1)^n/n}{(x-1)^{n-1}/(n-1)} \right] = -(x-1) = 1-x.$$

This will be less than unity if $x$ lies between 0 and 1, but if $x = 0$, the text fails. However, if $x = 0$, the series is the same as the harmonic series except for the sign, and thus diverges. The interval of convergence is $0 < x \le 2$.

We summarize the behavior of the Taylor series representation of the previous example: There is a point at which the logarithm function is not analytic, at $x = 0$. The function is analytic to the right of this point, and the series equals the function for positive values arbitrarily close to $x = 0$. Beyond $x = 2$, the series diverges. The interval of convergence is centered on the point about which the function is expanded, which is $x = 1$ in this case. That is, the distance from $x = 1$ to the left end of the interval of convergence is 1 unit, and the distance from $x = 1$ to the right end of the interval of convergence is also 1 unit. This distance is called the *radius of convergence*. Even though the function is defined for values of $x$ beyond $x = 2$, the series does not converge and cannot represent the function for values of $x$ beyond $x = 2$. Another Taylor series expanded about a value of $x$ larger than $x = 1$ can represent the logarithm function beyond $x = 2$.

### Problem 6.11

Find the series for $\ln(x)$, expanding about $x = 2$, and show that the radius of convergence for this series is equal to 2, so that the series can represent the function up to and including $x = 4$.

The behavior of the Taylor series representation of the logarithm function is typical. In general, the interval of convergence is centered on the point about which we are expanding. The radius of convergence is the distance from the point about which we are expanding to the closest point at which the function is not analytic, and the interval of convergence extends by this distance in either direction. If a function is defined on both sides of a point at which the function is not analytic, it is represented on the two sides by different series.

### Example 6.10

Find the interval of convergence for the series representing the exponential function in Eq. (6.25).

### Solution

We apply the ratio test,

$$r = \lim_{n \to \infty} \frac{t_n}{t_{n-1}} = \lim_{n \to \infty} \frac{x^n/n!}{x^{n-1}/(n-1)!} = \lim_{n \to \infty} \frac{x}{n}.$$

This limit vanishes for any finite value of $x$, so the series converges for any finite value of $x$, and the radius of convergence is infinite.

### Problem 6.12

a. Find the series for $1/(1 - x)$, expanding about $x = 0$. What is the interval of convergence?
b. Find the interval of convergence for the series for $\sin(x)$ and for $\cos(x)$.

Unfortunately, the situation is not always so simple as in the examples we have been discussing. A power series can represent a function for complex values of the independent variable, and points in the complex plane at which the function is not analytic can determine the radius of convergence of the function. In fact, a Taylor series converges to the function in a circle in the complex plane with radius equal to the radius of convergence. The radius of convergence is the distance from the point about which we expand to the closest point in the complex plane at which the function is not analytic.

For example, if we wanted to construct a Maclaurin series for the function

$$f(x) = \frac{1}{1 + x^2}$$

the radius of convergence would be determined by discontinuities at $x = i$ and $x = -i$, even though there are no discontinuities for real values of $x$. The radius of convergence equals unity, the distance from the origin to $x = \pm i$ in the Argand plane. We do not discuss the behavior of power series in the complex plane, but you can read more about this topic in the book by Churchill or the book by Kreyszig listed at the end of the chapter.

In physical chemistry, there are a number of applications of power series, but in most applications, a partial sum is used to approximate the series. For example, the behavior of a nonideal gas is often described by use of the *virial series*, or *virial equation of state*,

$$\frac{P\bar{V}}{RT} = 1 + \frac{B_2}{\bar{V}} + \frac{B_3}{\bar{V}^2} + \frac{B_4}{\bar{V}^3} + \cdots, \tag{6.28}$$

where $P$ is the pressure, $\bar{V}$ is the molar volume of the gas, $T$ is the Kelvin temperature, and $R$ is the ideal gas constant. The coefficients $B_2$, $B_3$, etc., are called *virial coefficients*.

If all the virial coefficients were known for a particular gas, the virial series would represent exactly the volumetric behavior of that gas at all values of $1/\bar{V}$. However, only the first few virial coefficients can be determined experimentally or theoretically, so a partial sum must be used. For many purposes, the two-term truncated equation is adequate:

$$\frac{P\bar{V}}{RT} \approx 1 + \frac{B_2}{\bar{V}}. \tag{6.29}$$

There is another commonly used series equation of state, sometimes called the *pressure virial equation of state*:

$$P\bar{V} = RT + A_2 P + A_3 P^2 + A_4 P^3 + \cdots. \tag{6.30}$$

This is a Maclaurin series in $P$. It is also truncated for practical use.

---

## Example 6.11

Show that the coefficient $A_2$ in Eq. (6.30) is equal to the coefficient $B_2$ in Eq. (5.29).

**Solution**

We multiply Eq. (6.28) on both sides by $RT/\bar{V}$ to obtain

$$P = \frac{RT}{\bar{V}} + \frac{RTB_2}{\bar{V}^2} + \frac{RTB_3}{\bar{V}^3} + \cdots. \tag{6.31}$$

This must be equal to

$$P = \frac{RT}{\bar{V}} + \frac{A_2 P}{\bar{V}} + \frac{A_3 P^2}{\bar{V}} + \cdots. \tag{6.32}$$

We convert the second series into a series in $1/\bar{V}$ by substituting the first series into the right-hand side wherever a $P$ occurs. When the entire series on the right-hand side of Eq. (6.31) is squared, every term will have a least a $\bar{V}^2$ in the denominator [see Eq. (11) of Appendix 3 for the square of a series]. Therefore, Eq. (6.32) becomes

$$P = \frac{RT}{\bar{V}} + \frac{A_2}{\bar{V}} \left( \frac{RT}{\bar{V}} + \frac{RTB_2}{\bar{V}} + \cdots \right) + O\left( \frac{1}{\bar{V}} \right)^3, \tag{6.33}$$

where the symbol $O(1/\bar{V})^3$ stands for terms of degree $1/\bar{V}^3$ or higher (containing no powers of $1/\bar{V}$ lower than the third power).

Equation (6.33) is thus

$$P = \frac{RT}{\bar{V}} + \frac{RTA_2}{\bar{V}^2} + O\left( \frac{1}{\bar{V}} \right)^3. \tag{6.34}$$

We now use a fact about series [Eq. (9) of Appendix 6]: *If two power series in the same independent variable are equal to each other for all values of the independent variable, then any coefficient in one series is equal to the corresponding coefficient of the other series.*

Comparison of Eq. (6.34) with Eq. (6.31) shows that

$$B_2 = A_2.$$

---

Another application of a power series in physical chemistry is in the discussion of colligative properties (freezing-point depression, boiling-point elevation, and osmotic pressure). For example, if $X_1$ is the mole fraction of solvent, $\Delta H_v$ is the molar heat of vaporization, $T_0$ is the pure solvent's boiling temperature, and $T$ is the solution's boiling temperature,

$$-\ln(X_1) = \frac{\Delta H_v}{R} \left( \frac{1}{T_0} - \frac{1}{T} \right). \tag{6.35}$$

If there is only one solute (component other than the solvent), then its mole fraction, $X_2$, is given by

$$X_2 = 1 - X_1. \tag{6.36}$$

The left-hand side of Eq. (6.35) is then (see Example 6.8 and Problem 6.9)

$$-\ln(X_1) = -\ln(1 - X_2) = X_2 + \frac{1}{2}X_2^2 + \cdots. \tag{6.37}$$

If $X_2$ is not too large, we can write

$$X_2 \approx \frac{\Delta H_v}{R} \left( \frac{1}{T_0} - \frac{1}{T} \right). \tag{6.38}$$

### Problem 6.13

Determine how large $X_2$ can be before the truncation of Eq. (6.37) that was used in Eq. (6.38) is inaccurate by more than 1%.

## SECTION 6.3. FOURIER SERIES

If we want to produce a series that will converge rapidly, so that we can approximate it fairly well with a partial sum containing only a few terms, it is good to choose basis functions that have as much as possible in common with the function to be represented.

The basis functions in Fourier series[3] are sine and cosine functions, which are periodic functions. Fourier series are therefore used to represent periodic functions. A Fourier series that represents a periodic function of period $2L$ is

$$f(x) = a_0 + \sum_{n=1}^{\infty} a_n \cos\left(\frac{n\pi x}{L}\right) + \sum_{n=1}^{\infty} b_n \sin\left(\frac{n\pi x}{L}\right). \tag{6.39}$$

### Problem 6.14

Show that the basis functions in the series in Eq. (6.39) are periodic with period $2L$. That is, show for arbitrary $n$ that

$$\sin\left[\frac{n\pi(x+2L)}{L}\right] = \sin\left(\frac{n\pi x}{L}\right)$$

and

$$\cos\left[\frac{n\pi(x+2L)}{L}\right] = \cos\left(\frac{n\pi x}{L}\right).$$

Fourier series occur in various physical theories involving waves, because waves often behave sinusoidally. For example, Fourier series can represent the constructive and destructive interference of standing waves in a vibrating string.[4] This fact provides a useful way of thinking about Fourier series. A periodic function of arbitrary shape is represented by adding up sine and cosine functions with shorter and shorter wavelengths, having different amplitudes adjusted to represent the function correctly.

---

[3] The Fourier series is named for its inventor, Jean Baptiste Joseph Fourier, 1768–1730, famous French mathematician and physicist.

[4] Robert G. Mortimer, *Physical Chemistry*, pp. 338–340, Benjamin–Cummings, Redwood City, CA, 1993.

This is analogous to the constructive and destructive interference of waves resulting from the addition of their displacements.

There are some important mathematical questions about Fourier series, including the convergence of a Fourier series and the completeness of the basis functions. A set of basis functions is said to be *complete* for representation of a set of functions if a series in these functions can accurately represent any function from the set. We do not discuss the mathematics, but state the facts that were proved by Fourier: (1) any Fourier series in $x$ is uniformly convergent for all real values of $x$; (2) the set of sine and cosine basis functions in Eq. (6.39) is a complete set for the representation of periodic functions of period $2L$. In many cases of functional series, the completeness of the set of basis functions has not been proved, but most people assume completeness and proceed.

### Finding the Coefficients of a Fourier Series—Orthogonality

In a power series, we found the coefficients by demanding that the function and the series have equal derivatives at the point about which we were expanding. In a Fourier series, we use a different procedure, utilizing a property of the basis functions that is called *orthogonality*. This property is expressed by the three equations:

$$\int_{-L}^{L} \cos\left(\frac{m\pi x}{L}\right)\cos\left(\frac{n\pi x}{L}\right)dx = L\delta_{mn} = \begin{cases} L & \text{if } m = n \\ 0 & \text{if } m \neq n \end{cases} \tag{6.40}$$

$$\int_{-L}^{L} \cos\left(\frac{m\pi x}{L}\right)\sin\left(\frac{n\pi x}{L}\right)dx = 0 \tag{6.41}$$

$$\int_{-L}^{L} \sin\left(\frac{m\pi x}{L}\right)\sin\left(\frac{n\pi x}{L}\right)dx = L\delta_{mn}. \tag{6.42}$$

The quantity $\delta_{\mathrm{mn}}$ is called the *Kronecker delta*. It is equal to unity if its two indices are equal and is equal to zero otherwise.

Equations (6.40) and (6.42) do not apply if $m$ and $n$ are both equal to zero. The integral in Eq. (6.40) is equal to $2L$ if $m = n = 0$ and the integral in Eq. (6.42) is equal to zero if $m = n = 0$.

Two different functions that yield zero when multiplied together and integrated are said to be *orthogonal* to each other. Equations (6.40), (6.41), and (6.42) indicate that all the basis functions for the Fourier series of period $2L$ are orthogonal to each other. This terminology is analogous to that used with vectors. If two vectors are at right angles to each other, they are said to be orthogonal to each other, and their scalar product is zero (see Chapter 2). An integral such as those in the three equations above is sometimes called a *scalar product* of the two functions. Since each of the basis functions is orthogonal to the others, its scalar product with a different basis function vanishes.

To find $a_m$, where $m \neq 0$ we multiply both sides of Eq. (6.39) by $\cos(m\pi x/L)$ and integrate from $-L$ to $L$.

$$\int_{-L}^{L} f(x)\cos\left(\frac{m\pi x}{L}\right)dx = \sum_{n=0}^{\infty} a_n \int_{-L}^{L} \cos\left(\frac{n\pi x}{L}\right)\cos\left(\frac{m\pi x}{L}\right)dx$$

$$+ \sum_{n=0}^{\infty} b_n \int_{-L}^{L} \sin\left(\frac{n\pi x}{L}\right)\cos\left(\frac{m\pi x}{L}\right)dx. \tag{6.43}$$

We have incorporated the $a_0$ term into the first sum, using the fact that $\cos(0) = 1$. We have also used the fact that the integral of a sum is equal to the sum of the integrals of the terms if the series is uniformly convergent.

We now apply the orthogonality facts, Eqs. (6.40)–(6.42), to find that all of the integrals on the right-hand side of Eq. (6.43) vanish except for the term with two cosines in which $n = m$. The result is

$$\int_{-L}^{L} f(x) \cos\left(\frac{m\pi x}{L}\right) dx = a_m L. \tag{6.44}$$

This is a formula for finding all of the $a$ coefficients except for $a_0$.

To find $a_0$, we use the fact that

$$\int_{-L}^{L} \cos(0) \cos(0) \, dx = \int_{-L}^{L} dx = 2L \tag{6.45}$$

which leads to our working equations for the $a$ coefficients:

$$\boxed{a_0 = \frac{1}{2L} \int_{-L}^{L} f(x) \, dx} \tag{6.46}$$

$$\boxed{a_n = \frac{1}{L} \int_{-L}^{L} f(x) \cos\left(\frac{n\pi x}{L}\right) dx}. \tag{6.47}$$

A similar procedure consisting of multiplication by $\sin(m\pi x/L)$ and integration from $-L$ to $L$ yields

$$\boxed{b_n = \frac{1}{L} \int_{-L}^{L} f(x) \sin\left(\frac{n\pi x}{L}\right) dx}. \tag{6.48}$$

---

**Problem 6.15**

Show that Eq. (6.48) is correct.

---

A function does not have to be analytic, or even continuous, in order to be represented by a Fourier series. It is only necessary that the function be integrable. As mentioned in Chapter 4, this permits a function to have step discontinuities, as long as the step in the function is finite. At a step discontinuity, a Fourier series will converge to a value halfway between the value just to the right of the discontinuity and the value just to the left of the discontinuity.

We can represent a function that is not necessarily periodic by a Fourier series if we are only interested in representing the function in the interval $-L < x < L$. The Fourier series will be periodic with period $2L$, and the series will be equal to the function inside the interval, but not necessarily equal to the function outside the interval.

If the function $f(x)$ is an even function, all of the $b_n$ coefficients will vanish, and only the cosine terms will appear in the series. Such a series is called a *Fourier cosine series*. If $f(x)$ is an odd function, only the sine terms will appear, and the series is called a *Fourier sine series*.

If we want to represent a function only in the interval $0 < x < L$ we can regard it as the right half of an odd function or the right half of an even function, and can therefore represent it either with a sine series or a cosine series. These two series would have the same value in the interval $0 < x < L$ but would be the negatives of each other in the interval $-L < x < 0$.

---

### Example 6.12

Find the Fourier series to represent the function $f(x) = x$ for the interval $-L < x < L$.

---

### Solution

The function is odd in the interval $(-L, L)$ so the series will be a sine series. Although our function is defined only for the interval $(-L, L)$ the series will be periodic, and will be the "sawtooth wave" that is shown in Fig. 6.2.

The coefficients are obtained from Eq. (6.48). Since the integrand is the product of two odd functions, it is an even function, and the integral is equal to twice the integral from 0 to $L$:

$$b_n = \frac{2}{L} \int_0^L x \sin\left(\frac{n\pi x}{L}\right) dx = \frac{2}{L} \left(\frac{L}{n\pi}\right)^2 \int_0^{n\pi} y \sin(y)\, dy = \frac{2L}{n\pi}(-1)^{n-1}.$$

The series is

$$f(x) = \sum_{n-1}^{\infty} \frac{2L}{n\pi}(-1)^{n-1} \sin\left(\frac{n\pi x}{L}\right).$$

---

### Problem 6.16

a. Show that the $a_n$ coefficients for the series representing the function in Example 6.12 all vanish.

b. Show that the series equals zero at $x = -L, x = L, x = 3L$, etc., rather than equaling the function at this point.

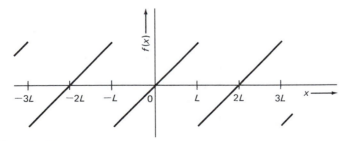

**FIGURE 6.2**   The "sawtooth wave" periodic function of Example 6.12.

c. Find the Fourier cosine series for the even function

$$f(x) = |x| \qquad \text{for } -L < x < L.$$

Draw a graph of the periodic function represented by the series.

---

It is a necessary condition for the convergence of Fourier series that the coefficients become smaller and smaller and approach zero as $n$ becomes larger and larger. If a Fourier series is convergent, it will be uniformly convergent for all values of $x$. If convergence is fairly rapid, it might be possible to approximate a Fourier series by one of its partial sums. Figure 6.3 shows three different partial sums of the series that represents the "square-wave" function

$$f(x) = \begin{cases} +1 & \text{for } 0 < x < L \\ -1 & \text{for } -L < x < 0. \end{cases}$$

Only the right half of one period is shown. The first partial sum only vaguely resembles the function, but $S_{10}$ is a better approximation. Notice the little "spike" or overshoot near the discontinuity. This is a typical behavior and is known as the "Gibbs phenomenon."[5] Even though $S_{100}$ fits the function more closely away from the discontinuity, it has a spike near the discontinuity that is just as high as that of $S_{10}$, although much narrower.

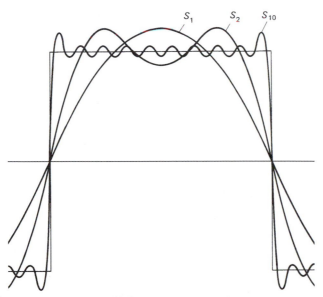

**FIGURE 6.3** The square wave function approximated by $S_1$, $S_2$, and $S_{10}$.

---

[5]Named for Josiah Willard Gibbs, 1839–1903, a prominent American physicist who made important contributions to mathematics and chemistry as well to physics.

### Fourier Series with Complex Exponential Basis Functions

The sine and cosine basis functions are closely related to complex exponential functions, as shown in Eqs. (2.124) and (2.125). One can write

$$b_n \sin\left(\frac{n\pi x}{L}\right) + a_n \cos\left(\frac{n\pi x}{L}\right) = \frac{1}{2}(a_n - ib_n)e^{in\pi x/L} + \frac{1}{2}(a_n + ib_n)e^{in\pi x/L}.$$
(6.49)

It is therefore possible to rewrite Eq. (6.39) as an *exponential Fourier series*:

$$f(x) = \sum_{n=-\infty}^{\infty} c_n e^{in\pi x/L}.$$
(6.50)

We have incorporated the terms with negative exponents into the same sum with the other terms by allowing the summation index to take on negative as well as positive values. The function being represented by a Fourier series does not have to be a real function. However, if it is a real function, the coefficients $a_n$ and $b_n$ will be real, so that $c_n$ will be complex.

### Other Functional Series with Orthogonal Basis Sets

Fourier series are just one example of series using orthogonal basis sets. For example, in quantum mechanics it is found that the eigenfunctions of quantum mechanical operators form orthogonal sets of functions, and these can be used as basis functions for series. It is generally assumed that such a set of functions is complete for representation of functions that obey the same boundary conditions as the basis functions. Boundary conditions are discussed in Chapter 7 in connection with differential equations.

Assume that we have a complete set of orthogonal functions, called $\phi_1$, $\phi_2$, $\phi_3$, etc., and that these functions have been *normalized*. This means that the functions have been multiplied by appropriate constants so that the scalar product of any one of the functions with itself is unity:

$$\int \phi_n^* \phi_m \, dx = \delta_{nm} = \begin{cases} 1 & \text{if } n = m \\ 0 & \text{if } n \neq m. \end{cases}$$
(6.51)

The basis functions for Fourier sine and cosine series are real functions, but in case the basis functions are complex, the scalar product is defined as the integral of the complex conjugate of the first function times the second function, as in Eq. (6.51).

Since the set of functions is assumed to be complete, we can expand an arbitrary function, $\psi$, in terms of the $\phi$ functions:

$$\psi = \sum_n c_n \phi_n.$$
(6.52)

The sum in this equation will include one term for each function in the complete set and can have infinitely many terms.

In order to find the coefficients $c_1$, $c_2$, $c_3$, etc., we multiply by the complex conjugate of $\phi_m$ and integrate. With Eq. (6.51), our result is

$$\int \phi_m^* \psi \, dx = \sum_n c_n \int \phi_m^* \phi_n \, dx = \sum_n c_n \delta_{nm} = c_m.$$
(6.53)

Equations (6.46), (6.47), and (6.48) are special cases of this equation.

### Problem 6.17

Write the formula for finding the coefficients for an exponential Fourier series. Is there any difference in the formulas for odd functions, even functions, or functions that are neither odd nor even? What conditions must the function obey to be represented by an exponential Fourier series?

## SECTION 6.4. MATHEMATICAL OPERATIONS ON SERIES

Carrying out mathematical operations such as integration or differentiation on a functional series with a finite number of terms is straightforward, since no questions of convergence arise. However, carrying out such operations on an infinite series presents a few difficulties. For example, the question arises whether differentiating each term and then summing the result gives the same result as first summing the series and then differentiating. Although we do not prove it, the principal fact is that *if a series is uniformly convergent the result of operating on the series is the same as the result of operating on the individual terms and then summing the resulting series.* For example, if

$$f(x) = \sum_{n=0}^{\infty} a_n g_n(x) \tag{6.54}$$

and if the series is uniformly convergent, then

$$\frac{df}{dx} = \frac{d}{dx}\left[\sum_{n=0}^{\infty} a_n g_n(x)\right] = \sum_{n=0}^{\infty} a_n \frac{dg_n}{dx}. \tag{6.55}$$

This amounts to interchange of the operations of summing and differentiating.

Similarly, for a uniformly convergent series,

$$\int_a^b f(x)\,dx = \int_a^b \left[\sum_{n=0}^{\infty} a_n g_n(x)\right] dx = \sum_{n=0}^{\infty} a_n \int_a^b g_n(x)\,dx. \tag{6.56}$$

This amounts to interchange of the operations of summing and integrating.

We have already used Eq. (6.56) in the previous section in deriving the formula for the coefficients in the Fourier series, without commenting on the fact that the series must be uniformly convergent to justify this procedure.

### Example 6.13

Find the Maclaurin series for $\cos(x)$ from the Maclaurin series for $\sin(x)$, using the fact that $d[\sin(x)]/dx = \cos(x)$.

### Solution

The series is uniformly convergent for all values of $x$. From Eq. (6.29), we have

$$\sin(x) = x - \frac{x^3}{3!} + \frac{x^5}{5!} - \frac{x^7}{7!} + \cdots$$

so that

$$\frac{d[\sin(x)]}{dx} = 1 - \frac{x^2}{2!} + \frac{x^4}{4!} - \frac{x^6}{6!} + \cdots.$$

---

### Problem 6.18

From the Taylor series for $\ln(x)$ expanded about $x = 1$ given in Eq. (6.27), find the Taylor series for $1/x$ about $x = 1$, using the fact that

$$\frac{d[\ln(x)]}{dx} = \frac{1}{x}.$$

---

## SECTION 6.5. INTEGRAL TRANSFORMS

Integral transforms are closely related to functional series. However, instead of a sum with each term consisting of a coefficient multiplying a basis function, we have an integral in which the summation index is replaced by an integration variable. The basis functions are multiplied by a function of this integration variable, and integration over this variable yields a representation of the function. This function of the integration variable is called the *transform* of the given function.

As with a functional series, the transform encodes the information about the original function in a different way from the original representation. The transform is a function of the integration variable, in the same way as the coefficients in a functional series depend on the value of the summation index. You can think of a transform as encoding the same information as in the original representation of the function, but with a different independent variable.

There are several kinds of integral transforms, including Mellin transforms, Hankel transforms, etc.,[6] but the principal kinds of transforms encountered by physical chemists are Fourier transforms and Laplace transforms.

### Fourier Transforms (Fourier Integrals)

Although Fourier transforms were once important only to mathematicians and some theoretical scientists, they are now widely used in spectroscopy, because instruments have been designed that produce a superposition of wave-like signals such that the spectrum is the Fourier transform of the detected signal.[7]

Fourier series are designed to represent periodic functions with period $2L$. If we allow $L$ to become larger and larger without bound, the values of $n\pi x/L$ become closer and closer together. We let

$$k = \frac{n\pi}{L}. \tag{6.57}$$

---

[6]A. Erdelyi Ed., *Tables of Integral Transforms*, Vols. I and II, McGraw–Hill, New York, 1954; A. G. Marshall, Ed., *Fourier, Hadamard, and Hilbert Transforms in Chemistry*, Plenum, New York, 1982.

[7]L. Glasser, *J. Chem. Educ.* **64**, A228(1987); *J. Chem. Educ.* **64**, A260(1987); and *J. Chem. Educ.* **64**, A306(1987).

As the limit $L \to \infty$ is taken, $k$ becomes a continuously variable quantity if $n \to \infty$ in the proper way. In this limit, an exponential Fourier series becomes an integral, which is called a *Fourier integral* or a *Fourier transform*,

$$f(x) = \frac{1}{\sqrt{2\pi}} \int_{-\infty}^{\infty} F(k)e^{ikx} \, dk, \tag{6.58}$$

where the coefficient $c_n$ in Eq. (6.50) has become a function of $k$, denoted by $F(k)$.

The equation for determining $F(k)$ is analogous to Eqs. (6.46), (6.47), and (6.48)

$$F(k) = \frac{1}{\sqrt{2\pi}} \int_{-\infty}^{\infty} f(x)e^{-ikx} \, dx. \tag{6.59}$$

We have introduced a factor of $1/\sqrt{2\pi}$ in front of the integral in Eq. (6.58) in order to have the same factor in front of this integral and the integral in Eq. (6.59).

The function $F(k)$ is called the *Fourier transform* of $f(x)$ and the function $f(x)$ is also called the Fourier transform of $F(k)$. The function $f(x)$ is no longer required to be periodic, because the period $2L$ has been allowed to become infinite.

Since we now have improper integrals, the functions $f(x)$ and $F(k)$ must have properties such that the integrals converge. For the integral of Eq. (6.59) to converge, the following integral must converge:

$$\int_{-\infty}^{\infty} |f(x)|^2 \, dx.$$

We say that the function $f(x)$ must be *square integrable*. The function $f(x)$ must approach zero as $x \to -\infty$ and as $x \to \infty$ to be square integrable. If the Fourier transform $F(k)$ exists, it will also be square integrable.

---

## Example 6.14

Find the Fourier transform of the Gaussian function

$$f(x) = e^{-ax^2}.$$

---

## Solution

$$F(k) = \frac{1}{\sqrt{2\pi}} \int_{-\infty}^{\infty} e^{-ax^2} e^{-ikx} \, dx.$$

From Eq. (2.112),

$$F(k) = \frac{1}{\sqrt{2\pi}} \int_{-\infty}^{\infty} e^{-ax^2} \cos(kx) \, dx - \frac{i}{\sqrt{2\pi}} \int_{-\infty}^{\infty} e^{-ax^2} \sin(kx) \, dx.$$

The second integral in this equation is equal to zero because its integrand is an odd function. The first integral is twice the integral of Eq. (47) of Appendix 6,

$$F(k) = \frac{2}{\sqrt{2\pi}} \int_{0}^{\infty} e^{-ax^2} \cos(kx) \, dx = \frac{2}{\sqrt{2\pi}} \frac{1}{2} \sqrt{\frac{\pi}{a}} e^{-k^2/4a}$$

$$= \frac{1}{\sqrt{2a}} e^{-k^2/4a}.$$

The Fourier transform of a Gaussian function is another Gaussian function.

---

In Example 6.15, the transform integral was expressed in two parts, one containing a cosine function and one containing a sine function. If the function $f(x)$ is an even function as in Example 6.14, its Fourier transform is a *Fourier cosine transform*:

$$F(k) = \sqrt{\frac{1}{2\pi}} \int_{-\infty}^{\infty} f(x) \cos(kx)\, dx = \sqrt{\frac{2}{\pi}} \int_{0}^{\infty} f(x) \cos(kx)\, dx \qquad (f \text{ even}).$$
(6.60)

The second version of the transform is called a one-sided cosine transform.

If $f(x)$ is an odd function, its Fourier transform is a *Fourier sine transform*:

$$F(k) = \sqrt{\frac{1}{2\pi}} \int_{-\infty}^{\infty} f(x) \sin(kx)\, dx = \sqrt{\frac{2}{\pi}} \int_{0}^{\infty} f(x) \sin(kx)\, dx \qquad (f \text{ odd}).$$
(6.61)

There is a useful theorem for the Fourier transform of a product of two functions, called the *convolution theorem* or the *Faltung theorem* (Faltung is German for "folding"). The *convolution* of two functions $f(x)$ and $g(x)$ is defined as the integral

$$\frac{1}{\sqrt{2\pi}} \int_{-\infty}^{\infty} f(y) g(x - y)\, dy.$$
(6.62)

This integral is a function of $x$, and its Fourier transform is equal to $F(k)G(k)$ where $F(k)$ is the Fourier transform of $f(x)$ and $G(k)$ is the Fourier transform of $g(x)$.[8] Since the Fourier transform is nearly the same going in both directions, the analogous convolution

$$\frac{1}{\sqrt{2\pi}} \int_{-\infty}^{\infty} F(l) G(k - l)\, dl$$
(6.63)

has as its Fourier transform the product $f(x)g(x)$.

---

### Problem 6.19

Take the two Gaussian functions

$$f(x) = e^{-ax^2} \qquad \text{and} \qquad g(x) = e^{-bx^2}.$$

Find the Fourier transform of the product of the functions using the convolution theorem. Show that this transform is the same as that obtained by multiplying the functions together and computing the transform in the usual way.

---

The two principal applications of Fourier transformation for chemists are in infrared spectroscopy and nuclear magnetic resonance spectroscopy. In both cases, a spectrum as a function of frequency is to be obtained, and the instrument takes raw data as a function of time.

---

[8] Philip M. Morse and Herman Feshbach, *Methods of Theoretical Physics*, Part I, pp. 464ff, McGraw–Hill, New York, 1953.

In a Fourier transform infrared instrument, an interferometer varies the intensities of radiation of various frequencies as a function of time, and a detector determines the intensity as a function of time, producing an interferogram, which is the Fourier transform of the desired spectrum. The Fourier transformation is carried out by a computer program for predetermined values of the frequency, so that the spectrum is obtained only for a discrete set of frequencies. Numerical Fourier transformation is usually a fairly slow process, demanding a lot of computer time, but a "fast Fourier transform" (FFT) algorithm has been developed that makes the process practical for routine usage.[9]

In a Fourier transform NMR instrument, a signal called the free induction decay signal is obtained as a function of time, and the Fourier transform of this signal is the desired NMR spectrum. The FFT algorithm is used to carry out the transformation.

## Laplace Transforms

The Laplace transform $F(s)$ of the function $f(t)$ is defined by

$$\boxed{F(s) = \int_0^\infty f(t)e^{-st}dt}. \tag{6.64}$$

We use the same notation as with the Fourier transform, denoting a Laplace transform by a capital letter and the function by a lower-case letter. You will have to tell from the context whether we are discussing a Fourier transform or a Laplace transform. We use the letter $t$ for the independent variable of the function, since Laplace transforms are commonly applied to functions of the time. The letter $x$ could also have been used.

The Laplace transform is similar to a one-sided Fourier transform, except that it has a real exponential instead of the complex exponential of the Fourier transform. If we consider complex values of the variables, the two transforms become different cases of the same transform, and their properties are related.[10] The integral that is carried out to invert the Laplace transform is carried out in the complex plane, and we do not discuss it. Fortunately, it is often possible to apply Laplace transforms without carrying out such an integral.[11] We will discuss the use of Laplace transforms in solving differential equations in Chapter 7.

The Laplace transform and its inverse are often denoted in the following way:

$$F(s) = \mathcal{L}\{f(t)\} \tag{6.65a}$$

$$f(t) = \mathcal{L}^{-1}\{F(s)\}. \tag{6.65b}$$

[9] J. W. Cooley and J. W. Tukey, *Math. Computation*, **19**, 297–301(1965). See the book by James under Additional Reading.

[10] Philip M. Morse and Herman Feshbach, *op. cit.*, pp. 467ff.

[11] Erwin Kreyszig, *Advanced Engineering Mathematics*, 3rd ed., pp. 147ff, Wiley, New York, 1972.

**TABLE 6.1　Laplace Transforms**

| $F(s)$ | $f(t)$ |
| --- | --- |
| $1/s$ | $1$ |
| $1/s^2$ | $t$ |
| $n!/s^{n+1}$ | $t^n$ |
| $\dfrac{1}{s-a}$ | $e^{at}$ |
| $\dfrac{s}{s^2+k^2}$ | $\cos(kt)$ |
| $\dfrac{k}{s^2+k^2}$ | $\sin(kt)$ |

Table 6.1 gives a few common Laplace transforms.

### Example 6.15

Find the Laplace transform of the function $f(t) = t^2$.

### Solution

$$F(s) = \int_0^\infty t^2 e^{-st}\, dt = \frac{1}{s^3} \int_0^\infty u^2 e^{-u}\, du = \frac{2!}{s^3},$$

where we have used Eq. (1) of Appendix 6 to obtain the value of the definite integral.

### Problem 6.20

Find the Laplace transform of the function $f(t) = e^{at}$ where $a$ is a constant.

There are several theorems that are useful in obtaining Laplace transforms of various functions.[12] The first is the *shifting theorem*:

$$\boxed{\mathcal{L}\{e^{at} f(t)\} = F(s-a)}. \tag{6.66}$$

### Example 6.16

Use the theorem of Eq. (6.66) to obtain the Laplace transform of the function

$$f(t) = e^{at} \cos(kt).$$

---

[12] Ibid.

**Solution**

We transcribe the entry for $\cos(kx)$ from Table 6.1, replacing $s$ by $s-a$, obtaining

$$F(s) = \frac{s-a}{(s-a)^2 + k^2}.$$

---

**Problem 6.21**

Find the Laplace transform of the function

$$f(t) = t^n e^{at}.$$

---

The next useful theorem is the *derivative theorem*,

$$\boxed{\mathcal{L}\{f'\} = s\mathcal{L}\{f\} - f(0)}, \tag{6.67}$$

where we use the notation $f'$ for the first derivative of $f$. This theorem can be applied to the solution of first-order differential equations and can be applied repeatedly to obtain the extended version,

$$\boxed{\mathcal{L}\{f^{(n)}\} = s^n \mathcal{L}\{f\} - s^{n-1} f(0) - s^{n-2} f''(0) - \cdots - f^{(n-1)}(0)}, \tag{6.68}$$

where we use the notation $f^{(n)}$ for the $n$th derivative of $f$, etc.

---

**Problem 6.22**

Derive the version of Eq. (6.68) for $n = 2$.

---

The next theorem is for the Laplace transform of an integral of a given function $f$,

$$\boxed{\mathcal{L}\left\{\int_0^t f(u)\, du\right\} = \frac{1}{s}\mathcal{L}\{f(t)\}}. \tag{6.69}$$

These theorems can be used to construct the Laplace transforms of various functions, and to find inverse transforms without carrying out an integral in the complex plane.

---

**Example 6.17**

Find the inverse Laplace transform of

$$\frac{1}{s(s-a)}.$$

**Solution**

From Table 5.1, we recognize $1/(s - a)$ as the Laplace transform of $e^{ax}$. From the theorem of Eq. (6.69),

$$\mathcal{L}\left\{\int_0^x e^{au}\,du\right\} = \frac{1}{s}\mathcal{L}\{e^{at}\} = \frac{1}{s}\frac{1}{s - a}.$$

Therefore, the inverse transform is

$$\mathcal{L}^{-1}\left\{\frac{1}{s(s - a)}\right\} = \int_0^x e^{ay}\,dy = \frac{1}{a}(e^{ax} - 1).$$

**Problem 6.23**

Find the inverse Laplace transform of

$$\frac{1}{s(s^2 + k^2)}.$$

## SUMMARY OF THE CHAPTER

In this chapter we introduced mathematical series and mathematical transforms. A finite series is a sum of a finite number of terms, and an infinite series is a sum of infinitely many terms. A constant series has terms that are constants, and a functional series has terms that are functions. The two important questions to ask about a constant series are whether the series converges, and if so, what value it converges to. We presented several tests that can be used to determine whether a series converges. Unfortunately, there appears to be no general method for finding the value to which a convergent series converges.

A functional series is one way of representing a function. Such a series consists of terms, each one of which is a basis function times a coefficient. A power series uses powers of the independent variable as basis functions and represents a function as a sum of the appropriate linear function, quadratic function, cubic function, etc. We discussed Taylor series, which contain powers of $x - h$, where $h$ is a constant, and also Maclaurin series, which are Taylor series with $h = 0$. Taylor series can represent a function of $x$ only in a region of convergence centered on $h$ and reaching no further than the closest point at which the function is not analytic. We found the general formula for determining the coefficients of a power series.

The other functional series that we discussed was the Fourier series, in which the basis functions are sine and cosine functions. This type of series is best suited for representing periodic functions and represents the function as a sum of the sine and cosine functions with the appropriate coefficients. The method of determining the coefficients to represent any particular function was given.

Integral transforms were discussed, including Fourier and Laplace transforms. Fourier transforms are the result of allowing the period of the function to be represented by a Fourier series to become larger and larger, so that the series approaches an integral in the limit. Fourier transforms are usually written with complex exponential basis functions, but sine and cosine transforms also occur. Laplace transforms are related to Fourier transforms, with real exponential basis functions. We presented several theorems that allow the determination of some kinds of inverse Laplace transforms and that allow later applications to the solution of differential equations.

## ADDITIONAL READING

R. V. Churchill, *Introduction to Complex Variables and Applications*, McGraw–Hill, New York, 1948. This book is an introduction to the calculus of complex variables. Chapter 6 of the book is devoted to power series of a complex variable, and questions of convergence in the complex plane are discussed. Much of the information about series of complex variables is important for series of a real variable.

R. V. Churchill, *Fourier Series and Boundary Value Problems*, 2nd ed., McGraw–Hill, New York, 1963. This book concerns the use of Fourier series in the solution of partial differential equations occurring in physics. Chapters 4 through 6 concern the principal properties of Fourier series and integrals.

Herbert B. Dwight, *Tables of Integrals and Other Mathematical Data*, 4th ed., Macmillan Co., New York, 1961. This little book contains quite a large number of constant series and power series, as well as tables of integrals.

A. Erdélyi, Ed., *Tables of Integral Transforms*, Vols. I and II, McGraw–Hill, New York, 1954. This set of two volumes contains a brief introduction of several types of integral transforms, with extensive tables of transforms of specific functions.

Gerald B. Folland, *Fourier Analysis and Its Applications*, Wadsworth and Brooks/Cole, Pacific Grove, CA, 1992. This book is designed for advanced undergraduates in mathematics and physics and attempts to steer a middle course between pure mathematics and applied mathematics viewed as a tool for applications.

J. F. James, *A Student's Guide to Fourier Transforms*, Cambridge Univ. Press, Cambridge, UK, 1995. This small book is designed to teach the subject to a student without previous knowledge of Fourier transforms. It contains a description of the fast Fourier transform method and a computer program in BASIC to carry out the transformation.

L. B. W. Jolley, *Summation of Series*, 2nd ed., Dover, New York, 1961. This collection of series contains a large number of series that the author has collected from many sources. It is a paperback book that is well worth the small price. There is also a fairly large section of infinite products.

Wilfred Kaplan, *Advanced Calculus*, Addison–Wesley, Reading, MA, 1952. This is a text for a calculus course beyond the first year. Chapter 6 concerns infinite series, and Chapter 7 concerns Fourier series.

Konrad Knopp, *Infinite Sequences and Series*, translated by F. Bagemihl, Dover, New York, 1956. This is a small but authoritative paperback book that discusses most of the important mathematics of sequences and series, including the theory of convergence. A very short chapter on the evaluation of series is included.

Erwin Kreyszig, *Advanced Engineering Mathematics*, 3rd ed., Wiley, New York, 1972. This book is meant for engineers, but engineers use mathematics in much the same way as chemists, and this book emphasizes applications rather than mathematical theory in a way that is useful to chemists.

Philip M. Morse and Herman Feshbach, *Methods of Theoretical Physics*, Vols. I and II, McGraw–Hill, New York, 1953. This two-volume work includes all of the mathematics commonly used in theoretical science, including a relatively rigorous discussion of Fourier and Laplace transforms.

David L. Powers, *Boundary Value Problems*, Academic Press, New York, 1972. This book covers the same area as the book by Churchill, and includes a 40-page chapter on Fourier series and integrals.

Angus E. Taylor and Robert W. Mann, *Advanced Calculus*, 2nd ed., Ginn, Lexington, MA, 1972. This book is similar to the book by Kaplan.

G. P. Tolstov, *Fourier Series*, Prentice–Hall, Englewood Cliffs, NJ, 1962. This is a good introduction to Fourier series and integrals.

In addition to the foregoing works, particular series can be found in various tables of integrals, such as the work by Gradshteyn and Rhyzhik listed in Chapter 4. You can also read about Taylor and Maclaurin series in almost any elementary calculus textbook.

## ADDITIONAL PROBLEMS

### 6.24

By use of the Maclaurin series already obtained in this chapter, prove the identity

$$e^{ix} = \cos(x) + i\sin(x).$$

### 6.25

a. Show that no Maclaurin series

$$f(x) = a_0 + a_1 x + a_2 x^2 + \cdots$$

can be formed to represent the function $f(x) = \sqrt{x}$. Why is this?

b. Find the first few coefficients for the Maclaurin series for the function

$$f(x) = \sqrt{1+x}.$$

### 6.26

If $f(x) = \tan(x)$, find the first few terms of the Taylor series

$$f(x) = a_0 + a_1\left(x - \frac{\pi}{4}\right) + a_2\left(x - \frac{\pi}{4}\right)^2 + \cdots,$$

where $x$ is measured in radians. What is the radius of convergence of the series?

### 6.27

The sine of $\pi/4$ radians (45°) is $\sqrt{2}/2 = 0.70710678\ldots$. How many terms in the series

$$\sin(x) = x - \frac{x^3}{3!} + \frac{x^5}{5!} - \frac{x^7}{7!} + \cdots$$

must be taken to achieve 1% accuracy at $x = \pi/4$?

### 6.28

Estimate the largest value of $x$ that allows $e^x$ to be approximated to 1% accuracy by the partial sum

$$e^x \approx 1 + x.$$

### 6.29

Find two different Taylor series to represent the function

$$f(x) = \frac{1}{x^2}$$

such that one series is

$$f(x) = a_0 + a_1(x-1) + a_2(x-1)^2 + \cdots$$

and the other is

$$f(x) = b_0 + b_1(x-2) + b_3(x-2)^2 + \cdots.$$

Show that $b_n = a_n/2^n$ for any value of $n$. Find the interval of convergence for each series (the ratio test may be used). Which series must you use in the vicinity of $x = 3$? Why?

### 6.30

A certain electronic circuit produces the following sawtooth wave,

$$f(t) = \begin{cases} a(-T-t), & -T < t < -T/2 \\ at, & -T/2 < t < T/2 \\ a(T-t), & -T/2 < t < T, \end{cases}$$

where $a$ and $T$ are constants and $t$ represents the time. Find the Fourier series that represents this function. The definition of the function given is for only an interval of length $2T$, but the Fourier series will be periodic. Make a graph of the function and of the first two partial sums.

### 6.31

Find the Fourier series that represents the square wave

$$A(t) = \begin{cases} 0, & -T/2 < t < 0 \\ A_0 & 0 < t < T/2, \end{cases}$$

where $A_0$ is a constant and $T$ is the period.

**6.32**

Using the Maclaurin series for $e^x$, show that the derivative of $e^x$ is equal to $e^x$.

**6.33**

Find the Maclaurin series that represents $\tan(x)$. What is its radius of convergence?

**6.34**

Find the Maclaurin series that represents $\cosh(x)$. What is its radius of convergence?

**6.35**

Find the Fourier transform of the function $a \exp(-(x - x_0)^2/b)$ where $a$ and $b$ are constants.

**6.36**

Find the Fourier transform of the function $ae^{-b|x|}$.

**6.37**

a. Find the Laplace transform of the function $a/(b^2 + t^2)$ where $a$ and $b$ are constants.
b. Find the Fourier transform of the same function.

**6.38**

a. Show that $\mathcal{L}\{t \cos(kt)\} = (s^2 - k^2)/(s^2 + k^2)^2$.

# 7
# DIFFERENTIAL EQUATIONS

## Preview

A differential equation contains one or more derivatives of an unknown function, and solving a differential equation means finding what that function is. One important class of differential equations consists of classical equations of motion, which come from Newton's second law of motion. We will discuss the solution of several kinds of ordinary differential equations, including linear differential equations, in which the unknown function and its derivatives enter only to the first power, and exact differential equations, which can be solved by a line integration. We will also discuss partial differential equations, in which partial derivatives occur and in which there are two or more independent variables. We will also discuss the solution of differential equations by use of Laplace transformation.

## Principal Facts and Ideas

1. The solution of a differential equation is a function whose derivative or derivatives satisfy the differential equation.

2. An equation of motion is a differential equation obtained from Newton's second law of motion, $\mathbf{F} = m\mathbf{a}$.

3. In principle, an equation of motion can be solved to give the position and velocity as a function of time for every particle in a system governed by Newton's laws of motion.

4. A homogeneous linear differential equation with constant coefficients can be solved by use of an exponential trial solution.

5. An inhomogeneous linear differential equation can be solved if a particular solution can be found.

6. An exact differential equation can be solved by a line integration.

7. Some inexact differential equations can be converted to exact differential equations by multiplication by an integrating factor.

8. Some partial differential equations can be solved by separation of variables.

9. A differential equation can be transformed into an algebraic equation by a Laplace transformation. Solution of this equation followed by inverse transformation provides a solution to the differential equation.

10. Differential equations can be solved numerically by a variety of methods.

## Objectives

After studying this chapter, you should be able to:

1. construct an equation of motion for a particle from Newton's second law;
2. solve a linear homogeneous differential equation with constant coefficients;
3. solve a differential equation whose variables can be separated;
4. solve an exact differential equation;
5. use an integrating factor to solve an inexact differential equation;
6. solve a simple partial differential equation by separation of variables;
7. solve a differential equation by use of Laplace transforms;
8. write and run computer programs in BASIC to implement Euler's method and the Runge–Kutta method for numerically solving simple differential equations.

## SECTION 7.1. DIFFERENTIAL EQUATIONS AND NEWTON'S LAWS OF MOTION

Our first examples of differential equations are equations of motion, obtained from Newton's second law of motion. These equations are used to determine the time dependence of the position and velocity of particles. The position of a particle is given by the position vector $\mathbf{r}$ with cartesian components $x$, $y$, and $z$. The velocity $\mathbf{v}$ of a particle is the rate of change of its position vector,

$$\mathbf{v} = \frac{d\mathbf{r}}{dt} = \mathbf{i}\frac{dx}{dt} + \mathbf{j}\frac{dy}{dt} + \mathbf{k}\frac{dz}{dt} = \mathbf{i}v_x + \mathbf{j}v_y + \mathbf{k}v_z, \tag{7.1}$$

where $\mathbf{i}$, $\mathbf{j}$, and $\mathbf{k}$ are the unit vectors defined in Chapter 2. The acceleration $\mathbf{a}$ is the rate of change of the velocity:

$$\mathbf{a} = \frac{d^2\mathbf{r}}{dt^2} = \mathbf{i}\frac{d^2x}{dt^2} + \mathbf{j}\frac{d^2y}{dt^2} + \mathbf{k}\frac{d^2z}{dt^2} = \mathbf{i}a_x + \mathbf{j}a_y + \mathbf{k}a_z. \tag{7.2}$$

Consider a particle that moves in the $z$ direction only, so that $v_x$, $v_y$, $a_x$, and $a_y$ vanish. If $a_z$ is known as a function of time,

$$a_z = a_z(t) \tag{7.3}$$

we can write an equation by equating the time derivative of the velocity to this known function:

$$\frac{dv_z}{dt} = a_z(t). \tag{7.4}$$

This is a differential equation for the velocity, since it contains the derivative of the velocity. Whereas the solution to an algebraic equation is generally a number or a set of numbers, solving this equation means finding the velocity as a function of time.

To solve Eq. (7.4), we multiply both sides by $dt$ and perform a definite integration

$$v_z(t_1) - v_z(0) = \int_0^{t_1} \left(\frac{dv_z}{dt}\right) dt = \int_0^{t_1} a_z(t)\,dt. \tag{7.5}$$

The result of this integration gives $v_z$ as a function of time, so that the position obeys a second differential equation

$$\frac{dz}{dt} = v_z(t). \tag{7.6}$$

A second integration gives the position as a function of time:

$$z(t_2) - z(0) = \int_0^{t_2} \left(\frac{dz}{dt_1}\right) dt_1 = \int_0^{t_2} v_z(t_1)\,dt_1. \tag{7.7}$$

There are inertial navigation systems used on submarines and space vehicles which determine the acceleration as a function of time and perform two numerical integrations in order to determine the position of the vehicle.

---

### Problem 7.1

At time $t = 0$, a certain particle has $z = 0$ and $v_z = 0$. Its acceleration is given as a function of time by

$$a_z = a_0 e^{-t/b},$$

where $a_0$ and $b$ are constants. Find $z$ as a function of time.

   a. Find the speed and the position of the particle at $t = 30$ s if $a_0 = 10\,\mathrm{m\,s^{-2}}$.
   b. Find the limiting value of the speed as $t \to \infty$.

---

Unfortunately, the acceleration is very seldom known as a function of time, so the simple integration of an equation like Eq. (7.3) cannot often be used to find how a particle moves. Instead, we must obtain the acceleration of the particle from knowledge of the force on it, using Newton's second law.

## Newton's Laws of Motion

These laws were deduced by Isaac Newton[1] from his analysis of observations of the motions of actual objects, including apples and celestial bodies. The laws can be

---

[1] Isaac Newton, 1642–1727, was a great English physicist and mathematician who discovered the law of gravity and the laws of motion and who helped invent calculus.

stated:

1. A body on which no forces act does not accelerate.
2. A body acted on by a force **F** accelerates according to

$$\mathbf{F} = m\mathbf{a}, \tag{7.8}$$

 where $m$ is the mass of the object.
3. Two bodies exert forces of equal magnitude and opposite direction on each other.

*Classical mechanics* is primarily the study of the consequences of these laws. The first law is just a special case of the second, and the third law is primarily used to obtain forces for the second law, so Newton's second law is the most important equation of classical mechanics.

If the force on a particle can be written as a function of its position alone, we have a differential equation for the position of that particle, which is called its *equation of motion*. If the force on a particle depends on the positions of other particles, the equations of motion of the particles are coupled together and must be solved simultaneously. The equations of motion cannot be solved exactly for a system of more than two interacting particles.

The simplest equation of motion is for a particle which can move in only one direction. From Newton's second law,

$$F_z(z) = m \frac{d^2 z}{dt^2}. \tag{7.9}$$

A force which depends only on position can be derived from a potential energy function, as in Eq. (5.62). Equation (7.9) becomes

$$-\frac{d\mathcal{V}}{dz} = m \frac{d^2 z}{dt^2}, \tag{7.10}$$

where $\mathcal{V}$ is the potential energy function.

## SECTION 7.2. THE HARMONIC OSCILLATOR: LINEAR DIFFERENTIAL EQUATIONS WITH CONSTANT COEFFICIENTS

Consider an object of mass $m$ attached to the end of a coil spring whose other end is rigidly fastened. Let the object move only in the $z$ direction, the direction in which the spring is stretched or compressed. Define the $z$ coordinate so that $z = 0$ when the spring has its equilibrium length. To a good approximation, the force on the object due to the spring is given by *Hooke's law*,[2]

$$F_z = -kz, \tag{7.11}$$

where $k$ is a constant called the *spring constant*. The negative sign produces a negative force (downward) when $z$ is positive and vice versa, so that the force pushes the mass toward its equilibrium position.

The *harmonic oscillator* is an idealized model of this system. That is, it is a hypothetical system (existing only in our minds) which has some properties in common

---

[2] After Robert Hooke, 1635–1703, one of Newton's contemporaries and rivals.

with the real system, but it is enough simpler to allow exact mathematical analysis. Our model is defined by saying that the spring has no mass and that Eq. (7.11) is exactly obeyed, even if $z$ has large magnitudes.

Replacement of $F$ by $md^2z/dt^2$ according to Newton's second law gives the equation of motion for our harmonic oscillator:

$$\frac{d^2z}{dt^2} + \left(\frac{k}{m}\right) z = 0. \tag{7.12}$$

This differential equation has the properties:

1. It is *linear*, which means that the dependent variable $z$ and its derivatives enter only to the first power.
2. It is *homogeneous*, which means that there are no terms that do not contain $z$.
3. It is *second order*, which means that the highest order derivative in the equation is a second derivative.
4. It has *constant coefficients*, which means that the quantities which multiply $z$ and its derivatives are constants.

There are two important facts about linear homogeneous differential equations:

1. If $z_1(t)$ and $z_2(t)$ are two functions that satisfy the equation, then the *linear combination* $z_3(t)$ is also a solution, where

$$z_3(t) = c_1 z_1(t) + c_2 z_2(t) \tag{7.13}$$

and $c_1$ and $c_2$ are constants.
2. If $z(t)$ satisfies the equation, then $cz(t)$ is also a solution, where $c$ is a constant.

---

### Problem 7.2

A linear differential equation of the $n$th order can be written

$$f_n(t)\frac{d^n z}{dt^n} + f_{n-1}(t)\frac{d^{n-1}z}{dt^{n-1}} + \cdots + f_1(t)\frac{dz}{dt} + f_0(t)z(t) = g(t). \tag{7.14}$$

For the case that $g(t) = 0$ (homogeneous equation), prove the two facts given above.

---

A linear homogeneous differential equation with constant coefficients can be solved by the following routine method:

1. Assume the trial solution

$$z(t) = e^{\lambda t}. \tag{7.15}$$

A trial solution is what the name implies. We try it by substituting it into the equation and produce an algebraic equation in $\lambda$ called the *characteristic equation*.
2. Find the values of $\lambda$ that satisfy the characteristic equation. These cause the trial solution to satisfy the equation. Call these $\lambda_1, \lambda_2, \ldots, \lambda_n$.

3. Use fact (1) to write a solution

$$z(t) = c_1 e^{\lambda_1 t} + c_2 e^{\lambda_2 t} + \cdots + c_n e^{\lambda_n t}. \tag{7.16}$$

---

## Example 7.1

Show that the differential equation

$$a_3 \left( \frac{d^3 y}{dx^3} \right) + a_2 \left( \frac{d^2 y}{dx^2} \right) + a_1 \left( \frac{dy}{dx} \right) + a_0 y = 0 \tag{7.17}$$

can be satisfied by a trial solution $y = e^{\lambda x}$.

---

## Solution

We substitute the trial solution $y = e^{\lambda x}$ into Eq. (7.17).

$$a_3 \lambda^3 e^{\lambda x} + a_2 \lambda^2 e^{\lambda x} + a_1 \lambda^2 e^{\lambda x} + a_1 \lambda e^{\lambda x} + a_0 e^{\lambda x} = 0. \tag{7.18}$$

If $x$ remains finite, we can divide by $e^{\lambda x}$ to obtain the characteristic equation:

$$a_3 \lambda^3 + a_2 \lambda^2 + a_1 \lambda + a_0 = 0. \tag{7.19}$$

This is an equation that can be solved for three values of $\lambda$ which cause the trial solution to satisfy Eq. (7.17).

---

## Example 7.2

Solve the differential equation

$$\frac{d^2 y}{dx^2} + \frac{dy}{dx} - 2y = 0. \tag{7.20}$$

---

## Solution

Substitution of the trial solution $y = e^{\lambda x}$ gives the characteristic equation

$$\lambda^2 + \lambda - 2 = 0.$$

The solutions to this equation are

$$\lambda = 1, \qquad \lambda = -2.$$

The solution to the differential equation is thus

$$y(x) = c_1 e^x + c_2 e^{-2x}. \tag{7.21}$$

The solution in Eq. (7.21) satisfies Eq. (7.20) no matter what values $c_1$ and $c_2$ have. The solution is a *family* of functions, one function for each set of values for $c_1$ and $c_2$.

---

A solution to a linear differential equation of order $n$ that contains $n$ arbitrary constants is known to be a general solution. A *general solution* is a family of functions which includes almost every solution to the differential equation.

A solution to a differential equation that contains no arbitrary constants is called a *particular solution*. A particular solution is usually one of the members of the general solution, but it might possibly be another function. We do not discuss the question of when particular solutions occur that are not included in the general solution. In the differential equations encountered in physical chemistry, you should assume that the general solution includes all solutions.

In the case of a linear homogeneous differential equation with constant coefficients, there will generally be $n$ values of $\lambda$ that satisfy the characteristic equation if the differential equation is of order $n$. Therefore, the solution of Eq. (7.16) is a general solution, since it contains two arbitrary constants. There is only one general solution to a differential equation. If you find two general solutions for the same differential equation that appear to be different, there must be some mathematical manipulations that will reduce both to the same form.

We are not finished with a problem when we find a general solution to a differential equation. We usually have additional information that will enable us to pick a particular solution out of the family of solutions. Such information consists of knowledge of boundary conditions and initial conditions. *Boundary conditions* arise from physical requirements on the solution, such as necessary conditions that apply to the boundaries of the region in space where the solution applies, or the requirement that the value of a physically measurable quantity must be a real number, etc. *Initial conditions* arise from knowledge of the state of the system described by the solution at some initial time.

We now solve the equation of motion for the harmonic oscillator, Eq. (7.12). We begin by finding the characteristic equation.

---

**Problem 7.3**

Show that the characteristic equation for Eq. (7.12) is

$$\lambda^2 + \frac{k}{m} = 0. \tag{7.22}$$

---

The solution of the characteristic equation for the harmonic oscillator is

$$\lambda = \pm i \left( \frac{k}{m} \right)^{1/2}, \tag{7.23}$$

where $i = \sqrt{-1}$, the imaginary unit.

The general solution to Eq. (7.12) is therefore

$$z = z(t) = c_1 \exp\left[ +i \left( \frac{k}{m} \right)^{1/2} t \right] + c_2 \exp\left[ -i \left( \frac{k}{m} \right)^{1/2} t \right], \tag{7.24}$$

where $c_1$ and $c_2$ are arbitrary constants.

Our principal boundary condition is that the solution be real, because imaginary and complex numbers cannot represent physically measurable quantities like the position of the oscillator. From Eq. (2.112), we can write

$$z = c_1[\cos(\omega t) + i \sin(\omega t)] + c_2[\cos(\omega t) - i \sin(\omega t)], \tag{7.25}$$

where

$$\omega = \left(\frac{k}{m}\right)^{1/2}.$$

If we let $c_1 + c_2 = b_1$ and $i(c_1 - c_2) = b_2$, then

$$z = b_1 \cos(\omega t) + b_2 \sin(\omega t). \tag{7.26}$$

Although the solutions in Eq. (7.25) and Eq. (7.26) look different, they are equivalent to each other. Since $z$ is a physically measurable quantity, it must be real. We can eliminate complex solutions by requiring that $b_1$ and $b_2$ be real.

---

**Problem 7.4**

Show that the function of Eq. (7.26) satisfies Eq. (7.12).

---

Our new general solution applies to a particular harmonic oscillator if we use that oscillator's values of $k$ and $m$ values to calculate the value of $\omega$. We can apply initial conditions to make our solution describe a particular case of its motion. Say that we have the initial conditions at $t = 0$:

$$z(0) = 0 \tag{7.27a}$$
$$v_z(0) = v_0 = \text{constant}. \tag{7.27b}$$

We require one initial condition to evaluate each arbitrary constant, so these two initial conditions will enable us get a particular solution for the case at hand. Knowledge of the position at time $t = 0$ without knowledge of the velocity at that time would not suffice, nor would knowledge of the velocity without knowledge of the position.

For our initial conditions, $b_1$ must vanish:

$$z(0) = b_1 \cos(0) + b_2 \sin(0) = b_1 = 0. \tag{7.28}$$

The position is therefore given by

$$z(t) = b_2 \sin(\omega t). \tag{7.29}$$

The expression for the velocity is obtained by differentiation,

$$v_z(t) = \frac{dz}{dt} = b_2\omega \cos(\omega t) \tag{7.30}$$

so

$$v_z(0) = b_2\omega \cos(0) = b_2\omega$$

which gives

$$b_2 = \frac{v_0}{\omega} \tag{7.31}$$

and gives us our particular solution

$$z(t) = \left(\frac{v_0}{\omega}\right)\sin(\omega t). \tag{7.32}$$

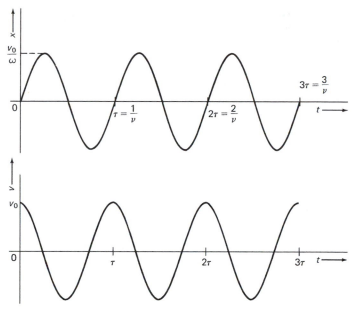

**FIGURE 7.1**    The position and velocity of a harmonic oscillator as functions of time.

The motion given by this solution is called *uniform harmonic motion*. It is a sinusoidal oscillation in time with a fixed frequency of oscillation. Figure 7.1 shows the position and the velocity of the suspended mass as a function of time.

The motion is *periodic*, repeating itself over and over. During one period, the argument of the sine changes by $2\pi$, so that if $\tau$ is the period (the length of time required for one cycle of the motion),

$$2\pi = \omega\tau = \left(\frac{k}{m}\right)^{1/2}\tau. \tag{7.33}$$

Thus,

$$\tau = 2\pi \left(\frac{m}{k}\right)^{1/2}. \tag{7.34}$$

The reciprocal of the period is called the *frequency*, denoted by $\nu$. (This is the Greek letter nu. Try not to confuse it with the letter vee.)

$$\nu = \frac{1}{2\pi}\sqrt{\frac{k}{m}}. \tag{7.35}$$

---

**Problem 7.5**

The vibration of a diatomic molecule resembles that of a harmonic oscillator. Since both nuclei move, the mass must be replaced by the *reduced mass*,

$$\mu = \frac{m_1 m_2}{m_1 + m_2},$$

where $m_1$ is the mass of one nucleus and $m_2$ the mass of the other nucleus. Calculate the frequency of vibration of a hydrogen chloride molecule. The force constant $k$ is equal to $481 \, \text{N} \, \text{m}^{-1} = 481 \, \text{J} \, \text{m}^{-2}$.

---

The kinetic energy of the harmonic oscillator is

$$\mathcal{K} = \frac{1}{2}mv^2 = \frac{1}{2}mv_z^2. \tag{7.36}$$

In order for the force to be the negative derivative of the potential energy as in Eq. (5.62), the potential energy of the harmonic oscillator must be

$$\mathcal{V} = \frac{1}{2}kz^2. \tag{7.37}$$

The total energy is the sum of the kinetic energy and the potential energy

$$E = \mathcal{K} + \mathcal{V} = \frac{1}{2}mv_0^2 \cos^2(\omega t) + \frac{1}{2}k\left(\frac{v_0}{\omega}\right)^2 \sin^2(\omega t) \tag{7.38}$$

$$= \frac{1}{2}mv_0^2,$$

where we have used the identity of Eq. (7) of Appendix 5.

As an undisturbed frictionless harmonic oscillator moves, the total energy remains constant. As the kinetic energy rises and falls, the potential energy changes so that the sum remains constant. We say that the energy is *conserved*, and that the system is *conservative*. If the forces on the particles of a system can be obtained from a potential energy function, the system will be conservative.

## The Damped Harmonic Oscillator—A Nonconservative System

We now discuss a damped harmonic oscillator, which is subject to an additional force that is proportional to the velocity, such as a frictional force due to fairly slow motion of an object through a fluid,

$$\mathbf{F}_f = -\zeta \mathbf{v} = -\zeta \frac{d\mathbf{r}}{dt}, \tag{7.39}$$

where $\zeta$ is called the friction constant. Since this force cannot be derived from a potential energy, the system is not conservative and its energy will change with time.

The equation of motion is

$$F_z = -\zeta \frac{dz}{dt} - kz = m\left(\frac{d^2z}{dt^2}\right). \tag{7.40}$$

Equation (7.40) is a linear homogeneous equation with constant coefficients, so a trial solution of the form of Eq. (7.15) will work. The characteristic equation is

$$\lambda^2 + \frac{\zeta \lambda}{m} + \frac{k}{m} = 0. \tag{7.41}$$

The solutions of this quadratic equation are

$$\lambda_1 = -\frac{\zeta}{2m} + \frac{\sqrt{(\zeta/m)^2 - 4k/m}}{2} \tag{7.42a}$$

$$\lambda_2 = -\frac{\zeta}{2m} - \frac{\sqrt{(\zeta/m)^2 - 4k/m}}{2} \tag{7.42b}$$

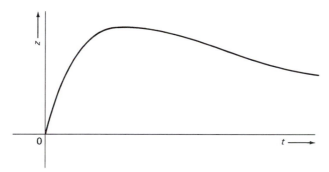

**FIGURE 7.2** The position of a greater than critically damped harmonic oscillator as a function of time.

and the general solution to the differential equation is

$$z(t) = c_1 e^{\lambda_1 t} + c_2 e^{\lambda_2 t}. \tag{7.43}$$

### Problem 7.6

Show that Eq. (7.41) is the correct characteristic equation, that Eq. (7.42) gives the correct solutions to the characteristic equation, and that the function of Eq. (7.43) does satisfy Eq. (7.40).

There are three cases. In the first case, the quantity inside the square root in Eq. (7.42) is positive, so that $\lambda_1$ and $\lambda_2$ are both real. This corresponds to a relatively large value of the friction constant $\zeta$, and the case is called *greater than critical damping*. In this case, the mass at the end of the spring does not oscillate, but returns smoothly to its equilibrium position of $z = 0$ if disturbed from this position.

Figure 7.2 shows the position of a greater than critically damped oscillator as a function of time for a particular set of initial conditions.

### Problem 7.7

From the fact that $\zeta, k$, and $m$ are all positive, show that $\lambda_1$ and $\lambda_2$ are both negative in the case of greater than critical damping, and from this fact, show that

$$\lim_{t \to \infty} z(t) = 0. \tag{7.44}$$

The next case is that of small values of $\zeta$, or *less than critical damping*. If

$$\left(\frac{\zeta}{m}\right)^2 < \frac{4k}{m}$$

the quantity inside the square root in Eq. (7.42) is negative, and $\lambda_1$ and $\lambda_2$ are complex quantities,

$$\lambda_1 = -\frac{\zeta}{2m} + i\omega \tag{7.45a}$$

$$\lambda_2 = -\frac{\zeta}{2m} - i\omega, \tag{7.45b}$$

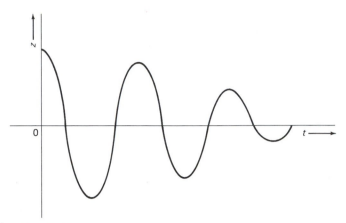

**FIGURE 7.3**   The position of a less than critically damped oscillator as a function of time for the initial conditions $z(0) = z_0$, $v_z(0) = 0$.

where

$$\omega = \sqrt{\frac{k}{m} - \left(\frac{\zeta}{2m}\right)^2}. \tag{7.46}$$

The solution thus becomes

$$z(t) = (c_1 e^{i\omega t} + c_2 e^{i\omega t})e^{-\zeta t/2m} \tag{7.47}$$

which can also be written in the form, similar to Eq. (7.26),

$$z(t) = [b_1 \cos(\omega t) + b_2 \sin(\omega t)]e^{-\zeta t/2m}. \tag{7.48}$$

This shows $z(t)$ to be an oscillatory function times an exponentially decreasing function, giving the "ringing" behavior shown in Fig. 7.3.

---

### Problem 7.8

If the position of the oscillator at time 0 is a particular value $z(0)$ and if the velocity at time zero is a particular value $v(0)$, express the constants $b_1$ and $b_2$ in terms of these values.

---

The final case is that of *critical damping*, in which the quantity inside the square root in Eq. (7.42) exactly vanishes. This case is not likely to happen by chance, but it is possible to construct an oscillating object such as a galvanometer mirror or a two-pan balance beam that is critically damped. The condition for critical damping is

$$\left(\frac{\zeta}{2m}\right)^2 = \frac{k}{m}. \tag{7.49}$$

An interesting thing happens to the solution of Eq. (7.43) in the case of critical damping. The values of $\lambda$ are equal to each other,

$$\lambda_1 = \lambda_2 = -\frac{\zeta}{2m} \tag{7.50}$$

so that Eq. (7.43) becomes

$$z(t) = (c_1 + c_2)e^{\lambda_1 t} = ce^{\lambda_1 t}. \tag{7.51}$$

This is not a general solution, since a general solution for a second order linear equation must contain two arbitrary constants, and a sum of two constants does not constitute two separate constants.

The problem with our two solutions, $e^{\lambda_1/t}$ and $e^{\lambda_2 t}$, is called *linear dependence*. With two functions, linear dependence means that the functions are proportional to each other, so that they are not distinct solutions. If we have several solutions, they are linearly dependent if one or more of the solutions equals a linear combination of the others.

Since we do not have a general solution, there must be another family of solutions not included in the solution of Eq. (7.51). One way to find it is by attempting additional trial functions until we find one that works. The one that works is

$$z(t) = te^{\lambda t}. \tag{7.52}$$

---

**Problem 7.9**

Substitute the trial solution of Eq. (7.52) into Eq. (7.40), using the condition of Eq. (7.49) to restrict the discussion to critical damping, and show that the equation is satisfied.

---

Our general solution is now

$$z(t) = (c_1 + c_2 t)e^{\lambda_1 t}. \tag{7.53}$$

For any particular set of initial conditions, we can find the appropriate values of $c_1$ and $c_2$. The behavior of a critically damped oscillator is much the same as that of Fig. 7.2.

## The Forced Harmonic Oscillator: Inhomogeneous Linear Differential Equations

If the function $g(t)$ in Eq. (7.14) is not zero, it is an *inhomogeneous term* and the equation is said to be an *inhomogeneous equation*. If an external force exerted on a harmonic oscillator depends only on the time, the equation of motion is an inhomogeneous equation,

$$\frac{d^2 z}{dt^2} + \left(\frac{k}{m}\right)z = \frac{F(t)}{m}, \tag{7.54}$$

where $F(t)$ is the external force.

A method for solving such an equation is:

*Step 1.* Solve the equation obtained by deleting the inhomogeneous term $F/m$. This homogeneous equation is called the *complementary equation*, and the general solution to this equation is called the *complementary function*.

*Step 2.* Find a particular solution to the inhomogeneous equation by whatever means may be necessary. The general solution to the inhomogeneous equation is the sum of the complementary function and this particular solution.

**TABLE 7.1   Particular Trial Solutions for the Variation of Parameters Method***

| Inhomogeneous term | Trial solution† | Forbidden characteristic root |
|---|---|---|
| $1$ | $A$ | $0$ |
| $t^n$ | $A_0 + A_1 t + A_2 t^2 + \cdots + A_n t^n$ | $0$ |
| $e^{\alpha t}$ | $A e^{\alpha t}$ | $\alpha$ |
| $t^n e^{\alpha t}$ | $e^{\alpha t}(A_0 + A_1 t + A_2 t^2 + \cdots + A_n t^n)$ | $\alpha$ |
| $e^{\alpha t} \sin(\beta t)$ | $e^{\alpha t}[A\cos(\beta t) + B\sin(\beta t)]$ | $\alpha, \beta$ |
| $e^{\alpha t} \cos(\beta t)$ | $e^{\alpha t}[A\cos(\beta t) + B\sin(\beta t)]$ | $\alpha, \beta$ |

*Source: M. Morris and O. E. Brown, *Differential Equations, 3rd ed.,* Prentice–Hall, Englewood Cliffs, NJ, 1952.

†The trial solution given will not work if the characteristic equation for the complementary differential equation has a root equal to the entry in this column. If such a root occurs with multiplicity $k$, multiply the trial solution by $t^k$ to obtain a trial solution that will work.

Note. $A$, $B$, $A_0$, $A_1$, etc., are parameters to be determined. $\alpha$ and $\beta$ are constants in the differential equation to be solved.

---

**Problem 7.10**

If $z_c(t)$ is a general solution to the complementary equation and $z_p(t)$ is a particular solution to the inhomogeneous equation, show that $z_c + z_p$ is a solution to the inhomogeneous equation of Eq. (7.14).

---

There is a method for finding a particular solution to a linear inhomogeneous equation, known as the *variation of parameters*. If the inhomogenous term is a power of $t$, an exponential, a sine, a cosine, or a combination of these functions, this method can be used. One proceeds by taking a suitable trial function that contains parameters (constants whose values need to be determined). This is substituted into the inhomogeneous equation and the values of the parameters are found so that the inhomogeneous equation is satisfied. Table 7.1 gives a list of suitable trial functions for various inhomogeneous terms.

Let us assume that the external force in Eq. (7.54) is

$$F(t) = F_0 \sin(\alpha t). \tag{7.55}$$

Use of Table 7.1 and determination of the parameters gives the particular solution

$$z_p(t) = \frac{F_0}{m(\omega^2 - \alpha^2)} \sin(\alpha t). \tag{7.56}$$

The complementary function is the solution given in Eq. (7.26), so the general solution is

$$z(t) = b_1 \cos(\omega t) + b_2 \sin(\omega t) + z_p(t). \tag{7.57}$$

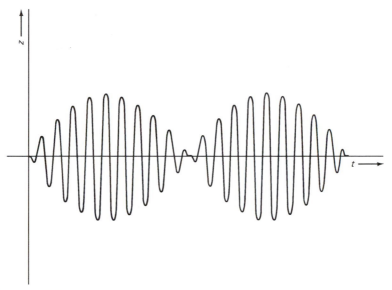

**FIGURE 7.4**   The position of a forced harmonic oscillator as a function of time for the case $\alpha = 1.1\omega$.

---

### Problem 7.11

Verify Eqs. (7.56) and (7.57).

---

The motion of the forced harmonic oscillator shows some interesting features. The solution is a linear combination of the natural motion and a motion proportional to the external force. If the frequencies of these are not very different, a motion such as shown in Fig. 7.4, known as "beating," can result.

## SECTION 7.3. DIFFERENTIAL EQUATIONS WITH SEPARABLE VARIABLES

In this section, we discuss equations that can be manipulated algebraically into the form

$$g(y)\left(\frac{dy}{dx}\right) = f(x), \tag{7.58}$$

where $g(y)$ is some integrable function of $y$ and $f(x)$ is some integrable function of $x$. To solve Eq. (7.58), we multiply both sides of the equation by $dx$ and use Eq. (3.18)

$$\left(\frac{dy}{dx}\right)dx = dy. \tag{7.59}$$

We now have

$$g(y)\,dy = f(x)\,dx. \tag{7.60}$$

If we have manipulated the equation into the form of Eq. (7.60), we say that we have *separated the variables*, because we have no $x$ dependence in the left-hand side of the equation and no $y$ dependence in the right-hand side. We can perform an indefinite integration on both sides of this equation to obtain

$$\int g(y)\,dy = \int f(x)\,dx + C, \tag{7.61}$$

where $C$ is a constant of integration. We can alternatively do a definite integration

$$\int_{y_1}^{y_2} g(y)\,dy = \int_{x_1}^{x_2} f(x)\,dx, \tag{7.62}$$

where

$$y_1 = y(x_1)$$
$$y_2 = y(x_2).$$

---

### Example 7.3

In a first-order chemical reaction with no back reaction, the concentration of the reactant is governed by

$$-\frac{dc}{dt} = kc, \tag{7.63}$$

where $c$ is the concentration of the single reactant, $t$ is the time, and $k$ is a function of temperature called the rate constant. Solve the equation to find $c$ as a function of $t$.

---

### Solution

We divide by $c$ and multiply by $dt$ to separate the variables:

$$\frac{1}{c}\frac{dc}{dt}\,dt = \frac{1}{c}dc = -k\,dt.$$

We perform an indefinite integration

$$\int \frac{1}{c}\,dc = \ln(c) = -k\int dt + C = -kt + C,$$

where $C$ is a constant of integration. Although each indefinite integration would require a constant of integration, we include only one constant, since the second constant of integration could be moved to the other side of the equation, giving the difference of two constants, which equals a constant.

We take the exponential of each side of Eq. (7.73) to obtain

$$e^{\ln(c)} = c = e^C e^{-kt} = c(0)e^{-kt}. \tag{7.64}$$

In the last step, we recognized that $e^C$ had to equal the concentration at time $t = 0$.

A definite integration can be carried out instead of an indefinite integration:

$$\int_{c(0)}^{c(t_1)} \frac{1}{c}\,dc = \ln\left(\frac{c(t_1)}{c(0)}\right) = -k\int_0^{t_1} dt = -kt_1.$$

This equation is the same as Eq. (7.64) except that the time is now called $t_1$ instead of $t$. The limits on the two definite integrations must be done correctly. If the lower limit of the time integration is zero, the lower limit of the concentration integration must be the value of the concentration at zero time. The upper limit is similar.

---

### Problem 7.12

In a second-order chemical reaction involving one reactant and having no back reaction,

$$-\frac{dc}{dt} = kc^2.$$

Solve this differential equation by separation of variables. Do a definite integration from $t = 0$ to $t = t_1$.

---

If you are faced with a differential equation, and if you think that there is some chance that separation of variables will work, try the method.

## SECTION 7.4. EXACT DIFFERENTIAL EQUATIONS

Sometimes you might be faced with an equation that cannot be manipulated into the form of Eq. (7.60), but which can be manipulated into the somewhat similar *pfaffian form*:

$$M(x, y)\, dx + N(x, y)\, dy = 0. \tag{7.65}$$

In Chapter 5, we discussed differentials like the left-hand side of this equation. Some such differentials are *exact*, which means that they are differentials of functions. Others are *inexact*, which means that they are not differentials of functions.

The test for exactness is based on the Euler reciprocity relation, as in Eq. (5.33): If

$$\left(\frac{\partial M}{\partial y}\right)_x = \left(\frac{\partial N}{\partial x}\right)_y \tag{7.66}$$

then the differential is exact.

If the differential is exact, there is a function $f(x, y)$ such that

$$df = M(x, y)\, dx + N(x, y)\, dy = 0 \tag{7.67}$$

which implies that

$$f(x, y) = C, \tag{7.68}$$

where $C$ is a constant, because only a constant function has a differential that vanishes. Equation (7.68) provides a solution to the differential equation, because it can be solved for $y$ in terms of $x$.

In Section 5.6, we discussed the procedure for finding the function in Eq. (7.68), using a line integral,

$$f(x_1, y_1) = f(x_0, y_0) + \int_C df, \tag{7.69}$$

where $C$ is a curve beginning at $(x_0, y_0)$ and ending at $(x_1, y_1)$.

A convenient curve is the rectangular path from $(x_0, y_0)$ to $(x_1, y_0)$ and then to $(x_1, y_1)$. On the first part of this path, $y$ is constant at $y_0$, so the $dy$ integral vanishes and $y$ is replaced by $y_0$ in the $dx$ integral. On the second part of the path, $x$ is constant at $x_1$, so the $dx$ integral vanishes and $x$ is replaced by $x_1$ in the $dy$ integral:

$$f(x_1, y_1) = f(x_0, y_0) + \int_{x_0}^{x_1} M(x, y_0)\, dx + \int_{y_0}^{y_1} N(x_1, y)\, dy. \qquad (7.70)$$

Both integrals are now ordinary integrals, so we have a solution if we can perform the integrals. The solution will contain an arbitrary constant, because any constant can be added to both $f(x_1, y_1)$ and $f(x_0, y_0)$ in Eq. (7.70) without changing the equality.

---

### Example 7.4

Solve the differential equation

$$2xy\, dx + x^2\, dy = 0.$$

---

### Solution

The equation is exact, because

$$\frac{\partial}{\partial y}(2xy) = 2x \qquad \text{and} \qquad \frac{\partial}{\partial x}(x^2) = 2x.$$

We do a line integral from $(x_0, y_0)$ to $(x_1, y_0)$ and then to $(x_0, y_1)$, letting $f(x, y)$ be the function whose differential must vanish:

$$
\begin{aligned}
0 = f(x_1, y_1) - f(x_0, y_0) &= \int_{x_0}^{x_1} 2xy_0\, dx + \int_{y_0}^{y_1} x_1^2\, dy \\
&= y_0 x_1^2 - y_0 x_0^2 + x_1^2 y_1 - x_1^2 y_0 \\
&= x_1^2 y_1 - x_0^2 y_0.
\end{aligned}
$$

We regard $x_0$ and $y_0$ as constants so that $x_0^2 y_0 = C$. We drop the subscripts on $x_1$ and $y_1$,

$$f(x, y) = x^2 y = C,$$

where $C$ is a constant. Our solution is $y = C/x^2$.

---

### Problem 7.13

  a.  Show that the solution in Example 7.4 satisfies the equation.
  b.  Solve the equation $(4x + y)\, dx + x\, dy = 0$.

---

## SECTION 7.5. SOLUTION OF INEXACT DIFFERENTIAL EQUATIONS BY THE USE OF INTEGRATING FACTORS

If we have an inexact pfaffian differential equation

$$M(x, y)\, dx + N(x, y)\, dy = 0 \qquad (7.71)$$

we cannot use the method of Section 7.4. However, as mentioned in Section 5.5, some inexact differentials yield an exact differential when multiplied by an *integrating factor.* If the function $g(x, y)$ is an integrating factor for the differential in Eq. (7.71), then

$$g(x, y)M(x, y)\, dx + g(x, y)N(x, y)\, dy = 0 \qquad (7.72)$$

is an exact differential equation that can be solved by the method of Section 7.4. A solution for Eq. (7.72) will also be a solution for Eq. (7.71).

---

### Example 7.5

Solve the differential equation

$$\frac{dy}{dx} = \frac{y}{x}.$$

---

### Solution

We convert the equation to the pfaffian form, $y\, dx - x\, dy = 0$
Test for exactness:

$$\left(\frac{\partial y}{\partial y}\right)_x = 1$$

$$\left[\frac{\partial(-x)}{\partial x}\right]_y = -1.$$

The equation is not exact. We show that $1/x^2$ is an integrating factor. Multiplication by this factor gives

$$\left(\frac{y}{x^2}\right) dx - \left(\frac{1}{x}\right) dy = 0. \qquad (7.73)$$

This is exact:

$$\left[\frac{\partial(y/x^2)}{\partial y}\right]_x = \frac{1}{x^2}$$

$$\left[\frac{\partial(-1x)}{\partial x}\right]_y = \frac{1}{x^2}.$$

We can solve Eq. (7.73) by the method of Section 7.4:

$$0 = \int_{x_0}^{x_1} \left(\frac{y_0}{x^2}\right) dx - \int_{y_0}^{y_1} \left(\frac{1}{x_1}\right) dy$$

$$= -y_0\left(\frac{1}{x_1} - \frac{1}{x_0}\right) - \frac{1}{x_1}(y_1 - y_0) = \frac{y_0}{x_0} - \frac{y_1}{x_1}.$$

We regard $x_0$ and $y_0$ as constants, so that

$$\frac{y}{x} = \frac{y_0}{x_0} = C = \text{constant}$$

or

$$y = Cx.$$

This is a general solution, since the original equation was first order and the solution contains one arbitrary constant.

---

If an inexact differential has one integrating factor, it has an infinite number of integrating factors. Therefore, there can be other integrating factors for a differential such as the one in the preceding example. Unfortunately, there is no general procedure for finding an integrating factor except by trial and error.

---

### Problem 7.14

Show that $1/y^2$ and $1/(x^2 + y^2)$ are integrating factors for the equation in Example 7.5 and that they lead to the same solution.

---

## SECTION 7.6. WAVES IN A STRING: PARTIAL DIFFERENTIAL EQUATIONS

Differential equations that contain several independent variables and contain partial derivatives are called *partial differential equations*. Differential equations also occur that contain more than one dependent variable, but you must have one equation for each dependent variable and must solve them simultaneously. You can read about such equations in some of the books listed at the end of the chapter.

We discuss only a rather simple method of solving a partial differential equation, devoting most of this section to an example, the classical equation of motion of a flexible string. There are important similarities between this equation and the Schrödinger equation[3] of quantum mechanics. The flexible string that we discuss is a mathematical model. It resembles a real string, but is defined by the following: (1) It is completely flexible, so that no force is required to bend the string. (2) Its motion is restricted to small vibrations, so that the string is not appreciably stretched. (3) Both ends of the string are fixed in position. Figure 7.5 shows the string. We choose one end of the string as our origin of coordinates and use the equilibrium (straight) position of the string as our $x$ axis. The location of the string at a particular value of $x$ is given by a value of $y$ and a value of $z$ for that value of $x$. Since the string moves, $y$ and $z$ are functions of time as well as of $x$:

$$y = y(x, t) \tag{7.74a}$$
$$z = z(x, t). \tag{7.74b}$$

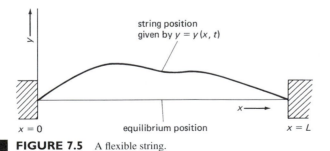

**FIGURE 7.5**   A flexible string.

---

[3] Named for its discoverer, Erwin Schrödinger, 1887–1961, joint 1933 Nobel Prize winner in physics with P. A. M. Dirac.

The equation of motion of the string is derived by writing Newton's second law for a small segment of the string and taking a mathematical limit as the length of the segment becomes infinitesimal.[4] The result is two partial differential equations,

$$\left(\frac{\partial^2 y}{\partial t^2}\right) = \frac{T}{\rho}\left(\frac{\partial^2 y}{\partial x^2}\right) = c^2\left(\frac{\partial^2 y}{\partial x^2}\right) \tag{7.75}$$

and a similar equation for $z$. In Eq. (7.75), $T$ is the magnitude of the tension force on the string and $\rho$ the mass of the string per unit length. The quantity $c$ turns out to be the speed of propagation of a wave along the string. Since the two equations are independent of each other, we can solve for $y$ and $z$ separately, and the two solutions will be identical except for the symbol used for the dependent variable.

## Solution by Separation of Variables

We do not seek a general solution for Eq. (7.75), but seek only a family of solutions that can be written as a product of factors, each of which depends on only one variable:

$$y(x, t) = \psi(x)\,\theta(t). \tag{7.76}$$

This is called a *solution with the variables separated*. We regard it as a trial solution and substitute it into the differential equation to see if it works. This method of separation of variables is slightly different from that of Section 7.3.

Since $\psi$ does not depend on $t$ and $\theta$ does not depend on $x$, the result is

$$\psi(x)\left(\frac{d^2\theta}{dt^2}\right) = c^2\theta(t)\left(\frac{d^2\psi}{dx^2}\right). \tag{7.77}$$

We write ordinary derivatives since we now have functions of only one variable.

We separate the variables by manipulating Eq. (7.77) into a form in which one term contains no $x$ dependence and the other term contains no $t$ dependence, just as we manipulated Eq. (7.58) into a form with only one variable in each term. The present separation of variables is different from that case, since we are now separating two independent variables from each other, instead of separating one independent variable from one dependent variable. The manipulations that we carry out are also different.

We divide both sides of Eq. (7.77) by the product $\psi(x)\,\theta(t)$. We also divide by $c^2$, but this is not essential.

$$\frac{1}{\psi(x)}\frac{d^2\psi}{dx^2} = \frac{1}{c^2\theta(t)}\frac{d^2\theta}{dt^2}. \tag{7.78}$$

The variables are now separated, since each term contains only one independent variable. We can use the fact that $x$ and $t$ are both independent variables to extract two ordinary differential equations from Eq. (7.78). If we temporarily keep $t$ fixed at some value, we can still allow $x$ to vary. Thus, even if $x$ varies, the function of $x$ on the left-hand side of Eq. (7.78) must be a constant function of $x$, because it equals a quantity that we can keep fixed.

---

[4]Robert G. Mortimer, *Physical Chemistry*, pp. 333ff, Benjamin–Cummings, Redwood City, CA, 1993.

Therefore

$$\frac{1}{\psi(x)} \frac{d^2\psi}{dx^2} = \text{constant} = -k^2. \tag{7.79}$$

For the same reason, the right-hand side is a constant function of $t$,

$$\frac{1}{c^2\theta(t)} \frac{d^2\theta}{dt^2} = -k^2, \tag{7.80}$$

where we denote the constant by the symbol $-k^2$. We now multiply Eq. (7.79) by $\psi(x)$ and multiply Eq. (7.80) by $c^2\theta(t)$,

$$\frac{d^2\psi}{dx^2} + k^2\psi = 0 \tag{7.81}$$

$$\frac{d^2\theta}{dt^2} + k^2c^2\theta = 0. \tag{7.82}$$

The separation of variables is complete, and we have two ordinary differential equations. Except for the symbols used, each of these equations are the same as Eq. (7.12). We transcribe the solution to that equation with appropriate changes in symbols:

$$\psi(x) = a_1 \cos(kx) + a_2 \sin(kx) \tag{7.83}$$
$$\theta(t) = b_1 \cos(kct) + b_2 \sin(kct). \tag{7.84}$$

These are general solutions to the ordinary differential equations, but we do not necessarily have a general solution to our partial differential equation, because there can be solutions that are not of the form of Eq. (7.76).

We are now ready to consider a specific case and to find values for the arbitrary constants that make our solution describe the specific case. We consider a string of length $L$ with the ends fixed, as in Fig. 7.5. We have the condition that $y = 0$ at $x = 0$ and $x = L$. These conditions are called *boundary conditions*, and literally arise from a condition at the boundaries of a region. If $y$ must vanish at $x = 0$ and at $x = L$, then $\psi$ must vanish at these points, since the factor $\theta$ does not necessarily vanish:

$$\psi(0) = 0 \tag{7.85}$$

and

$$\psi(L) = 0. \tag{7.86}$$

Equation (7.85) requires that

$$a_1 = 0 \tag{7.87}$$

since the cosine of zero is unity. Equation (7.86) requires that the argument of the sine function in Eq. (7.83) be equal to some integer times $\pi$ for $x = L$, because

$$\sin(n\pi) = 0 \qquad (n = 0, 1, 2, \ldots). \tag{7.88}$$

Therefore,

$$k = \frac{n\pi}{L} \qquad (n = 1, 2, 3, \ldots). \tag{7.89}$$

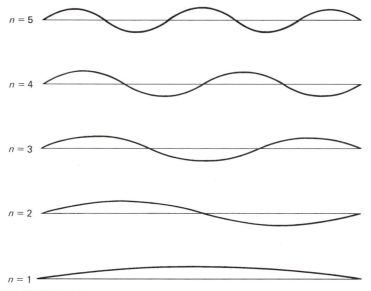

**FIGURE 7.6**  Standing waves in a flexible string.

We exclude $n = 0$, although this gives a legitimate solution corresponding to a stationary string at its equilibrium position.

$$\psi(x) = a_2 \sin\left(\frac{n\pi x}{L}\right). \tag{7.90}$$

Now we can apply *initial conditions* that make our solution apply to a particular case. For example, let us say that at $t = 0$, the string happens to be passing through its equilibrium position, which is $y = 0$ for all $x$. If so, then $b_1 = 0$, and we have

$$y(x, t) = A \sin\left(\frac{n\pi x}{L}\right) \sin\left(\frac{n\pi ct}{L}\right), \tag{7.91}$$

where we write $A = a_2 b_2$. The maximum amplitude is $A$, and another initial condition would be required to specify its value.

We have a set of solutions, one for each value of the integer $n$. Figure 7.6 shows the function $\psi(x)$ for several values of $n$. Each curve represents the shape of the string at an instant when $\theta = 1$. At other times, the string is vibrating between such a position and a position given by $-\psi(x)$. There are fixed points at which the string is stationary. These points are called *nodes*, and the number of nodes other than the two nodes at the ends of the string is $n - 1$. A wave with stationary nodes is called a *standing wave*.

If $\lambda$ is the wavelength, or the distance for the sine function in $\psi$ to go through a complete period, then

$$n\lambda = 2L. \tag{7.92}$$

The *period* of oscillation is the time required for the sine function in the factor $\theta$ to go through a complete oscillation and return the string to its original position and velocity, which requires the argument of the sine function to range through $2\pi$. If $\tau$

is the period,

$$\frac{n\pi c\tau}{L} = 2\pi; \qquad \tau = \frac{2L}{nc}. \tag{7.93}$$

The *frequency* $v$ is the reciprocal of the period:

$$v = \frac{nc}{2L} = \left(\frac{n}{2L}\right)\left(\frac{T}{\rho}\right)^{1/2}. \tag{7.94}$$

In musical acoustics, the oscillation corresponding to $n = 1$ is called the *fundamental*, that for $n = 2$ is the *first overtone*, etc. The fundamental is also called the first *harmonic*, the first overtone is called the *second harmonic*, etc.

---

### Problem 7.15

A certain violin string has a mass per unit length of $10.00 \text{ mg cm}^{-1}$ and a length of 50 cm. Find the tension force necessary to make it produce a fundamental tone of A above middle C (440 oscillations per second = 440 hertz).

---

When a string in a musical instrument is struck or bowed, it will usually not vibrate according to a single harmonic. The following Fourier series is a linear combination that satisfies Eq. (7.75) and can represent any possible motion of the string:

$$y(x, t) = \sum_{n=1}^{\infty} \sin\left(\frac{n\pi x}{L}\right)\left[a_n \cos\left(\frac{n\pi ct}{L}\right) + b_n \cos\left(\frac{n\pi ct}{L}\right)\right]. \tag{7.95}$$

The fact that a linear combination of solutions can be a solution to the equation is an example of the *principle of superposition*. We can regard the linear combination as a physical representation of constructive and destructive interference of the different harmonics. The strengths of the different harmonics are represented by the values of the coefficients $a_n$ and $b_n$. Different musical instruments have different relative strengths of different harmonics.

---

### Problem 7.16

Show that the function in Eq. (7.95) satisfies Eq. (7.75).

---

For a string of finite length with fixed ends, only standing waves can occur. For an infinitely long string, *traveling waves* can also occur. The following is a traveling wave:

$$y(x, t) = A \sin[k(x - ct)]. \tag{7.96}$$

---

### Problem 7.17

Show that the function of Eq. (7.96) satisfies Eq. (7.75).

---

The function in Eq. (7.96) is not a solution in which the variables are separated. However, using Eq. (14) of Appendix 2, we can show that

$$A \sin[k(x - ct)] = A[\sin(kx)\cos(kct) - \cos(kx)\sin(kct)]. \tag{7.97}$$

We can show that the function of Eq. (7.96) is a traveling wave by showing that a node in the wave moves along the string. When $t = 0$, there is a node at $x = 0$. At a later time, this node is located at a value of s such that $k(x - ct)$ is still equal to zero. At a time $t$, $x = ct$ at the node, so that the speed of the wave is equal to $c$.

---

### Problem 7.18

Find the speed of propagation of a traveling wave in an infinite string with the same mass per unit length and the same tension force as the violin string in Problem 7.15.

---

## SECTION 7.7. SOLUTION OF DIFFERENTIAL EQUATIONS WITH LAPLACE TRANSFORMS

Some differential equations can be solved by taking the Laplace transform of the equation, applying some of the theorems presented in Section 6.5 to obtain an expression for the Laplace transform of the unknown function, and then finding the inverse transform. We illustrate this procedure with the differential equation for the damped harmonic oscillator,[5] Eq. (7.40), which can be rewritten

$$\left(\frac{d^2z}{dt^2}\right) + \frac{\zeta}{m}\frac{dz}{dt} + \frac{k}{m}z = z'' + \frac{\zeta}{m}z' + \frac{k}{m}z = 0. \tag{7.98}$$

We introduce the notation $z''$ for the second derivative $d^2z/dt^2$ and $z'$ for the first derivative $dz/dt$. We take the Laplace transform of this equation, applying Eq. (6.67) and the $n = 2$ version of Eq. (6.68), to express the Laplace transforms of the first and second derivatives. We let $Z$ be the Laplace transform of $z$,

$$s^2 Z - sz(0) - z'(0) + \frac{\zeta}{m}(sZ - z(0)) + \frac{k}{m}Z = 0. \tag{7.99}$$

This algebraic equation is solved for $Z$:

$$Z = \frac{sz(0) + z'(0) + (\zeta/m)z(0)}{s^2 + (\zeta/m)s + k/m}. \tag{7.100}$$

When we find the inverse transform of this function, we will have our answer. We must carry out some algebraic manipulations before we can find the inverse transforms in Table 6.1. In order to match an expression for a transform in Table 6.1, we complete the square in the denominator:

$$Z = \frac{z(0)(s + \zeta/2m) + \zeta/2m + z'(0)}{(s^2 + \zeta/2m)^2 - \zeta^2/4m^2 + k/m}. \tag{7.101}$$

---

[5] Erwin Kreyszig, *Advanced Engineering Mathematics*, 3rd ed., pp. 156–157, Wiley, New York, 1972.

We have also expressed the numerator in terms of the quantity which is squared in the denominator. We now make the substitutions,

$$a = \frac{\zeta}{2m} \quad \text{and} \quad \omega^2 = \frac{k}{m} - \frac{\zeta^2}{4m^2}$$

so that Eq. (7.101) can be written

$$Z = \frac{z(0)(s + a)}{(s^2 + a)^2 + \omega^2} + \frac{z(0)a + z'(0)}{(s^2 + a)^2 + \omega^2}. \tag{7.102}$$

We assume the case of less than critical damping, so that $\omega^2$ is positive.

From Table 6.1, we have the inverse transforms,

$$\mathcal{L}^{-1}\left\{\frac{s}{s^2 + \omega^2}\right\} = \cos(\omega t)$$

$$\mathcal{L}^{-1}\left\{\frac{s}{s^2 + \omega^2}\right\} = \sin(\omega t)$$

and from the theorem of Eq. (6.66)

$$\mathcal{L}^{-1}\{e^{-at} f(t)\} = F(s + a) \tag{7.103}$$

so that

$$z(t) = \left[z(0)\cos(\omega t) + \frac{z(0)a + z'(0)}{\omega}\sin(\omega t)\right]e^{-at}. \tag{7.104}$$

---

### Problem 7.19

Substitute the function of Eq. (7.104) into Eq. (7.98) to show that it satisfies the equation.

---

### Problem 7.20

Obtain the solution of Eq. (7.98) in the case of critical damping, using Laplace transforms.

---

Our discussion of the Laplace transform method for solving differential equations suffices only to introduce the method. Chapter 4 of the book by Kreyszig or the book by Duffy is recommended for further study.

## SECTION 7.8. NUMERICAL SOLUTION OF DIFFERENTIAL EQUATIONS

Many differential equations occur for which no solution can be obtained with pencil and paper. A lot of these occur in the study of chemical reaction rates. With the use of programmable computers, it is now easy to obtain numerical approximations to the solutions of these equations to any desired degree of accuracy.

## Euler's Method

This is a method that is extremely simple to understand and implement. However, it is not very accurate and is not used in actual applications. Consider a differential equation for a variable $x$ as a function of time that can be schematically represented by

$$\frac{dx}{dt} = f(x, t) \tag{7.105}$$

with the initial condition that $x(0) = x_0$, a known value. A formal solution can be written

$$x(t') = x_0 + \int_0^{t'} f(x, t)\, dt. \tag{7.106}$$

Like any other formal solution, this cannot be used in practice, since the variable $x$ in the integrand function depends on $t$ in some way that we don't yet know.

Euler's method assumes that if $t'$ is small enough, the integrand function in Eq. (7.106) can be replaced by its value at the beginning of the integration. We replace $t'$ by the symbol $\Delta t$ and write

$$x(\Delta t) \approx x_0 + \int_0^{\Delta t} f(x_0, 0)\, dt = x_0 + \Delta t f(x_0, 0). \tag{7.107}$$

A small value of $\Delta t$ is chosen, and this process is repeated until the desired value of $t'$ is reached.

Let $x_i$ be the value of $x$ obtained after carrying out the process $i$ times, and let $t_i$ equal $i \Delta t$, the value of $t$ after carrying out the process $i$ times. We write

$$x_{i+1} \approx x_i + \Delta t f(x_i, t_i). \tag{7.108}$$

Euler's method is analogous to approximating an integral by the area under a bar graph, except that the height of each bar is obtained by starting with the approximate height of the previous bar and using the known slope of the tangent line.

---

## Example 7.6

The differential equation for a first-order chemical reaction without back reaction is

$$\frac{dc}{dt} = -kc,$$

where $c$ is the concentration of the single reactant and $k$ is the rate constant.

a. Write a computer program in the BASIC language to carry out Euler's method for this differential equation.

b. Run the program for initial concentration $1.000 \text{ mol L}^{-1}$, $k = 1.000 \text{ s}^{-1}$, for a time of 2.000 s and a value of $\Delta t$ equal to 0.00100 s.

---

### Solution

a. See Chapter 11 for information about writing and running a BASIC program using the processor TrueBASIC. Since the differential equation can be solved exactly (see Example 7.3), the program compares the numerical result with the exact solution.

```
100 REM PROGRAM TO INTEGRATE THE 1ST-ORDER KINETIC
EQUATION
110 REM BYEULER'S METHOD
120 OPTION NOLET
200 PRINT "INITIAL CONCENTRATION?"
210 INPUT C0
220 PRINT "RATE CONSTANT VALUE?"
230 INPUT K
240 PRINT "TIME OF REACTION (SECONDS)?"
250 INPUT T9
260 PRINT "TIME STEP VALUE (SECONDS)?"
270 INPUT H
280 PRINT "TIME INTERVAL FOR PRINTING CONCENTRATION?"
290 INPUT TI
300 C = C0
310 M = T9/TI
330 N = TI/H
340 FOR J = 1 TO M
400 FOR I = 1 TO N
410 F = -H*K*C
450 C = C + F
460 NEXT I
470 TP = TI*J
480 PRINT "TIME = ";TP; "CONCENTRATION = ";C
490 NEXT J
500 PRINT "FINAL CONCENTRATION = ";C
510 REM CALCULATE ACTUAL FINAL CONCENTRATION
520 CE = C0*EXP(-K*T9)
530 PRINT "EXACT FINAL CONCENTRATION = ";CE
999 END
```

b. Here is the output from running the program:

```
INITIAL CONCENTRATION?
? 1
RATE CONSTANT VALUE?
? 1
TIME OF REACTION (SECONDS)?
? 2
TIME STEP VALUE (SECONDS)?
? .001
TIME INTERVAL FOR PRINTING CONCENTRATION?
? .2
TIME = .2 CONCENTRATION = .818649
TIME = .4 CONCENTRATION = .670186
TIME = .6 CONCENTRATION = .548647
TIME = .8 CONCENTRATION = .449149
TIME = 1. CONCENTRATION = .367695
TIME = 1.2 CONCENTRATION = .301013
```

```
TIME = 1.4 CONCENTRATION = .246424
TIME = 1.6 CONCENTRATION = .201735
TIME = 1.8 CONCENTRATION =.16515
TIME = 2. CONCENTRATION = .1352
FINAL CONCENTRATION = .1352
EXACT FINAL CONCENTRATION = .135335
```

### Problem 7.21

Run the program of Example 7.6 with different values of $\Delta t$ (called H in the program) to see how the accuracy of the method varies with $\Delta t$.

## The Runge–Kutta Method

Since Euler's method is not accurate except for very small values of $\Delta t$, more sophisticated methods have been devised. One such widely used method is the Runge–Kutta method, which is somewhat analogous to using Simpson's method for a numerical integration, as discussed in Section 4.6.[6]

In the Runge–Kutta method, Eq. (7.108) is replaced by

$$x_{i+1} \approx x_i + \frac{1}{6}(F_1 + 2F_2 + 2F_3 + F_4), \tag{7.109}$$

where

$$F_1 = \Delta t f(x_i, t_i) \tag{7.110a}$$

$$F_2 = \Delta t f\left(x_i + \frac{1}{2}F_1, t_i + \frac{\Delta t}{2}\right) \tag{7.110b}$$

$$F_3 = \Delta t f\left(x_i + \frac{1}{2}F_2, t_i + \frac{\Delta t}{2}\right) \tag{7.110c}$$

$$F_4 = \Delta t f(x_i + F_3, t_i + \Delta t). \tag{7.110d}$$

We do not discuss the derivation of this method, which is discussed in the book by Burden, Faires, and Reynolds and the book by Hornbeck listed at the end of the chapter.

### Problem 7.22

Run the following BASIC program for various values of $\Delta t$ (called H in the program) and compare its performance with Euler's method.

```
100 REM PROGRAM TO INTEGRATE THE 1ST-ORDER KINETIC
EQUATION
110 REM BY THE RUNGE-KUTTA METHOD
120 OPTION NOLET
200 PRINT "INITIAL CONCENTRATION?"
```

---

[6]See the book by Burden, Faires, and Reynolds and the book by Hornbeck listed at the end of the chapter.

```
210 INPUT C0
220 PRINT "RATE CONSTANT VALUE?"
230 INPUT K
240 PRINT "TIME OF REACTION (SECONDS)?"
250 INPUT T9
260 PRINT "TIME STEP VALUE (SECONDS)?"
270 INPUT H
280 PRINT "TIME INTERVAL FOR PRINTING CONCENTRATION?"
290 INPUT TI
300 C = C0
310 M = T9/TI
330 N = TI/H
340 FOR J = 1 TO M
400 FOR I = 1 TO N
410 F1 = -H*K*C
420 F2 = -H*K*C - H*K*F1/2
430 F3 = -H*K*C - H*K*F2/2
440 F4 = -H*K*C - H*K*F3
450 C = C + (F1 + 2*F2 + 2*F3 + F4)/6
460 NEXT I
470 TP = TI*J
480 PRINT "TIME = ";TP; "CONCENTRATION = ";C
490 NEXT J
500 PRINT "FINAL CONCENTRATION = ";C
510 REM CALCULATE ACTUAL FINAL CONCENTRATION
520 CE = C0*EXP(-K*T9)
530 PRINT "EXACT FINAL CONCENTRATION = ";CE
999 END
```

---

There are also other numerical methods for solving differential equations, which we do not discuss. The numerical methods can be extended to sets of simultaneous differential equations such as occur in the analysis of chemical reaction mechanisms. Many of these sets of equations have a property called "stiffness" that makes them difficult to treat numerically. Techniques have been devised to handle this problem, which is beyond the scope of this book.[7]

## SUMMARY OF THE CHAPTER

A differential equation contains one or more derivatives, and its solution is a function that satisfies the equation. Classical equations of motion are differential equations based on Newton's laws of motion that when solved give the positions of particles as a function of time. We have presented the solution to several of these. These differential equations are deterministic. That is, given the equation of motion for a

---

[7]C. J. Aro, *Comput. Phys. Comm.* **97**, 304(1996).

given system and the initial conditions (position and velocity of every particle at some initial time), the positions and velocities are determined for all times.

Many homogeneous and inhomogeneous linear differential equations with constant coefficients can be solved by routine methods, which we discussed. An exact differential equation can also be solved in a routine way. Such an equation consists of an exact differential set equal to zero. Since a line integral of an exact differential (the differential of a function) is equal to the value of the function at the end of the integration minus the value of the function at the beginning of the integration, a line integration to a general ending point provides the formula for the function, solving the equation. Some inexact differential equations can be converted into exact equations by use of an integrating factor, and solution of the exact equation provides a solution to the inexact equation.

Many partial differential equations arising in physical problems can be solved by separation of variables. In this procedure, a trial solution consisting of factors depending on one variable each is introduced, and the resulting equation is manipulated until the variables occur only in separate terms. Setting these terms equal to constants gives one ordinary differential equation for each variable.

Some ordinary differential equations can be solved by using some theorems of Laplace transforms which transform a differential equation into an algebraic equation. If this equation can be solved for the transform of the unknown function, and if the inverse transform can be found, the equation is solved.

If a mathematical method for solving a differential equation cannot be found, numerical methods exist for generating numerical solutions to any desired degree of accuracy. Euler's method and the Runge–Kutta method were presented.

## ADDITIONAL READING

Ralph Palmer Agnew, *Differential Equations*, 2nd ed., McGraw–Hill, New York, 1960. This book puts some emphasis on underlying theory and contains interesting "remarks" which clarify some points.

Paul W. Berg and James L. McGregor, *Elementary Partial Differential Equations*, Holden–Day, San Francisco, 1966. This book discusses a number of physical problems, including classical wave equations and heat-flow equations.

Richard L. Burden, J. Douglas Faires, and Albert C. Reynolds, *Numerical Analysis*, 2nd ed., PWS–Kent, Boston, 1978. This is a standard text for an undergraduate course in numerical analysis. It includes a lot of useful material, including various algorithms for the numerical solution of differential equations.

Earl A. Coddington, *An Introduction to Ordinary Differential Equations*, Prentice–Hall, Englewood Cliffs, NJ, 1961. A general introductory treatment that assumes only a knowledge of elementary calculus.

Alice B. Dickinson, *Differential Equations, Theory and Use in Time and Motion*, Addison–Wesley, Reading, MA, 1972. This book puts special emphasis on the construction of differential equations for models of physical systems. It has a large chapter on solutions which are series.

Dean G. Duffy, *Transform Methods for Solving Partial Differential Equations*, CRC Press, Boca Raton, 1994. This book is a textbook for engineering students and focuses on practical applications.

Lester R. Ford, *Differential Equations*, 2nd ed., McGraw–Hill, New York, 1955. This book provides "cookbook" methods for solving quite a few kinds of differential equations.

Robert W. Hornbeck, *Numerical Methods*, Quantum Publishers, New York, 1975. This is a useful book that contains a lot of numerical methods, including the numerical solution of differential equations. It includes error analyses for the methods.

E. Kamke, *Differentialgleichungen—Lösungsmethoden und Lösungen (Differential Equations—Methods of Solution and Solutions)*, 2nd ed.,Vol. I, Akad. Verlagsgesellschaft, Leipzig, 1943 (reprinted by Edwards, Ann Arbor, MI, 1945). If you read German, you can read a summary of results, but even if you do not, you might be able to use the large compilation of special equations and their solutions which is included.

Wilfred Kaplan, *Ordinary Differential Equations*, Addison–Wesley, Reading, MA, 1958. An introductory book with numerous applications.

Erwin Kreyszig, *Advanced Engineering Mathematics*, 3rd ed., Wiley, New York, 1972. This book emphasizes practical applications of mathematics to engineering problems, which require much of the same mathematics as problems of physical chemistry. Its discussion of Laplace transforms is clear and is focused on applications rather than theory.

Max Morris and Orley E. Brown, *Differential Equations*, 4th ed., Prentice–Hall, Englewoods Cliffs, NJ, 1964. This book is a standard text for a beginning course in differential equations and is a fairly clear introduction to the subject. It contains a number of examples of differential equations in physical problems.

David L. Powers, *Boundary Value Problems*, Academic Press, New York, 1972. This book discusses some heat-flow equations and wave equations.

## ADDITIONAL PROBLEMS

### 7.23

An object moves through a fluid in the $x$ direction. The only force acting on the object is a frictional force that is proportional to the negative of the velocity:

$$F_x = -\zeta v_x = -\zeta\left(\frac{dx}{dt}\right).$$

Write the equation of motion of the object. Find the general solution to this equation, and obtain the particular solution that applies if $x(0) = 0$ and $v_x(0) = v_0 = $ constant. Draw a graph of the position as a function of time.

### 7.24

A particle moves along the $z$ axis. It is acted upon by a constant gravitational force equal to $-\mathbf{k}mg$, where $\mathbf{k}$ is the unit vector in the $z$ direction. It is also acted on by a frictional force given by

$$\mathbf{F}_f = -\mathbf{k}\left(\frac{dz}{dt}\right)\zeta.$$

Find the equation of motion and obtain a general solution. Find $z$ as a function of time if $z(0) = 0$ and $v_z(0) = 0$. Draw a graph of $z$ as a function of time.

### 7.25

An object sliding on a solid surface experiences a frictional force that is constant and in the opposite direction to the velocity if the particle is moving, and is zero it is not moving. Find the position of the particle as a function of time if it moves only in the $x$ direction and the initial position is $x(0) = 0$ and the initial velocity is $v_x(0) = v_0 =$ constant. Proceed as though the constant force were present at all times and then cut the solution off at the point at which the velocity vanishes. That is, just say that the particle is fixed after this time.

### 7.26

A tank contains a solution that is rapidly stirred, so that it remains uniform at all times. A solution of the same solute is flowing into the tank at a fixed rate of flow, and an overflow pipe allows solution from the tank to flow out at the same rate. If the solution flowing in has a fixed concentration that is different from the initial concentration in the tank, write and solve the differential equation that governs the number of moles of solute in the tank. The inlet pipe allows $A$ moles per hour to flow in and the overflow pipe allows $Bn$ moles per hour to flow out, where A and B are constants and $n$ is the number of moles of solute in the tank. Find the values of A and $B$ that correspond to a volume in the tank of 100 liters, an input of 1.000 L hour$^{-1}$ of a solution with 1.000 mol L$^{-1}$, and an output of 1.000 L hour$^{-1}$ of the solution in the tank. Find the concentration in the tank after 5.00 hour, if the initial concentration is zero.

### 7.27

An $n$th-order chemical reaction with one reactant obeys the differential equation

$$\frac{dc}{dt} = -kc^n,$$

where $c$ is the concentration of the reactant and $k$ is a constant. Solve this differential equation by separation of variables. If the initial concentration is $c_0$ moles per liter, find an expression for the time required for half of the reactant to react.

### 7.28

Find the solution to the differential equation

$$\left(\frac{d^3y}{dx^3}\right) - 2\left(\frac{d^2y}{dx^2}\right) - \left(\frac{dy}{dx}\right) + 2y = -xe^x.$$

### 7.29

Test the following equations for exactness, and solve those which are exact:

a. $(x^2 + xy + y^2)\,dx + (4x^2 - 2xy + 3y^2)\,dy = 0$
b. $ye^x dx + e^x\,dy = 0$
c. $[2xy - \cos(x)]\,dx + (x^2 - 1)\,dy = 0.$

**7.30**

Solve the differential equation

$$\left(\frac{dy}{dx}\right) + y\cos(x) = e^{-\sin(x)}.$$

**7.31**

Find a particular solution of

$$\left(\frac{d^2y}{dx^2}\right) - 4y = 2e^{3x} + \sin(x).$$

**7.32**

Radioactive nuclei decay according to the same differential equation which governs first-order chemical reactions, Eq. (7.63). In living matter, the isotope $^{14}C$ is continually replaced as it decays, but it decays without replacement beginning with the death of the organism. The half-life of the isotope (the time required for half of an initial sample to decay) is 5730 years. If a sample of charcoal from an archaeological specimen exhibits 0.97 disintegrations of $^{14}C$ per gram of carbon per minute and wood recently taken from a living tree exhibits 15.3 disintegrations of $^{14}C$ per gram of carbon per minute, estimate the age of the charcoal.

**7.33**

A pendulum of length $L$ oscillates in a vertical plane. Assuming that the mass of the pendulum is all concentrated at the end of the pendulum, show that it obeys the differential equation

$$L\left(\frac{d^2\theta}{dt^2}\right) = g\sin(\theta),$$

where $g$ is the acceleration due to gravity and $\theta$ the angle between the pendulum and the vertical. This equation cannot be solved exactly. For small oscillations such that

$$\sin(\theta) \approx \theta$$

find the solution to the equation. What is the period of the motion? What is the frequency? Evaluate these quantities if $L = 1.000$ m and if $L = 10.000$ m.

**7.34**

Obtain the solution for Eqs. (7.54) and (7.55) for the forced harmonic oscillator using Laplace transforms.

**7.35**

    a. Write and run a BASIC program to solve the differential equation of Problem 7.12 numerically, using Euler's method. Compare your numerical answer with the exact solution for your choice of initial concentrations and rate constant values.

    b. Repeat part a using the Runge–Kutta method.

# 8

# OPERATORS, MATRICES, AND GROUP THEORY

## Preview

A mathematical operator is a symbol standing for carrying out a mathematical operation or a set of operations. Operators are important in quantum mechanics, since each mechanical variable has a mathematical operator corresponding to it. Operator symbols can be manipulated symbolically in a way similar to the algebra of ordinary variables, but according to a different set of rules. An important difference between ordinary algebra and operator algebra is that multiplication of two operators is not necessarily commutative, so that if $\hat{A}$ and $\hat{B}$ are two operators, $\hat{A}\hat{B} \neq \hat{B}\hat{A}$ can occur.

A matrix is a list of quantities, arranged in rows and columns. We will introduce matrix algebra, which is a branch of algebra that has rules that are different from the algebra of ordinary variables, and which has similarities with operator algebra.

A group is a set of elements with defined properties, including a single operation which is not necessarily commutative. The elements of a group can represent symmetry operators, and group theory can provide useful information about quantum-mechanical wave functions for symmetrical molecules, spectroscopic transitions, etc.

## Principal Facts and Ideas

1. An operator is a symbol that stands for a mathematical operation. If an operator $\hat{A}$ operates on a function $f$ the result is a new function,
   $g : \hat{A}f = g$.
2. Operator algebra manipulates operator symbols according to rules slightly different from those of ordinary algebra.

**237**

3. An eigenvalue equation has the form $\hat{A}f = af$ where $f$ is an eigenfunction and $a$ is an eigenvalue.

4. Symmetry operators move points in space relative to a symmetry element. A symmetry operator which "belongs" to the nuclear framework of a molecule moves each nucleus to the former position of a nucleus of the same kind.

5. Symmetry operators can operate on functions as well as on points and can have eigenfunctions with eigenvalues equal to 1 or to $-1$. An electronic wave function of a molecule can be an eigenfunction of the symmetry operators which belong to the nuclear framework of a molecule.

6. Matrices can be manipulated according to the rules of matrix algebra, which are similar to the rules of ordinary algebra. One exception is that matrix multiplication is not necessarily commutative: if **A** and **B** are matrices, $\mathbf{AB} \neq \mathbf{BA}$ can occur.

7. The inverse of a matrix obeys $\mathbf{A}^{-1}\mathbf{A} = \mathbf{AA}^{-1} = \mathbf{E}$ where **E** is the identity matrix. The inverse of a given matrix can be obtained by the Gauss–Jordan elimination procedure.

8. A group is a set of elements obeying certain conditions, with a single operation combining two elements to give a third element of the group. This operation is called multiplication and is noncommutative.

9. The symmetry operators belonging to an symmetrical object such as the equilibrium nuclear framework of a molecule form a group.

10. A set of matrices obeying the same multiplication table as a group is a representation of the group.

11. Various theorems of group theory make it useful in studying the symmetry properties of molecules in quantum chemistry.

## Objectives

After studying this chapter, you should be able to:

1. perform the elementary operations of operator algebra;
2. identify and use symmetry operators associated with a symmetrical molecule;
3. perform the elementary operations of matrix algebra, including matrix multiplication and finding the inverse of a matrix;
4. identify a group of symmetry operators and construct a multiplication table for the group.

## SECTION 8.1. OPERATORS AND OPERATOR ALGEBRA

A mathematical operator is a symbol that stands for carrying out a mathematical operation on some function. For example, we can use the symbol $d/dx$ or the symbol $\hat{D}_x$ to stand for the operation of differentiating with respect to $x$. We will often assign a symbol to an operator that consists of a letter with a caret ($\hat{\ }$) over it. When an operator operates on a function, the result will generally be another function.

There are several types of operators that we will consider. Multiplication operators are operators that stand for multiplying a function either by a constant or by a specified function. The identity operator $\hat{E}$ stands for multiplication by unity,[1] in which case the resulting function is identical with the original function. Derivative operators stand for differentiating a function one or more times with respect to one or more independent variables. An operator can correspond to carrying out more than one operation, such as multiplication by a function followed by a differentiation.

---

### Example 8.1

Let the operator $\hat{A}$ be given by

$$\hat{A} = \hat{x} + \frac{d}{dx}. \tag{8.1}$$

Find $\hat{A}f$ if $f = a\sin(bx)$, where $a$ and $b$ are constants.

---

### Solution

$$\hat{A}a\sin(bx) = xa\sin(bx) + ab\cos(bx). \tag{8.2}$$

---

If the result of operating on a function with an operator is a function that is proportional to the original function, the function is called an *eigenfunction* of that operator, and the proportionality constant is called an *eigenvalue*.[2] If

$$\boxed{\hat{A}f = af} \tag{8.3}$$

then $f$ is an eigenfunction of $\hat{A}$ and $a$ is the eigenvalue corresponding to that eigenfunction. An equation like Eq. (8.3) is called an *eigenvalue equation*. The time-independent Schrödinger equation of quantum mechanics is an eigenvalue equation, and other eigenvalue equations are important in quantum mechanics.

---

### Example 8.2

Find the eigenfunctions and corresponding eigenvalues for the operator $d^2/dx^2$.

---

### Solution

We need to find a function $f(x)$ and a constant $a$ such that

$$\frac{d^2 f}{dx^2} = af.$$

This is a differential equation which is solved by the method of Section 7.3. The

---

[1]The letter $E$ stands for the German *Einheit*, meaning "unity."

[2]The word "eigenvalue" is a partial translation of the German *Eigenwert,* sometimes translated as "characteristic value." The word eigenfuction is a partial translation of the German *Eigenfunktion,* sometimes translated as "characteristic function."

general solution is

$$f(x) = A \exp(\sqrt{a}x) + B \exp(-\sqrt{a}x),$$

where $A$ and $B$ are constants. Since no boundary conditions were stated, the eigenvalue $a$ can take on any value, as can the constants $A$ and $B$.

---

### Problem 8.1

Find the eigenfunctions of the operator $-i\dfrac{d}{dx}$, where $i = \sqrt{-1}$.

---

## Mathematical Operations on Operators

Although a mathematical operator is a symbol that stands for the carrying out of an operation, we can define a kind of algebra in which we manipulate these symbols much as we manipulate ordinary variables and numbers.

We define the *sum of two operators* by

$$(\hat{A} + \hat{B})f = \hat{A}f + \hat{B}f, \tag{8.4}$$

where $\hat{A}$ and $\hat{B}$ are two operators and where $f$ is some function on which $\hat{A}$ and $\hat{B}$ can operate.

The *product of two operators* is defined as the successive operation of the operators, with the one on the right operating first. If

$$\hat{C} = \hat{A}\hat{B} \tag{8.5}$$

then

$$\hat{C}f = \hat{A}(\hat{B}f). \tag{8.6}$$

The result of $\hat{B}$ operating on $f$ is in turn operated on by $\hat{A}$ and the result is said to equal the result of operating on $f$ with the product $\hat{A}\hat{B}$. It is important that an operator operates on everything to its right in the same term.

Equation (8.5) is an *operator equation*. The two sides of the equation are equal in the sense that if each is applied to an arbitrary function, the two results are the same.

---

### Example 8.3

Find the operator equal to the operator product $\dfrac{d}{dx}\hat{x}$.

---

### Solution

We take an arbitrary differentiable function $f = f(x)$ and apply the operator product to it,

$$\frac{d}{dx}\hat{x}f = \hat{x}\frac{df}{dx} + f\frac{dx}{dx} = \left(\hat{x}\frac{d}{dx} + \hat{E}\right)f,$$

where $\hat{E}$ is the identity operator, the operator for multiplying by unity (or doing nothing). We can write the operator equation that is equivalent to this equation:

$$\frac{d}{dx}\hat{x} = \hat{x}\frac{d}{dx} + \hat{E}.$$

---

**Problem 8.2**

Find the operator equal to the operator product $\dfrac{d^2}{dx^2}\hat{x}$.

---

The difference of two operators is given by

$$(\hat{A} - \hat{B}) = \hat{A} + (-\hat{E})\hat{B}. \tag{8.7}$$

We now have an operator algebra in which we carry out the operations of addition and multiplication on the operators themselves. These operations have the following properties: Operator multiplication is *associative*. This means that if $\hat{A}$, $\hat{B}$, and $\hat{C}$ are operators, then

$$\boxed{(\hat{A}\hat{B})\hat{C} = \hat{A}(\hat{B}\hat{C})}. \tag{8.8}$$

Operator multiplication and addition are *distributive*. This means that if $\hat{A}$, $\hat{B}$, and $\hat{C}$ are operators

$$\boxed{\hat{A}(\hat{B} + \hat{C}) = \hat{A}\hat{B} + \hat{A}\hat{C}}. \tag{8.9}$$

Operator multiplication is not necessarily *commutative*. This means that in some cases

$$\boxed{\hat{A}\hat{B} \neq \hat{B}\hat{A} \quad \text{(possible)}}. \tag{8.10}$$

If the operator $\hat{A}\hat{B}$ is equal to the operator $\hat{B}\hat{A}$, then $\hat{A}$ and $\hat{B}$ are said to *commute*. The *commutator* of $\hat{A}$ and $\hat{B}$ is denoted by $[\hat{A}, \hat{B}]$ and defined by

$$\boxed{[\hat{A}, \hat{B}] = \hat{A}\hat{B} - \hat{B}\hat{A} \quad \text{(definition of the commutator)}}. \tag{8.11}$$

If $\hat{A}$ and $\hat{B}$ commute, then $[\hat{A}, \hat{B}] = \hat{0}$, where $\hat{0}$ is the *null operator*, equivalent to multiplying by zero.

---

**Example 8.4**

Find the commutator $\left[\dfrac{d}{dx}, \hat{x}\right]$.

**Solution**

We apply the commutator to an arbitrary function $f(x)$:

$$\left[\frac{d}{dx}, \hat{x}\right] f = \frac{d}{dx}(xf) - x\frac{df}{dx} = x\frac{df}{dx} + f - x\frac{df}{dx} = f. \qquad (8.12a)$$

Therefore,

$$\left[\frac{d}{dx}, \hat{x}\right] = \hat{E} = \hat{1}, \qquad (8.12b)$$

where the symbol $\hat{1}$ stands for multiplication by unity and is the same thing as $\hat{E}$.

---

**Problem 8.3**

Find the commutator $[\hat{x}^2, d^2/dx^2]$.

---

Here are a few facts that will predict in almost all cases whether two operators will commute:

1. An operator containing a multiplication by a function of $x$ and one containing $d/dx$ will not generally commute.
2. Two multiplication operators commute. If $g$ and $h$ are functions of the same or different independent variables or are constants, then

$$[\hat{g}, \hat{h}] = 0. \qquad (8.13)$$

3. Operators acting on different independent variables commute. For example,

$$\left[\hat{x}\frac{d}{dx}, \frac{d}{dy}\right] = 0. \qquad (8.14)$$

4. An operator for multiplication by a constant commutes with any other operator.

---

**Problem 8.4**

Show that Eq. (8.14) is correct, and that statement (4) is correct.

---

Since we have defined the product of two operators, we have a definition for the *powers of an operator*. An operator raised to the $n$th power is the operator for $n$ successive applications of the original operator:

$$\boxed{\hat{A}^n = \hat{A}\hat{A}\hat{A}\cdots\hat{A} \qquad (n \text{ factors})}.$$

---

**Example 8.5**

If the operator $\hat{A}$ is $\hat{x} + \dfrac{d}{dx}$, find $\hat{A}^3$.

## Solution

$$\hat{A}^3 = \left(\hat{x} + \frac{d}{dx}\right)\left(\hat{x} + \frac{d}{dx}\right)\left(\hat{x} + \frac{d}{dx}\right)$$

$$= \left(\hat{x} + \frac{d}{dx}\right)\left(\hat{x}^2 + \frac{d}{dx}\hat{x} + \hat{x}\frac{d}{dx} + \frac{d^2}{dx^2}\right) \tag{8.15}$$

$$= \hat{x}^3 + \hat{x}\frac{d}{dx}\hat{x} + \hat{x}^2\frac{d}{dx} + \hat{x}\frac{d^2}{dx^2} + \frac{d}{dx}\hat{x}^2 + \frac{d^2}{dx^2}\hat{x} + \frac{d}{dx}\hat{x}\frac{d}{dx} + \frac{d^3}{dx^3}.$$

The order of the factors in each term must be maintained because the two terms in the operator do not commute with each other.

## Problem 8.5

    a. For the operator $\hat{A}$ in Example 8.5, find $\hat{A}^3 f$ if $f(x) = \sin(ax)$.
    b. Find an expression for $\hat{B}^2$ if $\hat{B} = \hat{x}(d^2/dx^2)$ and find $\hat{B}^2 f$ if $f = bx^4$.

We do not define the operation of dividing by an operator. However, we define the *inverse of an operator* as that operator which "undoes" what the first operator does. The inverse of $\hat{A}$ is denoted by $\hat{A}^{-1}$, and

$$\boxed{\hat{A}^{-1}\hat{A} = \hat{E}} \tag{8.16}$$

$$\boxed{\hat{A}^{-1}\hat{A}f = \hat{E}f = f} . \tag{8.17}$$

Not all operators possess inverses. For example, there is no inverse for $\hat{0}$. The inverse of a nonzero multiplication operator is the operator for multiplication by the reciprocal of the original quantity.

Operator algebra can be used as a method of solving some differential equations.[3] A linear differential equation with constant coefficients can be written in operator notation and solved by operator algebra. The equation

$$\frac{d^2 y}{dx^2} - 3\frac{dy}{dx} + 2y = 0 \tag{8.18}$$

can be written as

$$\left(\hat{D}_x^2 - 3\hat{D}_x + 2\right)y = 0 \tag{8.19}$$

or as an operator equation

$$\left(\hat{D}_x^2 - 3\hat{D}_x + 2\right) = 0. \tag{8.20}$$

---

[3] See, for example, Max Morris and Orley Brown, *Differential Equations*, 3rd ed., pp. 86–89, Prentice–Hall, Englewood Cliffs, NJ, 1952.

Using operator algebra, we manipulate this equation as though it were an ordinary equation. We factor it to obtain

$$(\hat{D}_x - 2)(\hat{D}_x - 1) = 0. \tag{8.21}$$

The two roots are obtained from

$$\hat{D}_x - 2 = 0. \tag{8.22a}$$

and

$$\hat{D}_x - 1 = 0. \tag{8.22b}$$

Equation (8.22a) is the same as the equation

$$\frac{dy}{dx} - 2y = 0 \tag{8.23a}$$

and Eq. (8.22b) is the same as the equation

$$\frac{dy}{dx} - y = 0. \tag{8.23b}$$

The solution to Eq. (8.23a) is

$$y = e^{2x} \tag{8.24a}$$

and the solution to Eq. (8.23b) is

$$y = e^x. \tag{8.24b}$$

Since both of these must be solutions to the original equation, the general solution is

$$y = c_1 e^{2x} + c_2 e^x, \tag{8.25}$$

where $c_1$ and $c_2$ are arbitrary constants.

---

### Problem 8.6

Show that the function in Eq. (8.25) satisfies Eq. (8.18).

---

### Operators in Quantum Mechanics

One of the postulates of quantum mechanical theory is that for every mechanical quantity there is a mathematical operator. The eigenfunctions and eigenvalues of these operators play a central role in the theory. For example, the operator that corresponds to the mechanical energy is the Hamiltonian operator, and the time-independent Schrödinger equation is the eigenvalue equation for this operator.

## SECTION 8.2. SYMMETRY OPERATORS

Many common objects are said to be *symmetrical*. The most symmetrical object is a sphere, which looks just the same no matter which way it is turned. A cube, although less symmetrical than a sphere, is highly symmetrical, with 24 different orientations in which it looks the same. Many biological organisms have approximate *bilateral symmetry*, meaning that the left side looks like a mirror image of the right side. We will later define the symmetry of objects more carefully.

A *symmetry operator* acts to move a point in three-dimensional space to a new location. We will consider only *point symmetry operators*, a class of symmetry operators that do not move a point if it is at the origin. We denote the position of a point by its cartesian coordinates, keeping the cartesian coordinate axes fixed as the point moves. The action of the symmetry operator $\hat{O}$ is specified by writing

$$\hat{O}(x_1, y_1, z_1) = (x_2, y_2, z_2),$$ (8.26)

where $x_1, y_1, z_1$ are the coordinates of the original location of a point and $x_2, y_2, z_2$ are the coordinates of the location to which the operator moves the point.

Our first example of a symmetry operator is the *identity operator* $\hat{E}$, which leaves any point in its original location. We use the same symbol as for the multiplicative identity operator.

$$\hat{E}(x_1, y_1, z_1) = (x_1, y_1, z_1).$$ (8.27)

If $\mathbf{r}_1$ is the vector with components $(x_1, y_1, z_1)$, this equation can be written

$$\hat{E}\mathbf{r}_1 = \mathbf{r}_1.$$ (8.28)

The *inversion operator* is denoted by $\hat{i}$. It moves a point on a line from its original position through the origin to a location at the same distance from the origin as the original position:

$$\hat{i}(x_1, y_1, z_1) = (-x_1, -y_1, -z_1)$$ (8.29)

or

$$\hat{i}\mathbf{r}_1 = -\mathbf{r}_1.$$ (8.30)

For each symmetry operator, we define a *symmetry element*, which is a point, line, or plane relative to which the symmetry operation is performed. The symmetry element for a given symmetry operator is sometimes denoted by the same symbol as the operator, but without the caret (ˆ). For example, the symmetry element for the inversion operator is the origin. The symmetry element of any point symmetry operator must include the origin.

A *reflection operator* moves a point on a line perpendicular to a specified plane, through the plane to a location on the other side of the plane at the same distance from the plane as the original point. This motion is called *reflection through the plane*. The specified plane is the symmetry element and must pass through the origin if the operator is a point symmetry operator. There is a different reflection operator for each of the infinitely many planes passing through the origin.

The operator $\hat{\sigma}_h$ corresponds to reflection through the $x-y$ plane (the $h$ subscript stands for "horizontal"). Figure 8.1 shows the action of the $\hat{\sigma}_h$ operator.

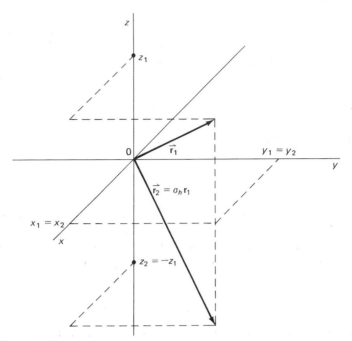

**FIGURE 8.1** The action of the reflection operator, $\hat{\sigma}_h$.

The action of $\hat{\sigma}_h$ corresponds to

$$\hat{\sigma}_h(x_1, y_1, z_1) = (x_1, y_1, -z_1)\,.$$ (8.31)

A reflection operator whose symmetry element is a vertical plane is denoted by $\hat{\sigma}_v$. You must separately specify which vertical plane is the symmetry element.

### Problem 8.7

Write an equation similar to Eq. (8.31) for the $\hat{\sigma}_v$ operator whose symmetry element is the $x-z$ plane, and one for the $\hat{\sigma}_v$ operator whose symmetry element is the $y-z$ plane.

Next we have a class of *rotation operators*. An ordinary rotation, in which a point moves as if it were part of a rigid object being rotated about an axis, is called a *proper rotation*. The axis of rotation is the symmetry element, and the action of the rotation operator is to move the point along an arc, staying at a fixed perpendicular distance from a point on the axis. The axis of rotation must pass through the origin if the rotation operator is a point symmetry operator. In addition to specifying the axis of rotation, one must specify the direction of rotation and the angle of rotation. By convention, the direction of rotation is taken as counterclockwise when viewed from the end of the axis that is designated as the positive end. We consider only angles

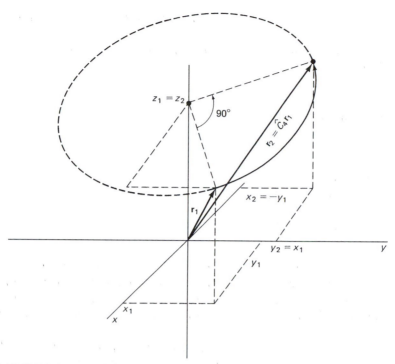

**FIGURE 8.2**   The action of a rotation operator, $\hat{C}_4$

of rotation such that $n$ applications of the rotation operator will produce exactly one complete rotation, where $n$ is a positive integer. Such a rotation operator is denoted by $\hat{C}_n$. The axis of rotation must be specified separately. For example, the operator for a rotation of $90°$ about the $z$ axis can be called $\hat{C}_4(z)$.

Figure 8.2 shows the action of the $\hat{C}_4(z)$ operator. For this operator,

$$\hat{C}_4(z)(x_1, y_1, z_1) = (-y_1, x_1, z_1) \tag{8.32}$$

so that

$$x_2 = -y_1, \qquad y_2 = x_1, \qquad z_2 = z_1. \tag{8.33}$$

---

**Problem 8.8**

Find the following:

a. $\hat{C}_4(z)(1, -4, 6)$
b. $C_2(x)(1, 2, -3)$.

---

An *improper rotation* is a proper rotation followed by a reflection through a plane perpendicular to the rotation axis. For this to be a point symmetry operation, both the rotation axis and the reflection plane must pass through the origin. The symbol for an improper rotation operator is $\hat{S}_n$, where the subscript $n$ has the same meaning as with a proper rotation. The symmetry element for an improper rotation is the axis of rotation.

The action of the operator for an improper rotation of $90°$ about the $z$ axis is given by

$$\hat{S}_4(z)(x_1, y_1, z_1) = (-y_1, x_1, -z_1). \tag{8.34}$$

An improper rotation operator $\hat{S}_2$ is the same as the inversion operator $\hat{i}$, and any $\hat{S}_1$ operator is the same as a reflection operator.

---

**Problem 8.9**

Find the following:

a. $\hat{S}_3(z)(1, 2, 3)$
b. $\hat{S}_2(y)(3, 4, 5)$.

---

We have defined all of the point symmetry operators. In addition, there are other symmetry operators and symmetry elements, such as translations, glide planes, screw axes, etc., which are useful in describing crystal lattices but which are not useful for molecules. We do not discuss these operators.

Symmetry operators can operate on a set of points as well as on a single point. For example, they can operate simultaneously on all the particles of a solid object or on all of the nuclei of a molecule, or on all of the electrons of a molecule. For example, a benzene molecule in its equilibrium conformation has the shape of a regular hexagon. If the center of mass is at the origin, the inversion operator moves each of the carbon nuclei to the original location of another carbon nucleus and each of the hydrogen nuclei to the original location of another hydrogen nucleus. It is customary to orient a symmetrical object such as a molecule so that the rotation axis of highest order (largest $n$) is on the $z$ axis.

If after a symmetry operation a molecule or other object is in the same conformation as before except for the exchange of identical particles, we say that the symmetry operators *belong* to the object. The symmetry properties of the object can be specified by listing all symmetry operators that belong to the object, or by listing their symmetry elements. Any object has a set of symmetry operators (or symmetry elements) that belong to it. An *unsymmetrical object* possesses only the identity operator, but a *symmetrical object* possesses at least one additional symmetry operator. We can describe the symmetry of a symmetrical object by listing either the symmetry operators belonging to it or the symmetry elements of these operators. It is found that the symmetry operators belonging to any rigid object form a mathematical group, to be discussed later in this chapter, and the symmetry of the object can be specified by giving the symbol assigned to the appropriate group.

A uniform spherical object is the most highly symmetrical object. If the center of the sphere is at the origin, every mirror plane, every rotation axis, every improper rotation axis, and the inversion center at the origin are symmetry elements of symmetry operators belonging to the sphere.

**Example 8.6**

List the symmetry elements of a uniform cube centered at the origin with its faces parallel to the coordinate planes.

────────

**Solution**

The symmetry elements are:

The inversion center at the origin.
Three $C_4$ axes coinciding with the coordinate axes.
Four $C_3$ axes passing through opposite corners of the cube.
Four $S_6$ axes coinciding with the $C_3$ axes.
Six $C_2$ axes connecting the midpoints of opposite edges.
Three mirror planes in the coordinate planes.
Six mirror planes passing through opposite edges.

**Problem 8.10**

a. List the symmetry elements of a right circular cylinder. The axis of the cylinder is placed on the $z$ axis and its center is at the origin. Since even an infinitesimal rotation belongs to the object, the $z$ axis is a $C_\infty$ axis.

b. List the symmetry elements of a uniform regular tetrahedron. It is possible to arrange the object so that its center is at the origin and the four corners are at alternate corners of a cube oriented as in Example 8.6.

**Example 8.7**

List the symmetry elements of the benzene molecule.

────────

**Solution**

Locate the molecule with its nuclei in their equilibrium conformation as shown in Fig. 8.3. The symmetry elements are:

The inversion center at the origin.
The $\sigma_h$ mirror plane containing the nuclei.
A $C_6$ axis and an $S_6$ axis on the $z$ axis.
Six vertical mirror planes, three through carbon nuclei and three that pass halfway between adjacent carbon nuclei.
Six $C_2$ axes located where the mirror planes intersect the $x$–$y$ plane. These are also $S_2$ axes.

Some of the symmetry elements are shown in Fig. 8.3. The symbols on the rotation axes identify them, with a hexagon labeling a sixfold axis, a square labeling a fourfold axis, etc.

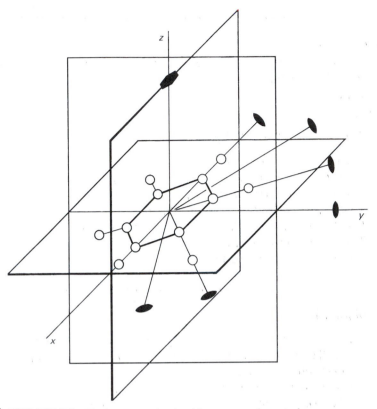

**FIGURE 8.3** The benzene molecule with symmetry elements shown.

**Problem 8.11**

List the symmetry elements for

a. $H_2O$ (bent)
b. $CH_4$ (tetrahedral)
c. $CO_2$ (linear).

The symmetry of molecules is exploited to gain information about the vibrations of the nuclei and also about the electronic wave functions. If the equilibrium nuclear configuration is symmetrical, the electronic and vibrational wave functions will have properties determined by this symmetry.

We have described the action of symmetry operators on points and on particles. We now define how they act on functions. If $\psi$ is some function and $\hat{O}$ is some symmetry operator, then

$$\hat{O}\psi = \phi, \tag{8.35}$$

where $\phi$ is a newly produced function. If $\hat{O}$ is the operator that carries a point from $(x_1, y_1, z_1)$ to $(x_2, y_2, z_2)$, we define $\phi$ to have the same value at $(x_2, y_2, z_2)$ that $\psi$

has at $(x_1, y_1, z_1)$:

$$\phi(x_2, y_2, z_2) = \psi(x_1, y_1, z_1). \tag{8.36}$$

This definition allows us to treat symmetry operators on an equal footing with other mathematical operators that operate on functions.

---

### Example 8.8

The unnormalized $2px$ wave function for the electron in a hydrogen atom is

$$\psi_{2px} = x \exp\left[\frac{-(x^2 + y^2 + z^2)^{1/2}}{2a_0}\right], \tag{8.37}$$

where $a_0$ is a constant called the Bohr radius. Find $\hat{C}_4(z)\psi_{2px}$.

---

**Solution**

The effect of this operator is given by Eq. (8.32):

$$\hat{C}_4(z)(x_1, y_1, z_1) = (x_2, y_2, z_2) = (-y_1, x_1, z_1).$$

The new function is thus

$$\phi(x_2, y_2, z_2) = \psi_{2px}(x_1, y_1, z_1) = y_2 \exp\left[\frac{-\left(y_2^2 + x_2^2 + z_2^2\right)^{1/2}}{2a_0}\right].$$

Thus

$$\hat{C}_4(z)\psi_{2px} = \psi_{2py}. \tag{8.38}$$

The symmetry operator has moved the region inside which the wave function is positive in the same way that it moves a rigid object.

Symmetry operators can have eigenfunctions, like any other operator. The result of operating on an eigenfunction of the operator is equal to a constant (the eigenvalue) times the original function. If a symmetry operator leaves a function unchanged, its eigenvalue is equal to unity. The only other possible eigenvalue for a symmetry operator is $-1$.

---

### Problem 8.12

Find $\hat{i}\psi_{2px}$. Show that $\psi_{2px}$ is an eigenfunction of the inversion operator, and find its eigenvalue.

---

The importance of symmetry operators in the study of electronic wave functions arises from the fact that two commuting operators can have a set of common eigenfunctions. In quantum mechanical theory, there is an operator corresponding to each mechanical variable. The most important quantum mechanical operator is the *Hamiltonian operator*, which corresponds to the energy. In the *Born–Oppenheimer approximation*, the electronic Hamiltonian operator operates on the coordinates of the electrons, but treats the nuclear coordinates as constants. This operator has a term in it that is the multiplication operator for multiplication by the potential energy as a function of the positions of the nuclei and electrons.

Symmetry operators are applied to molecules in two ways: they can operate on the nuclear coordinates or they can operate on the electronic wave functions. If a symmetry operator belongs to the nuclear framework it will leave the potential energy function unchanged if applied to the nuclei and not to the electrons. It will also leave the potential energy function unchanged if applied to the electrons and not to the nuclei. It will thus commute with the Born–Oppenheimer electronic Hamiltonian operator, so that electronic wave functions can occur which are eigenfunctions of the symmetry operators.

---

**Problem 8.13**

The potential energy of two charges $Q_1$ and $Q_2$ in a vacuum is

$$\mathcal{V} = \frac{Q_1 Q_2}{4\pi \varepsilon_0 r_{12}},$$

where $r_{12}$ is the distance between the charges and $\varepsilon_0$ is a constant called the permittivity of a vacuum. If a hydrogen molecule is placed so that the origin is midway between the two nuclei and the nuclei are on the $z$ axis, show that the inversion operator $\hat{\imath}$ and the reflection operator $\hat{\sigma}_h$ do not change the potential energy if applied to the electrons but not to the nuclei.

---

Full exploitation of the symmetry properties of electronic wave functions requires the use of group theory, which we briefly introduce in a later section of this chapter. However, we state some facts:

1. If a molecule has a permanent dipole moment, the dipole vector must lie along a proper rotation axis, and in a plane of symmetry.
2. A molecule with an improper rotation axis cannot be optically active.
3. A given object, such as a nuclear framework of a molecule, cannot possess a completely arbitrary collection of symmetry elements. Group theory can tell us which ones can belong together.

## SECTION 8.3. MATRIX ALGEBRA

A *matrix* is an array or list of numbers arranged in rows, columns, etc. Most of the matrices that you will encounter are two-dimensional arrays. That is, they have rows and columns. If the matrix **A** has $m$ rows and $n$ columns, it is called an $m$ by $n$ matrix and is written

$$\mathbf{A} = \begin{bmatrix} a_{11} & a_{12} & a_{13} & \cdots & a_{1n} \\ a_{21} & a_{22} & a_{23} & \cdots & a_{2n} \\ \cdots & & & & \\ \cdots & & & & \\ a_{m1} & a_{m2} & a_{m3} & \cdots & a_{mn} \end{bmatrix}. \tag{8.39}$$

The quantities that are entries in the two-dimensional list are called *elements* of the matrix. The brackets written on the left and right are included to show where the matrix starts and stops. If $m = n$, we say that the matrix is a *square matrix*. Two matrices are equal to each other if both have the same number of rows and the same number of columns and if every element of one is equal to the corresponding element of the other.

The sum of two matrices is defined by

$$\mathbf{C} = \mathbf{A} + \mathbf{B} \qquad \text{if and only if } c_{ij} = a_{ij} + b_{ij} \quad \text{for every } i \text{ and } j. \tag{8.40}$$

The product of a matrix and a scalar is defined by

$$\mathbf{B} = c\mathbf{A} \qquad \text{if and only if } b_{ij} = ca_{ij} \quad \text{for every } i \text{ and } j. \tag{8.41}$$

The product of two matrices is similar to the scalar product of two vectors. If we rename the components of two vectors $F_1, F_2, F_3, G_1, G_2,$ and $G_3$ instead of $F_x, F_y, F_z, G_x, G_y,$ and $G_z$, we can write Eq. (2.89) as

$$\mathbf{F} \cdot \mathbf{G} = F_1 G_1 + F_2 G_2 + F_3 G_3 = \sum_{i=1}^{3} F_i G_i. \tag{8.42}$$

We define matrix multiplication in a similar way. If $\mathbf{A}$, $\mathbf{B}$, and $\mathbf{C}$ are matrices such that $\mathbf{C}$ is the product $\mathbf{AB}$, we define

$$\boxed{c_{ij} = \sum_{k=1}^{n} a_{ik} b_{kj}} . \tag{8.43}$$

In this equation, $n$ is the number of columns in $\mathbf{A}$, which must equal the number of rows in the matrix $\mathbf{B}$. The matrix $\mathbf{C}$ will have as many rows as $\mathbf{A}$ and as many columns as $\mathbf{B}$.

We can think of the vector $\mathbf{F}$ in Eq. (8.42) as a matrix with one row and three columns (a *row vector*) and the vector $\mathbf{G}$ as being a matrix with three rows and one column (a *column vector*). Equation (8.42) is then a special case of Eq. (8.43):

$$F \cdot G = [F_1 \quad F_2 \quad F_3] \begin{bmatrix} G_1 \\ G_2 \\ G_3 \end{bmatrix} .$$

If $\mathbf{A}$ is a 2 by 3 matrix and $\mathbf{B}$ is a 3 by 3 matrix, we can write their matrix product as

$$\begin{bmatrix} a_{11} & a_{12} & a_{13} \\ a_{21} & a_{22} & a_{23} \end{bmatrix} \begin{bmatrix} b_{11} & b_{12} & b_{13} \\ b_{21} & b_{22} & b_{23} \\ b_{31} & b_{32} & b_{33} \end{bmatrix} = \begin{bmatrix} c_{11} & c_{12} & c_{13} \\ c_{21} & c_{22} & c_{23} \end{bmatrix} . \tag{8.44}$$

Each element in $\mathbf{C}$ is obtained in the same way as taking a scalar product of a row from $\mathbf{A}$ and a column from $\mathbf{B}$. For a particular element in $\mathbf{C}$, we take the row in $\mathbf{A}$ which is in the same position as the row in $\mathbf{C}$ containing the desired element, and the column in $\mathbf{B}$ which is in the same position as the column in $\mathbf{C}$ containing the desired element.

**Example 8.9**

Find the matrix product

$$
\begin{bmatrix} 1 & 0 & 2 \\ 0 & -1 & 1 \\ 0 & 0 & 1 \end{bmatrix}\begin{bmatrix} 0 & 0 & 2 \\ 3 & 0 & 1 \\ 1 & 1 & -1 \end{bmatrix}.
$$

**Solution**

$$
\begin{bmatrix} 1 & 0 & 2 \\ 0 & -1 & 1 \\ 0 & 0 & 1 \end{bmatrix}\begin{bmatrix} 0 & 0 & 2 \\ 3 & 0 & 1 \\ 1 & 1 & -1 \end{bmatrix}=\begin{bmatrix} 2 & 2 & 0 \\ -2 & 1 & -2 \\ 1 & 1 & -1 \end{bmatrix}.
$$

**Problem 8.14**

Find the two matrix products

$$
\begin{bmatrix} 1 & 2 & 3 \\ 3 & 2 & 1 \\ 1 & -1 & 2 \end{bmatrix}\begin{bmatrix} 1 & 3 & 2 \\ 2 & 2 & -1 \\ -2 & 1 & -1 \end{bmatrix} \quad \text{and} \quad \begin{bmatrix} 1 & 3 & 2 \\ 2 & 2 & -1 \\ -2 & -1 & -1 \end{bmatrix}\begin{bmatrix} 1 & 2 & 3 \\ 3 & 2 & 1 \\ 1 & -1 & 2 \end{bmatrix}.
$$

The left factor in one product is equal to the right factor in the other product, and vice versa. Are the two products equal?

As you can see, matrix multiplication with fairly large matrices can involve a lot of computation. Computer programs can be written to carry out the process, and such programs are typically included in BASIC processors such as TrueBASIC. It is possible to multiply two matrices that are already in the computer memory with a single statement, which we describe in Chapter 11.

If two matrices are square, as in Problem 8.14, they can be multiplied together in either order. However, the multiplication is not always commutative. It is possible that

$$\boxed{\mathbf{AB} \neq \mathbf{BA} \qquad \text{(in some cases)}}. \tag{8.45}$$

However, matrix multiplication is associative,

$$\boxed{\mathbf{A(BC)} = \mathbf{(AB)C}} \tag{8.46}$$

and matrix multiplication and addition are distributive,

$$\boxed{\mathbf{A(B + C)} = \mathbf{AB} + \mathbf{AC}}. \tag{8.47}$$

**Problem 8.15**

Show that the properties of Eqs. (8.46) and (8.47) are obeyed by the particular matrices

$$\mathbf{A} = \begin{bmatrix} 1 & 2 & 3 \\ 4 & 5 & 6 \\ 7 & 8 & 9 \end{bmatrix}, \qquad \mathbf{B} = \begin{bmatrix} 0 & 2 & 2 \\ -3 & 1 & 2 \\ 1 & -2 & -3 \end{bmatrix}, \qquad \mathbf{C} = \begin{bmatrix} 1 & 0 & 1 \\ 0 & 3 & -2 \\ 2 & 7 & -7 \end{bmatrix}.$$

The matrix multiplication that we have defined is similar to the operator multiplication defined in Section 8.1 in that both are associative and distributive but not necessarily commutative. In Section 8.1, we defined an identity operator, and we now define an identity matrix $\mathbf{E}$. We require

$$\boxed{\mathbf{EA} = \mathbf{AE} = \mathbf{A}}.$$

The fact that we require $\mathbf{E}$ to be the identity matrix when multiplied on either side of $\mathbf{A}$ requires both $\mathbf{A}$ and $\mathbf{E}$ to be square matrices. In fact, only with square matrices will we get a strict similarity between operator algebra and matrix algebra.

An identity matrix can have any number of rows and columns. It has the form

$$\mathbf{E} = \begin{bmatrix} 1 & 0 & 0 & \cdots & 0 \\ 0 & 1 & 0 & \cdots & 0 \\ 0 & 0 & 1 & \cdots & 0 \\ \cdots & \cdots & \cdots & \cdots & \cdots \\ 0 & 0 & 0 & \cdots & 1 \end{bmatrix}. \qquad (8.48)$$

The *diagonal elements* of any square matrix are those with both indices equal. The diagonal elements of $\mathbf{E}$ are all equal to 1 and are the only nonzero elements:

$$e_{ij} = \delta_{ij} = \begin{cases} 1 & \text{if } i = j \\ 0 & \text{if } i \neq j \end{cases}.$$

The quantity $\delta_{ij}$ is called the *Kronecker delta*.

**Problem 8.16**

Show by explicit multiplication that

$$\begin{bmatrix} 1 & 0 & 0 & 0 \\ 0 & 1 & 0 & 0 \\ 0 & 0 & 1 & 0 \\ 0 & 0 & 0 & 1 \end{bmatrix} \begin{bmatrix} a_{11} & a_{12} & a_{13} & a_{14} \\ a_{21} & a_{22} & a_{31} & a_{41} \\ a_{31} & a_{31} & a_{31} & a_{41} \\ a_{41} & a_{41} & a_{31} & a_{41} \end{bmatrix} = \begin{bmatrix} a_{11} & a_{12} & a_{13} & a_{14} \\ a_{21} & a_{21} & a_{31} & a_{41} \\ a_{31} & a_{31} & a_{31} & a_{41} \\ a_{41} & a_{41} & a_{31} & a_{41} \end{bmatrix}.$$

Just as in operator algebra, we do not define division by a matrix. In the operator algebra of Section 8.1, we defined an inverse operator, which undoes the effect of a given operator. We define now the *inverse of a matrix*. We denote the inverse of $\mathbf{A}$ by

$\mathbf{A}^{-1}$, so that

$$\boxed{\mathbf{A}^{-1}\mathbf{A} = \mathbf{A}\mathbf{A}^{-1} = \mathbf{E}}.\tag{8.49}$$

This multipliction of a matrix by its inverse is commutative, so that $\mathbf{A}$ is also the inverse of $\mathbf{A}^{-1}$.

From Eq. (8.43) we can write the second equality in Eq. (8.49) as

$$\sum_{k=1}^{n} a_{ik}(A^{-1})_{kj} = \delta_{ij},\tag{8.50}$$

where we write $(\mathbf{A}^{-1})_{kj}$ for the element of $\mathbf{A}^{-1}$. This is a set of simultaneous linear algebraic equations, one for each value of $i$ and each value of $j$, so that there are just enough equations to determine the elements of $\mathbf{A}^{-1}$. Finding the inverse of a matrix can involve a lot of computation, but BASIC processors such as TrueBASIC contain programs to carry out the calculation when called by a single statement, as described in Chapter 11.

One method for finding $\mathbf{A}^{-1}$ is called *Gauss–Jordan elimination*, which is a method of solving simultaneous linear algebraic equations. It consists of a set of operations to be applied to both sides of Eq. (8.50). The goal of the operations is to turn $\mathbf{A}$ into the identity matrix $\mathbf{E}$. These operations are not applied to $\mathbf{A}^{-1}$, but are applied to the right-hand side of Eq. (8.50), turning it into a matrix that we temporarily call $\mathbf{D}$. Equation (8.50) will be transformed into

$$\mathbf{E}(\mathbf{A}^{-1}) = \mathbf{D}\tag{8.51}$$

so that $\mathbf{A}^{-1}$ will be the same matrix as $\mathbf{D}$.

Equation (8.50) represents a matrix product, and is of the same form as Eq. (8.43). Equation (8.43) will still be valid if we multiply every element in a given row of $\mathbf{A}$ by some constant and multiply every element in the same row of $\mathbf{C}$ by the same constant, without changing $\mathbf{B}$. This is an example of a *row operation*.

Another row operation that we can apply is to subtract one row of $\mathbf{A}$ from another row of $\mathbf{A}$, element by element, while doing the same thing to $\mathbf{C}$ but not to $\mathbf{B}$. This amounts to subtracting the left-hand sides of pairs of equations and subtracting at the same time the right-hand sides of the equations, which produces valid equations. Successive application of these two row operations is sufficient to transform the left factor of a matrix product into the identity matrix. If we apply them in the appropriate way to Eq. (8.50), we can transform it into Eq. (8.51). We illustrate the procedure in the following example.

---

### Example 8.10

Find the inverse of the matrix

$$\mathbf{A} = \begin{bmatrix} 2 & 1 & 0 \\ 1 & 2 & 1 \\ 0 & 1 & 2 \end{bmatrix}.$$

## Solution

We carry out the row operations by writing the two matrices on which we operate side by side. The matrix that is not operated on, $\mathbf{A}^{-1}$, is not written. Our version of Eq. (8.50) is

$$\begin{bmatrix} 2 & 1 & 0 \\ 1 & 2 & 1 \\ 0 & 1 & 2 \end{bmatrix} \begin{bmatrix} (A^{-1})_{11} & (A^{-1})_{12} & (A^{-1})_{13} \\ (A^{-1})_{21} & (A^{-1})_{22} & (A^{-1})_{23} \\ (A^{-1})_{22} & (A^{-1})_{32} & (A^{-1})_{33} \end{bmatrix} = \begin{bmatrix} 1 & 0 & 0 \\ 0 & 1 & 0 \\ 0 & 0 & 1 \end{bmatrix}.$$

For our operations we represent this by a double matrix, omitting the matrix on which we do not operate. All operations are carried out on both sides of this double matrix,

$$\left[\begin{array}{ccc|ccc} 2 & 1 & 0 & 1 & 0 & 0 \\ 1 & 2 & 1 & 0 & 1 & 0 \\ 0 & 1 & 2 & 0 & 0 & 1 \end{array}\right].$$

It is usual to clear out the columns from left to right. We first want to get a zero in the place of $a_{21}$. We multiply the first row by $\frac{1}{2}$, obtaining

$$\left[\begin{array}{ccc|ccc} 1 & \frac{1}{2} & 0 & \frac{1}{2} & 0 & 0 \\ 1 & 2 & 1 & 0 & 1 & 0 \\ 0 & 1 & 2 & 0 & 0 & 1 \end{array}\right].$$

We subtract the first row from the second and replace the second row by this difference. The result is

$$\left[\begin{array}{ccc|ccc} 1 & \frac{1}{2} & 0 & \frac{1}{2} & 0 & 0 \\ 0 & \frac{3}{2} & 1 & -\frac{1}{2} & 1 & 0 \\ 0 & 1 & 2 & 0 & 0 & 1 \end{array}\right].$$

We have used the element $a_{11}$ as the *pivot element* in this procedure. The left column is now as we want it to be. We now use the $a_{22}$ element as the pivot element to clear the second column. We multiply the second row by $\frac{1}{3}$ and replace the first row by the difference of the first row and the second to obtain

$$\left[\begin{array}{ccc|ccc} 1 & 0 & -\frac{1}{3} & \frac{2}{3} & -\frac{1}{3} & 0 \\ 0 & \frac{1}{2} & \frac{1}{3} & -\frac{1}{6} & \frac{1}{3} & 0 \\ 0 & 1 & 2 & 0 & 0 & 1 \end{array}\right].$$

We now multiply the second row by 2, subtract this row, from the third row, and replace the third row by the difference. The result is

$$\left[\begin{array}{ccc|ccc} 1 & 0 & -\frac{1}{3} & \frac{2}{3} & -\frac{1}{3} & 0 \\ 0 & 1 & \frac{2}{3} & -\frac{1}{3} & \frac{2}{3} & 0 \\ 0 & 0 & \frac{4}{3} & \frac{1}{3} & -\frac{2}{3} & 1 \end{array}\right].$$

We now multiply the third row by $\frac{1}{2}$ in order to use the $a_{33}$ element as the pivot element. We subtract the third row from the second and replace the second row by the difference, obtaining

$$
\begin{bmatrix}
1 & 0 & -\frac{1}{3} & \vdots & \frac{2}{3} & -\frac{1}{3} & 0 \\
0 & 1 & 0 & \vdots & -\frac{1}{2} & 1 & -\frac{1}{2} \\
0 & 0 & \frac{2}{3} & \vdots & \frac{1}{6} & -\frac{1}{3} & \frac{1}{2}
\end{bmatrix}.
$$

We now multiply the third row by $\frac{1}{2}$, add it to the first row, and replace the first row by the sum. The result is

$$
\begin{bmatrix}
1 & 0 & 0 & \vdots & \frac{3}{4} & -\frac{1}{2} & \frac{1}{4} \\
0 & 1 & 0 & \vdots & -\frac{1}{2} & 1 & -\frac{1}{2} \\
0 & 0 & \frac{1}{3} & \vdots & \frac{1}{12} & -\frac{1}{6} & \frac{1}{4}
\end{bmatrix}.
$$

The final row operation is multiplication of the third row by 3 to obtain

$$
\begin{bmatrix}
1 & 0 & 0 & \vdots & \frac{3}{4} & -\frac{1}{2} & \frac{1}{4} \\
0 & 1 & 0 & \vdots & -\frac{1}{2} & 1 & -\frac{1}{2} \\
0 & 0 & 1 & \vdots & \frac{1}{4} & -\frac{1}{2} & \frac{3}{4}
\end{bmatrix}.
$$

This represents a matrix equation such that the left-hand side of the equation is $\mathbf{E}\mathbf{A}^{-1}$ and the right-hand side is the right half of the double matrix. Therefore, the right half of this double matrix is $\mathbf{A}^{-1}$:

$$
\mathbf{A}^{-1} =
\begin{bmatrix}
\frac{3}{4} & -\frac{1}{2} & \frac{1}{4} \\
-\frac{1}{2} & 1 & -\frac{1}{2} \\
\frac{1}{4} & -\frac{1}{2} & \frac{3}{4}
\end{bmatrix}.
$$

---

**Problem 8.17**

  a. Show that $\mathbf{A}\mathbf{A}^{-1} = \mathbf{E}$ and that $\mathbf{A}^{-1}\mathbf{A} = \mathbf{E}$ for the matrices of Example 8.10.
  b. Find the inverse of the matrix

$$
\mathbf{A} =
\begin{bmatrix}
1 & 2 \\
3 & 4
\end{bmatrix}
$$

and show that your result is the inverse of $\mathbf{A}$.

---

Only square matrices have inverses, but not all square matrices have inverses. Associated with each square matrix is a determinant, which we define in the next section. If the determinant of a square matrix vanishes, the matrix is said to be *singular*. A singular matrix has no inverse.

The inversion of a matrix by hand can be a tedious process. Computer programs can be written to accomplish the inversion, and such programs are built into BASIC processes such as TrueBASIC and can be called with a single statement. See Chapter 11 for details.

We conclude this section with the definition of several terms that apply to square matrices. The *trace* of a matrix is the sum of the diagonal elements of the matrix:

$$\mathrm{Tr}(\mathbf{A}) = \sum_{i=1}^{n} a_{ii}. \tag{8.52}$$

The trace is sometimes called the *spur*, from the German word *Spur*, which means track or trace. For example, the trace of the $n$ by $n$ identity matrix is equal to $n$.

A matrix in which all the elements below the diagonal elements vanish is called an *upper triangular matrix*. A matrix in which all the elements above the diagonal elements vanish is called a *lower triangular matrix*, and a matrix in which all the elements except the diagonal elements vanish is called a *diagonal matrix*. The matrix in which all of the elements vanish is called the *null matrix* or the *zero matrix*. The *transpose* of a matrix is obtained by replacing the first column by the first row, the second column by the second row of the original matrix, etc. The transpose of $\mathbf{A}$ is denoted by $\tilde{\mathbf{A}}$ (pronounced "A tilde"),

$$(\tilde{\mathbf{A}})_{ij} = \tilde{a}_{ij} = a_{ji}. \tag{8.53}$$

If a matrix is equal to its transpose, it is a *symmetric matrix*. The matrix in Example 8.10 is symmetric, and its inverse is also symmetric.

The *hermitian conjugate* of a matrix is obtained by taking the complex conjugate of each element and then taking the transpose of the matrix. If a matrix has only real elements, the hermitian conjugate is the same as the transpose. The hermitian conjugate is also called the *adjoint* (mostly by physicists) and the *associate* (mostly by mathematicians, who use the term "adjoint" for something else). The hermitian conjugate is denoted by $\mathbf{A}^{\dagger}$.

$$(\mathbf{A}^{\dagger})_{ij} = a_{ji}^{*}. \tag{8.54}$$

A matrix that is equal to its hermitian conjugate is said to be a *hermitian matrix*.

An *orthogonal matrix* is one whose inverse is equal to its transpose. If A is orthogonal, then

$$\mathbf{A}^{-1} = \mathbf{A} \qquad \text{(orthogonal matrix).} \tag{8.55}$$

A *unitary matrix* is one whose inverse is equal to its hermitian conjugate. If $\mathbf{A}$ is unitary, then

$$\mathbf{A}^{-1} = \mathbf{A}^{\dagger} = \tilde{\mathbf{A}}^{*} \qquad \text{(unitary matrix).} \tag{8.56}$$

---

**Problem 8.18**

Which of the following matrices are diagonal, symmetric, hermitian, orthogonal, or unitary?

$$\text{a.} \begin{bmatrix} 1 & 0 \\ 0 & 1 \end{bmatrix} \quad \text{b.} \begin{bmatrix} 1 & 0 \\ 1 & 1 \end{bmatrix} \quad \text{c.} \begin{bmatrix} i & i \\ 0 & 1 \end{bmatrix} \quad \text{d.} \begin{bmatrix} 0 & i \\ -i & 1 \end{bmatrix}.$$

---

## SECTION 8.4. DETERMINANTS

Associated with every square matrix is a quantity called a *determinant*. If the elements of the matrix are constants, the determinant is a single constant, defined as a certain sum of products of subsets of the elements. If the matrix has $n$ rows and columns, each term in the sum making up the determinant will have $n$ factors in it.

The determinant of a 2 by 2 matrix is defined by

$$\det(\mathbf{A}) = \begin{vmatrix} a_{11} & a_{12} \\ a_{21} & a_{22} \end{vmatrix} = a_{11}a_{22} - a_{12}a_{21} \,. \tag{8.57}$$

The determinant is written in much the same way as the matrix, except that straight vertical lines are used on the left and right.

---

**Example 8.11**

Find the value of the determinant

$$\begin{vmatrix} 3 & -17 \\ 1 & 5 \end{vmatrix}.$$

---

**Solution**

$$\begin{vmatrix} 3 & -17 \\ 1 & 5 \end{vmatrix} = (3)(5) - (-17)(1) = 15 + 17 = 32.$$

---

Finding the value of a large determinant is tedious. One way to do it is by *expanding by minors,* as follows:

1. Pick a row or a column of the determinant. Any row or column will do, but one with zeros in it will minimize the work.

2. The determinant is a sum of terms, one for each element in the row or column. Each term consists of an element of the chosen row or column times the *minor* of that element, with an assigned sign, either positive or negative. The minor of an element in a determinant is the determinant that is obtained by deleting the row and the column containing that element. It is a determinant with one less row and one less column than the original determinant.

3. Determine the sign assigned to a term as follows: Count the number of steps of one row or one column required to get from the upper left element to the element whose minor is desired. If the number of steps is odd, the sign is negative. If the number of steps is even (including zero), the sign is positive.

4. Repeat the entire process with each determinant in the expansion until you have a sum of 2 by 2 determinants, which can be evaluated by Eq. (8.57a).

The *cofactor* of an element in a determinant is the minor multiplied by the appropriate factor of 1 or $-1$, determined as in step 3. In addition to the minor which we have defined, other minors of different order are defined, in which two or more rows and columns are deleted. We do not need to use these and will not discuss them.

### Example 8.12

Expand the 3 by 3 determinant of the matrix **A** by minors.

### Solution

$$\begin{vmatrix} a_{11} & a_{12} & a_{13} \\ a_{21} & a_{22} & a_{23} \\ a_{31} & a_{32} & a_{33} \end{vmatrix} = a_{11} \begin{vmatrix} a_{22} & a_{23} \\ a_{32} & a_{33} \end{vmatrix} - a_{12} \begin{vmatrix} a_{21} & a_{23} \\ a_{31} & a_{33} \end{vmatrix} + a_{13} \begin{vmatrix} a_{21} & a_{22} \\ a_{31} & a_{32} \end{vmatrix}.$$

$$= a_{11}a_{22}a_{33} - a_{11}a_{23}a_{32} - a_{12}a_{21}a_{33} + a_{12}a_{23}a_{31}$$
$$+ a_{13}a_{21}a_{32} - a_{13}a_{22}a_{31}.$$

### Problem 8.19

Expand the following determinant by minors:

$$\begin{vmatrix} 3 & 2 & 0 \\ 7 & -1 & 5 \\ 2 & 3 & 4 \end{vmatrix}.$$

### Problem 8.20

a. Expand the 4 by 4 determinant by minors

$$\begin{vmatrix} a_{11} & a_{12} & a_{13} & a_{14} \\ a_{21} & a_{22} & a_{23} & a_{24} \\ a_{31} & a_{32} & a_{33} & a_{34} \\ a_{41} & a_{42} & a_{43} & a_{44} \end{vmatrix}.$$

b. Write a computer program in the BASIC language to evaluate a 4 by 4 determinant with constant elements.

Determinants have a number of important properties:

*Property 1.* If two rows of a determinant are interchanged, the result will be a determinant whose value is the negative of the original determinant. The same is true if two columns are interchanged.

*Property 2.* If two rows or two columns of a determinant are equal, the determinant has value zero.

*Property 3.* If each element in one row or one column of a determinant is multiplied by the same quantity $c$ the value of the new determinant is $c$ times the value of the original determinant. Therefore, if an $n$ by $n$ determinant has every element multiplied by $c$, the new determinant is $c^n$ times the original determinant.

*Property 4.* If every element in any one row or in any one column of a determinant is zero, the value of the determinant is zero.

*Property 5.* If any row is replaced, element by element, by that row plus a constant times another row, the value of the determinant is unchanged. The same is true for two columns. For example,

$$\begin{vmatrix} a_{11} + c a_{12} & a_{12} & a_{13} \\ a_{21} + c a_{22} & a_{22} & a_{13} \\ a_{31} + c a_{32} & a_{32} & a_{33} \end{vmatrix} = \begin{vmatrix} a_{11} & a_{12} & a_{13} \\ a_{21} & a_{22} & a_{23} \\ a_{31} & a_{32} & a_{33} \end{vmatrix}. \tag{8.58}$$

*Property 6.* The determinant of a triangular matrix (a *triangular determinant*) is equal to the product of the diagonal elements. For example,

$$\begin{vmatrix} a_{11} & 0 & 0 \\ a_{22} & a_{22} & 0 \\ a_{21} & a_{32} & a_{33} \end{vmatrix} = a_{11} a_{22} a_{33}. \tag{8.59}$$

*Property 7.* The determinant of a matrix is equal to the determinant of the transpose of that matrix.

$$\det(\tilde{\mathbf{A}}) = \det(\mathbf{A}). \tag{8.60}$$

These properties can be verified using the expansion of a determinant by minors.

---

**Problem 8.21**

    a. Find the value of the determinant

$$\begin{vmatrix} 3 & 4 & 5 \\ 2 & 1 & 6 \\ 3 & -5 & 10 \end{vmatrix}.$$

    b. Interchange the first and second columns and find the value of the resulting determinant.

    c. Replace the second column by the sum of the first and second columns and find the value of the resulting determinant.

    d. Replace the second column by the first, thus making two identical columns, and find the value of the resulting determinant.

---

There is an application of determinants in quantum chemistry which comes from Property 1. The electronic wave function of a system containing two or more electrons must change sign but keep the same magnitude if the positions of two of the electrons are interchanged (must be *antisymmetric*). For example, if $\mathbf{r}_1$ and $\mathbf{r}_2$ are the position vectors of two electrons and $\Psi$ is a multi-electron wave function, then the wave function must obey

$$\Psi(\mathbf{r}_1, \mathbf{r}_2, \mathbf{r}_3, \mathbf{r}_4, \ldots, \mathbf{r}_n) = -\Psi(\mathbf{r}_2, \mathbf{r}_1, \mathbf{r}_3, \mathbf{r}_4, \ldots, \mathbf{r}_n) \tag{8.61}$$

with similar equations for exchanging any other pair of electrons' coordinates. Many approximate multi-electron wave functions are constructed as a product of one-electron wave functions, or *orbitals*. If $\psi_1$, $\psi_2$, etc., are orbitals such a wave function might be, for $n$ electrons

$$\Psi = \psi_1(\mathbf{r}_1)\psi_2(\mathbf{r}_2)\psi_3(\mathbf{r}_3)\psi_4(\mathbf{r}_4) \cdots \psi_n(\mathbf{r}_n). \tag{8.62}$$

This wave function does not obey Eq. (8.61). A wave function that does obey this equation can be constructed as a *Slater determinant*.[4] The elements of this determinant are orbital functions, not constants, so the determinant is equal to a function, not a constant.

$$\Psi(\mathbf{r}_1, \mathbf{r}_2, \ldots, \mathbf{r}_n) = \frac{1}{\sqrt{n!}} \begin{vmatrix} \psi_1(\mathbf{r}_1) & \psi_1(\mathbf{r}_2) & \psi_1(\mathbf{r}_3) & \cdots & \psi_1(\mathbf{r}_n) \\ \psi_2(\mathbf{r}_1) & \psi_2(\mathbf{r}_2) & \psi_2(\mathbf{r}_3) & \cdots & \psi_2(\mathbf{r}_n) \\ \psi_3(\mathbf{r}_1) & \psi_3(\mathbf{r}_2) & \psi_3(\mathbf{r}_3) & \cdots & \psi_3(\mathbf{r}_n) \\ \cdots & \cdots & \cdots & \cdots & \cdots \\ \psi_n(\mathbf{r}_1) & \psi_n(\mathbf{r}_2) & \psi_n(\mathbf{r}_3) & \cdots & \psi_n(\mathbf{r}_n) \end{vmatrix}. \quad (8.63)$$

The factor $1/\sqrt{n!}$ in front is a normalizing factor, which is not important to us now. The Slater determinant obeys Eq. (8.61), since interchanging $\mathbf{r}_1$ and $\mathbf{r}_2$, for example, is the same as interchanging two columns, which changes the sign of the determinant.

Notice that if we attempt to write such a wave function with two electrons in the same orbital (two of the $\psi$ factors identical), then two rows of the determinant are identical, and the entire determinant vanishes by Property 2. This is the *Pauli exclusion principle,* which states that no two electrons in the same atom or molecule can occupy the same orbital.

## SECTION 8.5. AN ELEMENTARY INTRODUCTION TO GROUP THEORY

A *group* is a collection of elements with a single operation for combining two elements of the group. We call the operation multiplication in order to exploit the similarities of this operation with matrix and operator multiplication. The following requirements must be met:

1. If **A** and **B** are members of the group, and **F** is the product **AB**, then **F** must be a member of the group.
2. The group must contain the identity element, **E**, such that

$$\mathbf{AE} = \mathbf{EA} = \mathbf{A}. \quad (8.64)$$

3. The inverse of every element of the group must be a member of the group.
4. The associative law must hold:

$$\mathbf{A(BC)} = \mathbf{(AB)C}. \quad (8.65)$$

It is not necessary that the elements of the group commute with each other. That is, it is possible that

$$\mathbf{AB} \neq \mathbf{BA} \quad \text{(possible)}. \quad (8.66)$$

If all the members of the group commute, the group is called *abelian*.[5]

The set of symmetry operators which "belong" to a symmetrical object in the sense of Section 8.3 form a group if we define operator multiplication to be the

---

[4] After John C. Slater, 1900–1976, a prominent American physicist and chemist.

[5] After Niels Henrik Abel, 1802–1829, a great Norwegian mathematician, who was the first to show that a general fifth-degree algebraic equation cannot be solved by a radical expression.

a

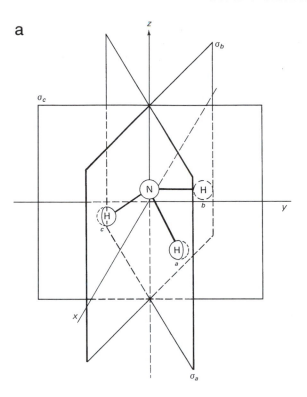

b

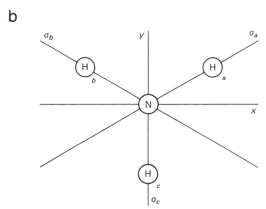

**FIGURE 8.4**   (a) The $NH_3$ molecule in its coordinate axes, with symmetry elements shown (after Levine). (b) The $NH_3$ molecule viewed from the positive $z$ axis (after Levine).

operation in the group. We will illustrate this fact for a particular molecule, $NH_3$, in its equilibrium conformation, a triangular pyramid.[6] Figure 8.4a shows the nuclear framework as viewed from the first octant of the coordinate system, and Fig. 8.4b shows the framework as viewed from the positive end of the $z$ axis. The molecule is placed in the coordinate system with the center of mass at the origin, and the rotation axis of highest order (largest value of $n$) along the $z$ axis.

[6]This discussion is adapted from Ira N. Levine, *Molecular Spectroscopy*, pp. 390ff, Wiley, New York, 1975.

**TABLE 8.1  Multiplication Table for the Symmetry Operators of the $NH_3$ Molecule**

|  | $\hat{E}$ | $\hat{C}_3$ | $\hat{C}_3^2$ | $\hat{\sigma}_a$ | $\hat{\sigma}_b$ | $\hat{\sigma}_c$ |
|---|---|---|---|---|---|---|
| $\hat{E}$ | $\hat{E}$ | $\hat{C}_3$ | $\hat{C}_3^2$ | $\hat{\sigma}_a$ | $\hat{\sigma}_b$ | $\hat{\sigma}_c$ |
| $\hat{C}_3$ | $\hat{C}_3$ | $\hat{C}_3^2$ | $\hat{E}$ | $\hat{\sigma}_c$ | $\hat{\sigma}_a$ | $\hat{\sigma}_b$ |
| $\hat{C}_3^2$ | $\hat{C}_3^2$ | $\hat{E}$ | $\hat{C}_3$ | $\hat{\sigma}_b$ | $\hat{\sigma}_c$ | $\hat{\sigma}_a$ |
| $\hat{\sigma}_a$ | $\hat{\sigma}_a$ | $\hat{\sigma}_b$ | $\hat{\sigma}_c$ | $\hat{E}$ | $\hat{C}_3$ | $\hat{C}_3^2$ |
| $\hat{\sigma}_b$ | $\hat{\sigma}_b$ | $\hat{\sigma}_c$ | $\hat{\sigma}_a$ | $\hat{C}_3^2$ | $\hat{E}$ | $\hat{C}_3$ |
| $\hat{\sigma}_c$ | $\hat{\sigma}_c$ | $\hat{\sigma}_a$ | $\hat{\sigma}_b$ | $\hat{C}_3$ | $\hat{C}_3^2$ | $\hat{E}$ |

The symmetry elements of the molecule are shown in the figure. The symmetry operators that belong to the molecule are $\hat{E}$, $\hat{C}_3$, $\hat{C}_3^2$ and the three reflection operators corresponding to vertical mirror planes passing through each of the three hydrogen nuclei, which we call $\hat{\sigma}_a$, $\hat{\sigma}_b$, and $\hat{\sigma}_c$. The square of the $\hat{C}_3$ operator is included, because we must include the inverse of all operators in the group.

We now satisfy ourselves that the four conditions to have a group are met:

*Condition 1.* The product of any two members of the group must be a member of the group. We show this by constructing a multiplication table, as shown in Table 8.1. The operators listed in the first column of the table are to be used as the left factor, and the operators listed in the first row of the table are to be used as the right factor in a product. We must specify which factor comes first because the operators do not necessarily commute. The entries in the table are obtained as in the example:

---

**Example 8.13**

Find the product $\hat{\sigma}_c \hat{C}_3$.

---

**Solution**

Both of these operators leave the nitrogen nucleus where it is. The $\hat{C}_3$ operator moves the hydrogen nucleus originally at the $\sigma_a$ plane to the $\sigma_b$ plane, the nucleus originally at the $\sigma_b$ plane to the $\sigma_c$ plane, and the nucleus originally at the $\sigma_c$ plane to the $\sigma_a$ plane. The $\hat{\sigma}_c$ operator reflects in the $\sigma_c$ plane, so that it exchanges the nuclei at the $\sigma_a$ and $\sigma_b$ planes. It thus returns the nucleus originally at the $\sigma_a$ plane to its original position and moves the nucleus originally at the $\sigma_c$ plane to the $\sigma_b$ plane. This is the same as the effect that the $\hat{\sigma}_a$ operator would have, so

$$\hat{\sigma}_c \hat{C} = \hat{\sigma}_a. \tag{8.67}$$

---

**Problem 8.22**

Verify several of the entries in Table 8.1.

---

Some pairs of operators in this group commute, whereas others do not. For example, $\hat{C}_3 \hat{\sigma}_c = \hat{\sigma}_b$, whereas $\hat{\sigma}_c \hat{C}_3 = \hat{\sigma}_a$.

*Condition 2.* The group does contain the identity operator, $\hat{E}$.

*Condition 3.* The inverse of every operator is in the group. Each reflection operator is its own inverse, and the inverse of $\hat{C}_3$ is $\hat{C}_3^2$.

*Condition 4.* The multiplication operation is associative, because operator multiplication is always associative.

A group that consists of point symmetry operators is called a *point group*. There is only a limited number of point groups that exist, and each is assigned a symbol, called a *Schoenflies symbol*. The point group of the $NH_3$ molecule is called the $C_{3v}$ group. This symbol is chosen because the principal rotation axis is a $C_3$ axis, and because there are vertical mirror planes. You can communicate what the symmetry properties of the $NH_3$ molecule are by saying that it has $C_{3v}$ symmetry. Flow charts have been constructed for the routine assignment of Schoenflies symbols.[7]

The $H_2O$ molecule belongs to the $C_{2v}$ point group, which contains the operators $\hat{E}, \hat{C}_2$, and two reflection operators, one whose mirror plane is the plane of the nuclei and one whose mirror plane bisects the angle between the bonds and is perpendicular to the first.

---

## Problem 8.23

Obtain the multiplication table for the $C_{2v}$ point group and show that it satisfies the conditions to be a group.

---

## Symmetry Operators and Matrices

Operator algebra and matrix algebra are quite similar, and matrices can be used to represent symmetry operators. Equation (8.26) represents the action of a symmetry operator, $\hat{A}$, on the location of a point. Let the original location of the point be given by the cartesian coordinates $(x, y, z)$, and the final coordinates be given by $(x', y', z')$:

$$\hat{A}(x, y, z) = (x', y', z'). \tag{8.68}$$

If we represent the position vectors by 3 by 1 matrices (column vectors) this can be written as a matrix equation:

$$\begin{bmatrix} a_{11} & a_{12} & a_{13} \\ a_{21} & a_{22} & a_{23} \\ a_{31} & a_{32} & a_{33} \end{bmatrix} \begin{bmatrix} x \\ y \\ z \end{bmatrix} = \begin{bmatrix} x' \\ y' \\ z' \end{bmatrix} \tag{8.69}$$

Equation (8.69) is the same as three ordinary equations:

$$a_{11}x + a_{12}y + a_{13}z = x' \tag{8.70a}$$

$$a_{21}x + a_{22}y + a_{23}z = y' \tag{8.70b}$$

$$a_{31}x + a_{32}y + a_{33}z = z'. \tag{8.70c}$$

We can obtain the elements of the matrix for $\hat{A}$ by comparing these equations with the equations obtained in Section 8.2 for various symmetry operators. For example, in the case of the identity operator, $x = x'$, $y = y'$, and $z = z'$, so that the matrix for

---

[7]See, for example, P. W. Atkins, *Physical Chemistry.* 6th ed. p. 433, Freeman, New York, 1998.

the identity symmetry operator is the 3 by 3 identity matrix.

$$\hat{E} \leftrightarrow \mathbf{E} = \begin{bmatrix} 1 & 0 & 0 \\ 0 & 1 & 0 \\ 0 & 0 & 1 \end{bmatrix}. \tag{8.71}$$

The double-headed arrow means that the symmetry operator $\hat{E}$ and the matrix $\mathbf{E}$ are equivalent. That is, the matrix product in Eq. (8.69) and the operator expression in Eq. (8.68) give the same result. We sometimes say that there is a one-to-one correspondence between this operator and this matrix.

---

### Example 8.14

Find the matrix equivalent to $\hat{C}_n(z)$.

---

### Solution

Let $\alpha = 2\pi/n$ radians, the angle through which the operator rotates a particle,

$$x' = \cos(\alpha)x - \sin(\alpha)y$$
$$y' = \sin(\alpha)x + \cos(\alpha)y$$
$$z' = z.$$

Comparison of this with Eq. (8.70) gives

$$\hat{C}_n(z) \leftrightarrow \begin{bmatrix} \cos(\alpha) & -\sin(\alpha) & 0 \\ \sin(\alpha) & \cos(\alpha) & 0 \\ 0 & 0 & 1 \end{bmatrix}. \tag{8.72}$$

---

### Problem 8.24

a. Verify Eqs. (8.71) and (8.72) by matrix multiplication.
b. Use Eq. (8.72) to find the matrix for $\hat{C}_2(z)$.
c. Find the matrices equivalent to $\hat{S}_3(z)$ and $\hat{\sigma}_h$.

---

The matrices that are equivalent to a group of symmetry operators have the same effect as the symmetry operators, so they must multiply together in the same way. We can show this by carrying out the matrix multiplications. The matrices for the $C_{3v}$ group are

$$\hat{E} \leftrightarrow \begin{bmatrix} 1 & 0 & 0 \\ 0 & 1 & 0 \\ 0 & 0 & 1 \end{bmatrix} = \mathbf{E}, \qquad \hat{C}_3 \leftrightarrow \begin{bmatrix} -1/2 & -\sqrt{3}/2 & 0 \\ \sqrt{3}/2 & -1/2 & 0 \\ 0 & 0 & 1 \end{bmatrix} = \mathbf{A}$$

$$\hat{C}_3^2 \leftrightarrow \begin{bmatrix} -1/2 & \sqrt{3}/2 & 0 \\ -\sqrt{3}/2 & -1/2 & 0 \\ 0 & 0 & 1 \end{bmatrix} = \mathbf{B}, \qquad \hat{\sigma}_a \leftrightarrow \begin{bmatrix} 1/2 & \sqrt{3}/2 & 0 \\ \sqrt{3}/2 & -1/2 & 0 \\ 0 & 0 & 1 \end{bmatrix} = \mathbf{C}$$

$$\hat{\sigma}_b \leftrightarrow \begin{bmatrix} 1/2 & -\sqrt{3}/2 & 0 \\ -\sqrt{3}/2 & -1/2 & 0 \\ 0 & 0 & 1 \end{bmatrix} = \mathbf{D}, \qquad \hat{\sigma}_c \leftrightarrow \begin{bmatrix} -1 & 0 & 0 \\ 0 & 1 & 0 \\ 0 & 0 & 1 \end{bmatrix} = \mathbf{F}, \tag{8.73}$$

where we have given each matrix an arbitrarily chosen letter symbol.

---

**Problem 8.25**

By transcribing Table 8.1 with appropriate changes in symbols, generate the multiplication table for the matrices in Eq. (8.73). Verify several of the entries by matrix multiplication.

---

Each of the matrices in Eq. (8.73) is equivalent to one of the symmetry operators in the $C_{3v}$ group, but it is not exactly identical to it, being only one possible way to represent the symmetry operator. The set of matrices forms another group of the same *order* (same number of members) as the $C_{3v}$ point group. The fact that it obeys the same multiplication table is expressed by saying that it is *isomorphic* with the group of symmetry operators. A group of matrices that is isomorphic with a group of symmetry operators is called a *faithful representation* of the group. Our group of matrices consists of 3 by 3 matrices and is said to be of *dimension 3*.

Any set of matrices that obeys the same multiplication table as a given group is called a representation of that group. There can be other representations of a symmetry group besides the group of matrices that are equivalent to the symmetry operators. The matrices in the representation do not have to have any physical interpretation, but they must multiply in the same way as do the symmetry operators, must be square, and all must have the same number of rows and columns. In some representations, called *unfaithful* or *homomorphic*, there are fewer matrices than there are symmetry operators, so that one matrix occurs in the places in the multiplication table where two or more symmetry operators occur.

Group representations are divided into two kinds, *reducible* and *irreducible*. In a *reducible representation*, the matrices are "block-diagonal" or can be put into block-diagonal form by a similarity transformation. A *block-diagonal matrix* is one in which all elements are zero except those in square regions along the diagonal. The following matrix has two 2 by 2 blocks and a 1 by 1 block

$$\begin{bmatrix} 1 & 2 & 0 & 0 & 0 \\ 3 & 2 & 0 & 0 & 0 \\ 0 & 0 & 4 & 3 & 0 \\ 0 & 0 & 3 & 3 & 0 \\ 0 & 0 & 0 & 0 & 2 \end{bmatrix}.$$

All the matrices in a reducible representation have to have the same size blocks in the same order. The representation of the group $C_{3v}$ given in Eq. (8.73) is reducible, since each matrix has a 2 by 2 block and a 1 by 1 block. When two block-diagonal matrices with the same size blocks are multiplied together, the result is a matrix that is block-diagonal with the same size blocks in the same order. This is apparent in the case of the matrices in Eq. (8.73), which produce only each other when multiplied together.

---

**Problem 8.26**

Show by matrix multiplication that two matrices with a 2 by 2 block and two 1 by 1 blocks produce another such matrix when multiplied together.

---

Because of the way in which block-diagonal matrices multiply, the 2 by 2 blocks in the matrices in Eq. (8.73) if taken alone form another representation of the $C_{3v}$ group. When a reducible representation is written with its matrices in block-diagonal form, the block submatrices form *irreducible representations*, and the reducible representation is said to be the *direct sum* of the irreducible representations. Both the representations obtained from the submatrices are irreducible. The 1 by 1 blocks form an unfaithful or homomorphic representation, in which every operator is represented by the 1 by 1 identity matrix. This one-dimensional representation is called the *totally symmetric representation*.

In this particular case we could not get three one-dimensional representations, because the $\hat{C}_3(z)$ operator mixes the $x$ and $y$ coordinates of a particle, preventing the matrices from being diagonal.

---

## Problem 8.27

Pick a few pairs of 2 by 2 submatrices from Eq. (8.73) and show that they multiply in the same way as the 3 by 3 matrices.

---

In any representation of a symmetry group, the trace of a matrix is called the *character* of the corresponding operator for that representation.

---

## Problem 8.28

Find the characters of the operators in the $C_{3v}$ group for the representation in Eq. (8.73).

---

The two irreducible representations of $C_{3v}$ that we have obtained thus far are said to be *nonequivalent*, since they have different dimensions. There are several theorems governing irreducible representations for a particular group.[8] These theorems can be used, for example, to determine that three irreducible representations of the $C_{3v}$ group occur, and that their dimensions are 2,1, and 1. The other one-dimensional representation is

$$\hat{E} \leftrightarrow 1 \quad \hat{C}_3 \leftrightarrow 1 \quad \hat{C}_3^2 \leftrightarrow 1$$
$$\hat{\sigma}_a \leftrightarrow -1 \quad \hat{\sigma}_b \leftrightarrow -1 \quad \hat{\sigma}_c \leftrightarrow -1. \tag{8.74}$$

---

## Problem 8.29

Show that the 1 by 1 matrices (scalars) in Eq. (8.74) obey the same multiplication table as does the group of symmetry operators.

---

Group theory can be applied to several different areas of molecular quantum mechanics, including the symmetry of electronic and vibrational wave functions and the study of transitions between energy levels.[9] There is also a theorem which says that

[8] Ira N. Levine, *Quantum Chemistry*, Vol. II, p. 389, Allyn & Bacon, Boston, 1970.

[9] See P. W. Atkins, *Physical Chemistry*, 5th ed., Chap. 15, Freeman, New York, 1994.

there is a correspondence between an energy level and some one of the irreducible representations of the symmetry group of the molecule, and that the degeneracy (number of states in the level) is equal to the dimension of that irreducible representation.

## SUMMARY OF THE CHAPTER

We presented three topics in this chapter: operator algebra, matrix algebra, and group theory. A mathematical operator is a symbol standing for the carrying out of a mathematical operation. Operating on a function with an operator produces a new function not necessarily the same as the old function. Operator symbols can be manipulated in a way similar to the algebra of ordinary variables, without reference to any function that might be operated on. We defined the sum and the product of two operators. The product of two operators was defined as successive operation with the operators. The quotient of two operators was not defined, but we defined the inverse of an operator, which undoes the effect of that operator. The principal difference between operator algebra and ordinary algebra is that multiplication of two operators is not necessarily commutative. We discussed symmetry operators, a useful class of operators which move points in space relative to a symmetry element.

A matrix is a list of quantities, arranged in rows and columns. Matrix algebra is similar to operator algebra in that multiplication of two matrices is not necessarily commutative. The inverse of a matrix is similar to the inverse of an operator. If $A^{-1}$ is the inverse of A, then $A^{-1}A = AA^{-1} = E$ where $E$ is the identity matrix. We presented the Gauss–Jordan method for obtaining the inverse of a nonsingular square matrix.

Group theory is a branch of mathematics that involves elements with defined properties and a single operation called multiplication of two elements. The symmetry operators belonging to a symmetrical object form a group, and the theorems of group theory can provide useful information about electronic wave functions for symmetrical molecules, spectroscopic transitions, etc.

## ADDITIONAL READING

Peter W. Atkins, *Physical Chemistry*, 5th ed, Chap. 15, Freeman, New York, 1994. This is a standard physical chemistry textbook. Chapter 15 discusses molecular symmetry, including group theory.

W. W. Bell, *Matrices for Scientists and Engineers*, Van Nostrand–Reinhold, New York, 1975. This book is a discussion of matrix algebra and its application to scientific problems.

Robert L. Carter, *Molecular Symmetry and Group Theory*, Wiley, New York, 1998. This is a paperback book that presents a complete introduction to molecular symmetry.

Robert H. Dicke and James P. Wittke, *Introduction to Quantum Mechanics*, Addison–Wesley, Reading, MA, 1960. This is a quantum mechanics text for physicists. It contains a discussion of operator algebra and the construction of quantum mechanical operators. There are many similar books, and you can read about this subject in almost any of them.

M. Hamermesh, *Group Theory*, Addison–Wesley, Reading, MA, 1962. This is an advanced work, carefully done, which is probably better suited to physicists than to chemists.

Robert W. Hornbeck, *Numerical Methods*, Quantum Publishers, New York, 1975. This book is a clear introduction to numerical analysis. It contains a discussion of matrix inversion, including a FORTRAN computer program to carry out the procedure, as well as a number of other useful things. It is available in an inexpensive paperback version.

H. H. Jaffe and Milton Orchin, *Symmetry in Chemistry*, Wiley, New York, 1965. This is a small book that gives an elementary introduction to the subject of symmetry for chemists.

Ira N. Levine, *Quantum Chemistry*, Allyn & Bacon, Boston, 1970. This work comes in two volumes. Volume I contains a discussion of operators and Volume II contains a very good discussion of symmetry, group theory, and group representations. The book is written at the senior/first year graduate level.

Ira N. Levine, *Molecular Spectroscopy*, Wiley, New York, 1975. This book is similar to Vol. II of the above work.

Milton Orchin and H. H. Jaffe, *Symmetry, Orbitals, and Spectra*, Wiley–Interscience, New York, 1971. This book contains a discussion of both quantum mechanics and the application of group theory to it. It includes a separate pamphlet with tables of characters for group representations.

D. S. Schonland, *Molecular Symmetry*, Van Nostrand, Princeton, NJ, 1965. This is a careful introduction to the subject, written for chemists.

## ADDITIONAL PROBLEMS

### 8.30

Find the following commutators: (a) $[\hat{D}_x, \sin(x)]$; (b) $[\hat{D}_x^3, x]$; (c) $[\hat{D}_x^2, f(x)]$.

### 8.31

Show that the $x$ and $z$ components of the angular momentum have quantum mechanical operators that do not commute:

$$\hat{L}_x = \frac{h}{i}\left(\hat{y}\frac{d}{dz} - \hat{z}\frac{d}{dy}\right), \qquad \hat{L}_z = \frac{h}{i}\left(\hat{x}\frac{d}{dy} - \hat{y}\frac{d}{dx}\right).$$

### 8.32

The Hamiltonian operator for a one-dimensional harmonic oscillator moving in the $x$ direction is

$$\hat{H} = -\frac{\hbar^2}{2m}\frac{d^2}{dx^2} + \frac{kx^2}{2}.$$

Find the value of the constant $a$ such that the function $e^{-ax^2}$ is an eigenfunction of the Hamiltonian operator. The quantity $k$ is the force constant, $m$ is the mass of the oscillating particle, and $\hbar$ is Planck's constant divided by $2\pi$.

**8.33**

In quantum mechanics, the *expectation value* of a mechanical quantity is given by

$$\langle A \rangle = \frac{\int \psi^* \hat{A} \psi \, dx}{\int \psi^* \psi \, dx},$$

where $\hat{A}$ is the operator for the mechanical quantity and $\psi$ is the wave function for the state of the system. The integrals are over all permitted values of the coordinates of the system. The expectation value is defined as the prediction of the mean of a large number of measurements of the mechanical quantity, given that the system is in the state corresponding to $\psi$ prior to each measurement.

For a particle moving in the $x$ direction only and confined to a region on the $x$ axis from $x = 0$ to $x = a$, the integrals are single integrals from 0 to $a$ and $\hat{p}_x$ is given by $(\hbar/i)\, \partial/\partial x$. Find the expectation value of $p_x$ and of $p_x^2$ if the wave function is

$$\psi = C \sin\left(\frac{\pi x}{a}\right),$$

where $C$ is a constant.

**8.34**

If $\hat{A}$ is the operator corresponding to the mechanical quantity $A$ and $\phi_n$ is an eigenfunction of $\hat{A}$, such that

$$\hat{A}\phi_n = a_n \phi_n$$

show that the expectation value of $A$ is equal to $a_n$ if the state of the system corresponds to $\phi_n$. See Problem 8.33 for the formula for the expectation value.

**8.35**

If $x$ is an ordinary variable, the Maclaurin series for $1/(1-x)$ is

$$\frac{1}{1-x} = 1 + x^2 + x^3 + x^4 + \cdots.$$

If $\hat{X}$ is some operator, show that the series

$$1 + \hat{X} + \hat{X}^2 + \hat{X}^3 + \hat{X}^4 + \cdots$$

is the inverse of the operator $1 - \hat{X}$.

**8.36**

Find the result of each operation on the given point (represented by cartesian coordinates):

    a. $\hat{C}_2(y)(1, 1, 1)$

    b. $\hat{C}_3(z)(1, 1, 1)$

    c. $\hat{S}_4(z)(1, 1, 1)$

d. $\hat{C}_2(z)\hat{i}\hat{\sigma}_h(1, 1, 1)$

e. $\hat{S}_2(y)\hat{\sigma}_h(1, 1, 0)$.

**8.37**

Find the 3 by 3 matrix that is equivalent in its action to each of the symmetry operators:

(a) $\hat{S}_2(z)$; (b) $\hat{C}_2(x)$; (c) $\hat{C}_8(x)$; (d) $\hat{S}_6(x)$.

**8.38**

Give the function that results if the given symmetry operator operates on the given function for each of the following:

| | Operator | Function |
|---|---|---|
| a. | $\hat{C}_4(z)$ | $x^2$ |
| b. | $\hat{\sigma}_h$ | $x\cos(x/y)$ |
| c. | $\hat{i}$ | $(x + y + z^2)$ |
| d. | $\hat{S}_4(x)$ | $(x + y + z)$ |

**8.39**

Find the matrix products:

a. $\begin{bmatrix} 0 & 1 & 2 \\ 4 & 3 & 2 \\ 7 & 6 & 1 \end{bmatrix} \begin{bmatrix} 1 & 2 & 3 \\ 6 & 8 & 1 \\ 7 & 4 & 3 \end{bmatrix}$

b. $\begin{bmatrix} 6 & 3 & 2 & -1 \\ -7 & 4 & 3 & 2 \\ 1 & 3 & 2 & -2 \\ 6 & 7 & -1 & -3 \end{bmatrix} \begin{bmatrix} 4 & 7 & -6 & -8 \\ 3 & -6 & 8 & -6 \\ 2 & 3 & -3 & 4 \\ -1 & 4 & 2 & 3 \end{bmatrix}$

c. $\begin{bmatrix} 3 & 2 & 1 & 4 \end{bmatrix} \begin{bmatrix} 1 & 2 & 3 \\ 0 & 3 & -4 \\ 1 & -2 & 1 \\ 3 & 1 & 0 \end{bmatrix}$

d. $\begin{bmatrix} 6 & 3 & -1 \\ 7 & 4 & -2 \end{bmatrix} \begin{bmatrix} 1 & 4 & -7 & 3 \\ 2 & 5 & 8 & -2 \\ 3 & 6 & -9 & 1 \end{bmatrix}$.

**8.40**

Show that $(\mathbf{AB})\mathbf{C} = \mathbf{A}(\mathbf{BC})$ for the example matrices:

$$\mathbf{A} = \begin{bmatrix} 0 & 1 & 2 \\ 3 & 1 & -4 \\ 2 & 3 & 1 \end{bmatrix}, \quad \mathbf{B} = \begin{bmatrix} 3 & 1 & 4 \\ -2 & 0 & 1 \\ 3 & 2 & 1 \end{bmatrix}, \quad \mathbf{C} = \begin{bmatrix} 0 & 3 & 1 \\ -4 & 2 & 3 \\ 3 & 1 & -2 \end{bmatrix}.$$

**8.41**

Show that $\mathbf{A}(\mathbf{B} + \mathbf{C}) = \mathbf{AB} + \mathbf{AC}$ for the example matrices in Problem 8.40.

**8.42**

Test the following matrices for singularity. Find the inverses of any that are nonsingular. Multiply the original matrix by its inverse to check your work.

a. $\begin{bmatrix} 0 & 1 & 2 \\ 2 & 3 & 1 \\ 2 & 4 & 3 \end{bmatrix}$

b. $\begin{bmatrix} 6 & 8 & 1 \\ 7 & 3 & 2 \\ 4 & 6 & -9 \end{bmatrix}$

c. $\begin{bmatrix} 3 & 2 & -1 \\ -4 & 6 & 3 \\ 7 & 2 & -1 \end{bmatrix}$

d. $\begin{bmatrix} 0 & 2 & 3 \\ 1 & 1 & 1 \\ 2 & 0 & 1 \end{bmatrix}$.

**8.43**

Find the matrix **P** which results from the similarity transformation

$$\mathbf{P} = \mathbf{X}^{-1}\mathbf{Q}\mathbf{X},$$

where

$$\mathbf{Q} = \begin{bmatrix} 1 & 2 \\ 2 & 1 \end{bmatrix}, \qquad \mathbf{X} = \begin{bmatrix} 2 & 3 \\ 4 & 3 \end{bmatrix}.$$

**8.44**

The $H_2O$ molecule belongs to the point group $C_{2v}$, which contains the symmetry operators $\hat{E}, \hat{C}_2, \hat{\sigma}_a$, and $\hat{\sigma}_b$, where the $C_2$ axis passes through the oxygen nucleus and midway between the two hydrogen nuclei, and where the $\sigma_a$ mirror plane contains the three nuclei and the $\sigma_b$ mirror plane is perpendicular to the $\sigma_a$ mirror plane.

a. Find the 3 by 3 matrix that is equivalent to each symmetry operator.
b. Show that the matrices obtained in part (a) have the same multiplication table as the symmetry operators, and that they form a group. The multiplication table for the group was to be obtained in Problem 8.23.

**8.45**

Permutation operators are operators that interchange objects. Three objects can be arranged in $3! = 6$ different ways, or permutations. From a given arrangement, all six permutations can be attained by application of the six operators: $\hat{E}$, the identity operator; $\hat{P}_{12}$, which interchanges objects 1 and 2; $\hat{P}_{23}$, which interchanges objects 2 and 3; $\hat{P}_{13}$, which interchanges objects 1 and 3; $\hat{P}_{23}\hat{P}_{12}$, which interchanges objects 1 and 2 and then interchanges objects 2 and 3, and $\hat{P}_{23}\hat{P}_{13}$, which interchanges objects 1 and 3 and then interchanges objects 2 and 3.

Satisfy yourself that each of these operators produces a different arrangement. Show that the six operators form a group. Construct a multiplication table for the group.

# 9

# THE SOLUTION OF ALGEBRAIC EQUATIONS

## Preview

If an equation is written in the form $f(x) = 0$, where $f$ is some function and $x$ is a variable, we seek to find those constant values of $x$ such that the equation is correct. These values are called solutions or roots of the equation. We discuss both algebraic and numerical methods for finding roots to algebraic equations.

If there are two variables in the equation, such as $F(x, y) = 0$, then the equation can be solved for $y$ as a function of $x$ or $x$ as a function of $y$, but in order to solve for constant values of both variables, a second equation, such as $G(x, y) = 0$, is required, and the two equations must be solved simultaneously. In general, if there are $n$ variables, $n$ independent and consistent equations are required. In this chapter, we discuss various methods for finding the roots to sets of simultaneous equations.

## Principal Facts and Ideas

1. The solution to an algebraic equation is generally a value or a set of values of the independent variable such that substitution of such a value into the equation produces an identically correct equation such as $0 = 0$.
2. For a single independent variable, one equation is required.
3. An algebraic equation in one variable can be solved algebraically, graphically, or numerically.
4. Polynomial equations through the fourth degree can be solved algebraically, but some equations of fifth and higher degree cannot be solved algebraically.

5. For $n$ variables, $n$ equations are required, but these equations must be independent and consistent.
6. Simultaneous equations can be solved with various techniques, including elimination, use of Cramer's formula, and by matrix inversion.
7. Linear homogeneous simultaneous equations have a nontrivial solution only when a certain dependence condition is met.

## Objectives

After studying this chapter, you should be able to:

1. solve any quadratic equation and determine which root is physically meaningful;
2. obtain an accurate numerical approximation to the roots of any single equation in one unknown;
3. solve any fairly simple set of two or three simultaneous linear equations by the method of elimination;
4. solve a set of linear inhomogeneous simultaneous equations by Cramer's method and by matrix inversion;
5. solve a set of linear homogeneous simultaneous equations using the dependence condition.

## SECTION 9.1. ALGEBRAIC METHODS FOR SOLVING ONE EQUATION WITH ONE UNKNOWN

If you have one algebraic equation containing only one variable, there will generally be a set of one or more constant values of that variable which make the equation valid. They are said to satisfy the equation, and the values in the set are the *roots* or *solutions* of the equation. Other values of the variable do not produce a valid equation and are said not to satisfy the equation.

## Polynomial Equations

A polynomial equation is written in the form

$$a_0 + a_1 x + a_2 x^2 + \cdots + a_n x^n = 0, \tag{9.1}$$

where $n$ is some positive integer. If $n = 1$, the equation is a *linear equation*. If $n = 2$, the equation is a *quadratic equation*. If $n = 3$, the equation is a *cubic equation*. If $n = 4$, it is a *quartic equation*, and so on. The integer $n$ is called the *degree* of the equation.

Generally, there are $n$ roots to an $n$th-degree polynomial equation, but two or more of the roots can be equal to each other. It is possible for some of the roots to be complex numbers. For most equations arising from physical and chemical problems, there will be only one root that is physically reasonable, and the others must be disregarded. For example, a concentration cannot be negative, and if a quadratic equation for a concentration produces a positive root and a negative root, the positive root must be chosen. Complex roots must ordinarily be disregarded for equations involving physically measurable quantities.

## Linear Equations

If

$$a_0 + a_1 x = 0 \tag{9.2}$$

then the single root of the equation is

$$x = \frac{a_0}{a_1}. \tag{9.3}$$

## Quadratic Equations

A quadratic equation is written in a standard form as

$$ax^2 + bx + c = 0. \tag{9.4}$$

Some quadratic expressions can be factored, which means that the equation can be written

$$a(x - x_1)(x - x_2) = 0, \tag{9.5}$$

where $x_1$ and $x_2$ are constants. In this case, $x_1$ and $x_2$ are the two roots of the equation. If a quadratic equation cannot be factored, you can apply the *quadratic formula*

$$\boxed{x = \frac{-b \pm \sqrt{b^2 - 4ac}}{2a}}. \tag{9.6}$$

This equation provides two roots, one when the positive sign in front of the square root is chosen and one when the negative sign is chosen. There are three cases: (1) if the *discriminant* $b^2 - 4ac$ is positive, the roots will be real and unequal; (2) if the discriminant is equal to zero, the two roots will be equal to each other; (3) if the discriminant is negative, the roots will be unequal and complex, since the square root of a negative quantity is imaginary. Each root will be the complex conjugate of the other root.

---

### Problem 9.1

Show by substituting Eq. (9.6) into Eq. (9.4) that the quadratic formula provides the roots to a quadratic equation.

---

A common application of a quadratic equation in elementary chemistry is the calculation of the hydrogen ion concentration in a solution of a weak acid. If activity coefficients are assumed to equal unity, the equilibrium expression in terms of molar concentrations is

$$K_a = \frac{([H^+]/c^\circ)([A^-]/c^\circ)}{([HA]/c^\circ)}, \tag{9.7}$$

where $[H^+]$ is the hydrogen-ion concentration, $[A^-]$ the acid-anion concentration, $[HA]$ is the concentration of the undissociated acid, $c°$ is defined to equal 1 mol $L^{-1}$, and $K_a$ is the acid ionization constant. It is true that the hydrogen ions are nearly all attached to water molecules or water molecule dimers, etc., so that we could write $[H_3O^+]$ instead of $[H^+]$, but this makes no difference in the calculation.

---

### Example 9.1

For acetic acid, $K_a = 1.754 \times 10^{-5}$ at 25°C. Find $[H^+]$ if 0.1000 mole of acetic acid is dissolved in enough water to make 1.000 L.

---

### Solution

Assuming that no other sources of hydrogen ions or acetate ions are present, $[H^+]/c° = [A^-]/c°$, which we denote by $x$,

$$K_a = \frac{x^2}{0.1000 - x}$$

or

$$x^2 + K_a x - 0.1000 K_a = 0.$$

From Eq. (9.6), our solution is

$$x = \frac{-K_a \pm \sqrt{K_a^2 - 0.4000 K_a}}{2}$$

$$= 1.316 \times 10^{-3} \quad \text{or} \quad -1.333 \times 10^{-3}$$

$$[H^+] = [A^-] = 1.316 \times 10^{-3} \, \text{mol} \, L^{-1}.$$

We must disregard the negative root, because a concentration cannot be negative.

---

### Problem 9.2

a. Express the answer to Example 9.1 in terms of pH, defined for our present purposes as

$$pH = -\log_{10}([H^+]/c°),$$

where $c°$ is defined to equal 1 mol $L^{-1}$ (exactly).

b. Find the pH of a solution formed from 0.075 mol of $NH_3$ and enough water to make 1.00 L of $NH_3$ solution. The ionization that occurs is

$$NH_3 + H_2O \longrightarrow NH_4^+ + OH^-.$$

The equilibrium expression in terms of molar concentrations is

$$K_b = \frac{([NH_4^+]/c°)([OH^-]/c°)}{([NH_3]/c°)},$$

where the water concentration is replaced by its mole fraction, which is very nearly equal to unity. $K_b$ equals $1.80 \times 10^{-5}$ for $NH_3$.

---

## Solutions to Approximate Equations

It is usually necessary to seek approximate roots to equations other than polynomial equations, such as equations containing sines, cosines, logarithms, exponentials, etc., which are called *transcendental equations*. If you have a cubic or higher-order polynomial equation, it is probably best to find approximations to the roots rather than attempting an exact solution, even though algebraic methods are available for cubic and quartic equations.[1]

The first method for finding approximate roots to an equation is to modify the equation by making simplifying assumptions. As an example, let us consider an equation for the hydrogen-ion concentration in a solution of a weak acid which is more nearly correct than Eq. (9.7). If we ignore activity coefficients, this equation is in terms of molar concentrations[2]

$$K_a = \frac{x(x - K_w/x)}{c/c^o - x + K_w/x},\tag{9.8}$$

where $c$ is the "gross" concentration of the acid (the concentration of acid that would occur if no ionization occurred), where $c^o$ is 1 mol L$^{-1}$, where $K_w$ is the ionization constant of water, equal to $1.00 \times 10^{-14}$ near 25°C, and where $x = [\text{H}^+]/c^o$. The difference between this equation and Eq. (9.7) is that this equation includes the hydrogen ions produced by the ionization of the water. If we multiply this equation out, we obtain the cubic equation

$$x^3 + K_a x2 - ((c/c^o)K_a + K_w)x - K_a K_w = 0.\tag{9.9}$$

In some cases, it is possible to simplify this equation by making approximations.

---

### Example 9.2

For the case of acetic acid with a gross acid concentration of 0.100 mol L$^{-1}$, convert Eq. (9.8) to a simpler approximate equation by discarding any negligibly small terms.

---

### Solution

Equation (9.8) contains two terms in the numerator and three in the denominator. If one term in a polynomial is much smaller than the other terms, it might be possible to neglect this term. In the numerator, we know that $x = [\text{H}^+]/c^o$ will lie somewhere between 0.1 and $10^{-7}$, the value for pure water. In fact, we know from our approximate solution in Example 9.1 that $[\text{H}^+]$ is near $10^{-3}$ mol L$^{-1}$. Since $K_w$ equals $1 \times 10^{-14}$ mol$^2$ L$^{-2}$, the second term must be near $10^{-11}$ mol L$^{-1}$, which is smaller than the first term by a factor of $10^8$. We therefore drop the term $K_w/x$. We also drop the same term in the denominator, and obtain the equation

$$K_a = \frac{x^2}{c/c^o - x}\tag{9.10}$$

---

[1] Older editions of *The Handbook of Chemistry and Physics* give methods for cubic equations. See, for example, the 33rd edition, pp. 272–273, Chem. Rubber Co., Cleveland, 1951–1952.

[2] Robert G. Mortimer, *Physical Chemistry*, p. 278, Benjamin–Cummings, Redwood City, CA, 1993.

TABLE 9.1    Results for the Hydrogen-Ion Concentration in Acetic Acid Solutions at 25°C from Different Equations at Different Concentrations

| c/mol L$^{-1}$ | [H$^+$]/mol L$^{-1}$ | | |
|---|---|---|---|
| | by Eq. (9.8) | by Eq. (9.10) | by Eq. (9.12) |
| 0.1000 | $1.31565 \times 10^{-3}$ | $1.31565 \times 10^{-3}$ | $1.324 \times 10^{-3}$ |
| $1.000 \times 10^{-3}$ | $1.23959 \times 10^{-4}$ | $1.23959 \times 10^{-4}$ | $1.324 \times 10^{-4}$ |
| $1.000 \times 10^{-5}$ | $0.711545 \times 10^{-5}$ | $0.711436 \times 10^{-5}$ | $1.324 \times 10^{-5}$ |
| $1.000 \times 10^{-7}$ | $0.161145 \times 10^{-6}$ | $0.099435 \times 10^{-6}$ | $1.324 \times 10^{-6}$ |

which is the same as Eq. (9.7), which was obtained with the assumption that $[H^+] = [A^-]$.

It is possible in some cases to make a further approximation on Eq. (9.10). If only a small fraction of the weak acid ionizes, $[H^+]$ will be small compared with $c$, so that $x$ can be neglected in the denominator. In the case of acetic acid and a gross acid concentration of $0.100 \, \text{mol L}^{-1}$, $[H^+]$ is approximately equal to $10^{-3} \, \text{mol L}^{-1}$, only about 1% as large as $c$. If we can tolerate an error of about 1%, we can further approximate Eq. (9.10) as

$$K_a = \frac{x^2}{c} \tag{9.11}$$

which has the solution

$$x = \sqrt{cK_a}. \tag{9.12}$$

However, as $c$ is made smaller, Eq. (9.12) quickly becomes a poor approximation, and for very small acid concentrations, Eq. (9.10) also becomes inaccurate. Table 9.1 shows the results from the three equations at different acid concentrations. Equation (9.10), the quadratic equation, remains fairly accurate down to $c = 10^{-5} \, \text{mol L}^{-1}$, but Eq. (9.12) is wrong by about 7% at $10^{-3} \, \text{mol L}^{-1}$, and much worse than that at lower concentrations.

In the case that approximations such as that of Eq. (9.11) are inaccurate, we can apply the method of *successive approximations*. In this method, one begins by solving an equation such as Eq. (9.11). The result of this approximation is used to approximate the term which was neglected in the first approximation and the solution is repeated. If needed, the result of this second approximation is used to replace the term that was originally neglected and the solution is repeated. This procedure is repeated (iterated) as many times as is necessary.

### Example 9.3

Solve the problem of Example 9.1 by successive approximation.

**Solution**

We write the equilibrium expression in the form

$$x^2 = K_a(0.1000 - x) \approx K_a(0.1000) = 1.754 \times 10^{-6}$$
$$x \approx \sqrt{1.754 \times 10^{-6}} = 0.01324.$$

This is the result that would be obtained from Eq. (9.12). Ths next approximation is obtained by replacing the $x$ in the right-hand side of the first equation by this value,

$$x^2 \approx (1.754 \times 10^{-5})(0.1000 - 0.0132) = 1.731 \times 10^{-6}$$
$$x \approx \sqrt{1.731 \times 10^{-6}} = 0.01316.$$

This result shows that the first approximation is acceptable. Further iterations would make even smaller changes, so we stop at this point.

**Problem 9.3**

Solve for the hydrogen ion concentration in solutions of acetic acid with gross molarities equal to

a. $0.00100 \, \text{mol L}^{-1}$
b. $0.0000100 \, \text{mol L}^{-1}$.

Use the method of successive approximations on Eq. (9.9) and on Eq. (9.10).

In the next example, we see another way in which an equation can be made into a tractable approximate equation, by linearizing a function.

**Example 9.4**

The Dieterici equation of state is

$$P e^{a/\bar{V}RT}(\bar{V} - b) = RT, \tag{9.13}$$

where $P$ is the pressure, $T$ is the temperature, $\bar{V}$ is the molar volume, and $R$ is the ideal gas constant. The constant parameters $a$ and $b$ are assigned values that depend on the identity of the gas being described. For carbon dioxide, $a = 0.468 \, \text{Pa} \, \text{m}^6 \, \text{mol}^{-2}$, $b = 4.63 \times 10^{-5} \, \text{m}^3 \, \text{mol}^{-1}$. Find the molar volume of carbon dioxide if $T = 298.15 \, \text{K}$ and $P = 10.000 \, \text{atm} = 1.01325 \times 10^6 \, \text{Pa}$.

**Solution**

We expand the exponential function in a Maclaurin series, using Eq. (6.25):

$$e^{a/\bar{V}RT} = 1 + a/\bar{V}RT + \frac{1}{2!}(a/\bar{V}RT)^2 + \cdots. \tag{9.14}$$

Rough calculation shows that $a/\bar{V}RT \approx 0.09$, and that $(a/\bar{V}RT)^2 \approx 0.008$. We therefore discard all of the terms past the $a/\bar{V}RT$ term and write to a fairly good approximation:

$$e^{a/\bar{V}RT} \approx 1 + a/\bar{V}RT.$$

We say that we have *linearized* the exponential function. We substitute this approximation into the original equation of state, obtaining

$$P(1 + a/\bar{V}RT)(\bar{V} - b) = RT$$

which can be written in the standard form for a quadratic equation:

$$P\bar{V}^2 + \left(\frac{Pa}{RT} - Pb - RT\right)\bar{V} - \frac{Pab}{RT} = 0.$$

We divide by $P$ and substitute the numerical values. We temporarily let $x = \bar{V}/(1\,\text{m}^3\,\text{mol}^{-1})$ and obtain

$$x^2 - (2.304 \times 10^{-3})x + 8.741 \times 10^{-9} = 0.$$

Applying the quadratic formula,

$$x = \frac{2.304 \times 10^{-3} \pm \sqrt{(0.002304)^2 - 4(8.741 \times 10^{-9})}}{2} = \begin{cases} 2.3 \times 10^{-3} \\ 4.000 \times 10^{-6}. \end{cases}$$

We disregard the second value as too small to correspond to the physical situation, and obtain

$$\bar{V} = 2.30 \times 10^{-3}\,\text{m}^3\,\text{mol}^{-1}.$$

The ideal gas equation of state gives $\bar{V} = 2.447 \times 10^{-3}\,\text{m}^3\,\text{mol}^{-1}$, for a difference of about 6%.

---

**Problem 9.4**

    a. Verify the prediction of the ideal gas equation of state given in Example 9.4.

    b. Substitute the value of the molar volume obtained in Example 9.4 and the given temperature into the Dieterici equation of state to calculate the pressure. Compare the calculated pressure with 10.00 atm, to check the validity of the linearization approximation used in Example 9.4.

---

## SECTION 9.2. GRAPHICAL SOLUTION OF EQUATIONS

Instead of approximating an equation and then solving the approximate equations exactly, we can seek approximate solutions to an exact equation. The *graphical method* is an old method for doing this. It is sometimes quite valuable, because you can see what you are doing. When you solve an equation numerically, as discussed later, it is often a good idea to make an approximate graphical solution first in order to make sure that the numerical method finds the root in which you are interested.

The equation to be solved is first written in the form

$$f(x) = 0, \tag{9.15}$$

where we denote the independent variable by $x$. If a graph of the function $f$ is drawn, any real roots to the equation can be obtained by reading off the values of $x$ where the curve crosses the $x$ axis.

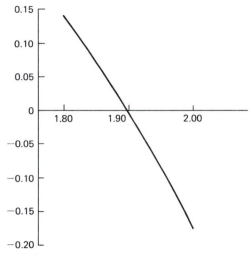

**FIGURE 9.1**  The graph used in Example 9.5.

---

### Example 9.5

By graphing, find a root of the equation

$$2\sin(x) - x = 0.$$

---

### Solution

A graph of the function for the interval $1.8 < x < 2.0$ is shown in Fig. 9.1. From the graph, it appears that the root is near $x = 1.895$.

---

### Problem 9.5

Find approximately the smallest positive root of the equation

$$\tan(x) - x = 0.$$

---

## SECTION 9.3. NUMERICAL SOLUTION OF EQUATIONS

Strictly speaking, one does not solve an equation numerically. One obtains an approximation to a root. However, with a computer or with a hand calculator, it is easy to obtain enough significant digits for almost any purpose.

### The Method of Trial and Error

If the equation is written in the form

$$f(x) = 0$$

one repeatedly evaluates the function $f$, choosing different values of $x$, until $f$ nearly vanishes. When using this method with a hand calculator, it is usually possible to adopt a strategy of finding two values of $x$ such that $f$ has different signs for the two values of $x$, and then to choose values of $x$ within this interval until $f \approx 0$. If the equation has more than one root, you must either find all of the roots, or must make sure that you have found the one that you want. If the equation has complex roots, you must vary both the real and imaginary parts of $x$, which is done conveniently by using a computer program that will carry out arithmetic with complex quantities.

---

### Example 9.6

Use the method of trial and error to solve the equation of Example 9.5.

---

### Solution

We let $f(x) = 2\sin(x) - x$, which vanishes at the root. We find quickly that $f(1) = 0.68294$, and that $f(2) = -0.1814$, so that there must be a root between $x = 1$ and $x = 2$. We find that $f(1.5) = 0.49499$, so the root lies between 1.5 and 2. We find that $f(1.75) = 0.21798$, so the root is larger than 1.75. However, $f(1.9) = -0.00740$, so the root is smaller than 1.9. We find that $f(1.89) = 0.00897$, so the root is between 1.89 and 1.90. We find that $f(1.895) = 0.000809$ and that $f(1.896) = -0.000829$. To five significant digits, the root is $x = 1.8955$.

---

### Problem 9.6

Use the method of trial and error to solve the equation of Problem 9.5.

---

## The Method of Bisection

This is a systematic variation of the method of trial and error and is what was done in Example 9.6. You start with two values of $x$ for which the function $f(x)$ has opposite signs, and then evaluate the function for the midpoint of the interval. If the function has the same sign at the midpoint as at the left end of the interval, the root is in the right half of the interval. The midpoint of the half of the original interval containing the root is taken, and it is determined which half of this new interval contains the root. The method is continued, repeating the process until the interval known to contain the root is as small as twice the error you are willing to tolerate. The middle of the last interval is then taken as the approximation to the root.

Occasionally, a root will occur at which the graph of the function does not cross the $x$ axis, but is tangent to it at the root. The method of bisection will not work for this kind of a root, but the method can be used on the first derivative of the function, which must change sign at such a root.

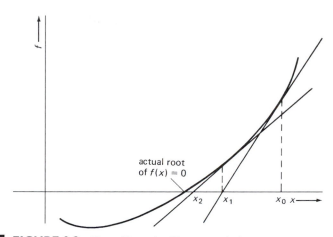

**FIGURE 9.2**   Figure illustrating Newtons method.

---

### Problem 9.7

The equation

$$x^3 - 2x^2 + x - 1 = 0$$

has one positive real root. Find it to five significant digits, using the method of bisection.

---

### Newton's Method

This method, which is also called the *Newton–Raphson method*, is an iterative procedure, as was the method of bisection. An iterative procedure is one that is repeated until the desired degree of accuracy is attained. The procedure is illustrated in Fig. 9.2.

We assume that we have an equation written as in Eq. (9.15). The process is as follows:

*Step 1.* Guess at a value, $x_0$, which is "not too far" from the actual root. A rough graph of the function $f(x)$ can help you to choose a good value for $x_0$.

*Step 2.* Find the value of $f(x)$ and the value of $df/dx$ at $x = x_0$.

*Step 3.* Using the value of $f(x)$ and $df/dx$, find the value of $x$ at which the tangent line to the curve at $x = x_0$ crosses the axis. This value of $x$, which we call $x_1$, is our next approximation to the root. It is given by

$$x_1 = x_0 - \frac{f(x_0)}{f'(x_0)}, \tag{9.16}$$

where we use the notation

$$f'(x_0) = \frac{df}{dx}\bigg|_{x=x_0}. \tag{9.17}$$

*Step 4.* Repeat the process until you are satisfied with the accuracy obtained. The $n$th approximation is given by

$$
\boxed{x_n = x_{n-1} - \frac{f(x_{n-1})}{f'(x_{n-1})}}. \tag{9.18}
$$

### Problem 9.8

Using the definition of the derivative, show that Eqs. (9.16) and (9.18) are correct.

You must decide when to stop your iteration. If the graph of the function $f(x)$ crosses the $x$ axis at the root, you can probably stop when the difference between $x_n$ and $x_{n+1}$ is smaller than the error you can tolerate. However, if the curve becomes tangent to the $x$ axis at the root, the method may converge very slowly near the root. It may be necessary to pick another trial root on the other side of the root and to compare the results from the two iterations. Another possibility is to take the derivative of the function, set that equal to zero, and solve for that root, since the first derivative changes sign if the function is tangent to the $x$ axis.

You should not demand too much of Newton's method. A poor choice of $x_0$ can make the method converge to the wrong root, especially if the function is oscillatory. If your choice of $x_0$ is near a local maximum or a local minimum, the first application of the procedure might give a value of $x_1$ that is nowhere near the desired root. These kinds of problems can be especially troublesome when you are using a computer to do the iterations, because you might not see the values of $x_1$, $x_2$, etc., until the program has finished execution. Remember the first law of computing: "Garbage in, garbage out." Carrying out an approximate graphical solution before you start is a good idea so that you can make a good choice for your first approximation to the root and so that you can know if you found the desired root instead of some other root.

### Example 9.7

Write a computer program in BASIC to carry out Newton's method on Eq. (9.8) and use it to obtain one of the results in Table 9.1.

### Solution

The program and the output follow. The program is written so that iteration stops if the difference between two approximations is less than one part in $10^6$ of the value of the root. If the root were at $x = 0$, this criterion could not be used. In the present case, this criterion should give about five significant digits in the answer.

```
 90 REM PROGRAM ACIDEQ
100 PRINT "THIS PROGRAM SOLVES FOR THE HYDROGEN ION
CONCENTRATION IN A"
110 PRINT "SOLUTION OF A WEAK ACID, USING NEWTON'S
METHOD ON THE CUBIC"
120 PRINT "EQUATION WHICH INCLUDES HYDROGEN IONS
FROM THE DISSOCIATION"
```

```
130 PRINT "OF THE WATER. ACTIVITY COEFFICIENTS ARE
ASSUMED TO EQUAL UNITY."
140 PRINT "ITERATION IS DISCONTINUED WHEN THE NEXT
ITERATION CHANGES THE"
150 PRINT "VALUE OF THE APPROXIMATION TO THE ROOT BY
LESS THAN ONE"
160 PRINT "MILLIONTH OF THE VALUE OF THE ROOT."
200 PRINT "VALUE OF THE ACID DISSOCIATION CONSTANT?"
220 PRINT "GROSS ACID CONCENTRATION IN MOLES/LITER?"
210 INPUT K1
230 INPUT C0
240 PRINT "VALUE OF THE TRIAL ROOT?"
250 INPUT X
255 K0=1.00E-14
260 FOR I=1 TO 100
270 Y=X^3/K1+X^2-(C0+K0/K1)*X - K0
280 Y1=3*X^2/K1+2*X - C0 - K0/K1
290 Z=Y/Y1
300 IF ABS(Z/X) < 1E-6 THEN 400
310 X=X - Z
320 NEXT I
400 PRINT "NUMBER OF ITERATIONS =";I
410 PRINT "HYDROGEN ION CONCENTRATION =";X
420 P= -LOG(X)/2.303
430 PRINT "pH =";P
999 END
```

The output follows:

```
THIS PROGRAM SOLVES FOR THE HYDROGEN ION CONCENTRATION
IN A SOLUTION OF A WEAK ACID, USING NEWTON'S METHOD ON
THE CUBIC EQUATION WHICH INCLUDES HYDROGEN IONS FROM
THE DISSOCIATION OF THE WATER. ACTIVITY COEFFICIENTS
ARE ASSUMED TO EQUAL UNITY. ITERATION IS DISCONTINUED
WHEN THE NEXT ITERATION CHANGES THE VALUE OF THE
APPROXIMATION TO THE ROOT BY LESS THAN ONE MILLIONTH
OF THE VALUE OF THE ROOT.
VALUE OF THE ACID DISSOCIATION CONSTANT?
? 1.754E-5
GROSS ACID CONCENTRATION IN MOLES/LITER?
? 1E-3
VALUE OF THE TRIAL ROOT?
? 1E-4
NUMBER OF ITERATIONS = 5
HYDROGEN ION CONCENTRATION = .123959E-03
pH = 3.90602
```

## SECTION 9.4. SIMULTANEOUS EQUATIONS: SOLUTION OF TWO EQUATIONS WITH TWO UNKNOWNS

We will discuss only algebraic methods for solving simultaneous equations. However, Newton's method can be modified to work on simultaneous equations.[3] The simplest set of simultaneous equations is

$$a_{11}x + a_{12}y = c_1 \tag{9.19a}$$
$$a_{21}x + a_{22}y = c_2, \tag{9.19b}$$

where the $a$'s and the $c$'s are constants. The set of equations in Eq. (9.19) is called *linear*, because the unknowns $x$ and $y$ enter only to the first power, and is called *inhomogeneous*, because there are constant terms, not containing $x$ or $y$. If certain conditions are met, such a set of equations can be solved for a solution set consisting of a single value of $x$ and a single value of $y$.

### The Method of Substitution

The first step of this method is to solve one equation to give one variable as a function of the other. This function then is substituted into the other equation to give an equation in one unknown which can be solved. The result is then substituted into either of the original equations, which is then solved for the second variable.

___

### Example 9.8

Use the method of substitution on Eq. (9.19).

___

### Solution

We solve Eq. (9.19a) for $y$ in terms of $x$:

$$y = \frac{-a_{11}x}{a_{12}} + \frac{c_1}{a_{12}}. \tag{9.20}$$

We substitute this into Eq. (9.22b):

$$a_{21}x + a_{22}\left(\frac{c_1}{a_{12}} - \frac{a_{11}x}{a_{12}}\right) = c_2. \tag{9.21}$$

This contains only $x$ and not $y$, so it can be solved for $x$ to give the root

$$x = \frac{c_1 a_{22} - c_2 a_{12}}{a_{11}a_{22} - a_{12}a_{21}}. \tag{9.22}$$

Equation (9.22) can be substituted into Eq. (9.19a) or (9.19b) to yield

$$y = \frac{c_2 a_{11} - c_1 a_{21}}{a_{11}a_{22} - a_{12}a_{21}}. \tag{9.23}$$

___

[3]S. D. Conte and Carl de Boor, *Elementary Numerical Analysis*, 2nd ed., pp. 86ff, McGraw–Hill, New York, 1972.

**Problem 9.9**

Do the algebraic manipulations to obtain Eq. (9.23).

The method of substitution is not limited to two equations and is not limited to linear equations. However, with several equations it is tedious, since you usually have to solve for one variable in terms of all the others, substitute, and then repeat this procedure with other variables.

**Problem 9.10**

Solve the pair of simultaneous equations by the method of substitution:

$$x^2 - 2xy - x = 0$$
$$\frac{1}{x} + \frac{1}{y} = 2.$$

Hint. Multiply the second equation by $xy$ before proceeding.

In Problem 9.10 there are two solution sets, since the first equation is quadratic in $x$. When it is solved after substituting to eliminate $y$, two values of $x$ are found to satisfy the equation, and there is a root in $y$ for each of these. Nonlinear equations are more difficult to handle than linear equations, so we consider only linear equations for the rest of this section.

## The Method of Elimination

This method is used with linear equations. It applies the process of subtracting one equation from another to obtain a simpler equation. That is, the left-hand side of the first equation is subtracted from the left-hand side of the second equation and the right-hand side of the first equation is subtracted from the right-hand side of the second to yield a new equation that is simpler.

**Example 9.9**

Solve the following pair of equations:

$$x + y = 3$$
$$2x + y = 0.$$

**Solution**

We subtract the first equation from the second to obtain

$$x = -3.$$

This is substituted into either of the original equations to obtain

$$y = 6.$$

If necessary to get a simpler equation, you can multiply one of the equations by a constant before taking the difference, and it is possible to add the equations instead of subtracting.

---

**Problem 9.11**

Solve the set of equations

$$3x + 2y = 40$$
$$2x - y = 10.$$

---

There are two common difficulties that can arise with pairs of simultaneous equations: These are (1) that the equations might be inconsistent, and (2) that the equations might not be independent. If two equations are *inconsistent*, there is no solution that can satisfy both of them, and if the equations are not *independent*, they express the same information, so that there is really only one equation, which can be solved for one variable in terms of the other, but cannot be solved to give a constant solution set.

---

**Example 9.10**

Show that the pair of equations is inconsistent:

$$2x + 3y = 15$$
$$4x + 6y = 45.$$

---

**Solution**

We attempt solution by elimination. We multiply the first equation by 2 and subtract the second from the first, obtaining

$$0 = -15$$

which is obviously not correct. The equations are inconsistent.

---

**Example 9.11**

Show that the equations are not independent:

$$3x + 4y = 7$$
$$6x + 8y = 14.$$

---

**Solution**

We attempt a solution by elimination, multiplying the first equation by 2. However, this makes the two equations identical, so that if we subtract one from the other, we obtain

$$0 = 0$$

which is correct, but not of any use to us. We have just one independent equation

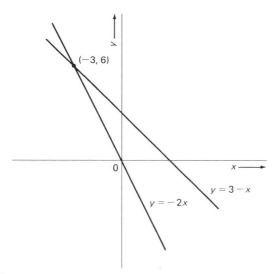

**FIGURE 9.3**   Graphical representation of the two consistent and linearly independent equations of Example 9.9.

instead of two, so that we could solve for $x$ in terms of $y$ or for $y$ in terms of $x$, but not for values of either $x$ or $y$.

We can understand consistency and independence in simultaneous equations by looking at the graphs of the equations. Each of the equations represents $y$ as a function of $x$. With linear equations, these functions are represented by straight lines. Figure 9.3 shows the two lines representing the two equations of Example 9.9. The two equations are consistent and independent, and the lines cross at the point whose coordinates represent the solution set, consisting of a value of $x$ and a value of $y$.

Figure 9.4 shows the lines representing the two equations of Example 9.10. Since the lines do not cross, there is no solution to this pair of inconsistent equations. A single line represents both of the equations of Example 9.11. Any point on the line satisfies both equations, which are *linearly dependent*.

The concept of linear dependence is important in the study of homogeneous linear equations. A pair of *homogeneous linear equations* is similar to those of Eq. (9.19) except that both $c_1$ and $c_2$ vanish. Consider the set of homogeneous linear equations:

$$a_{11}x + a_{12}y = 0 \qquad (9.24a)$$
$$a_{21}x + a_{22}y = 0. \qquad (9.24b)$$

We solve both of these equations for $y$ in terms of $x$:

$$y = \frac{-a_{11}x}{a_{12}} \qquad (9.25a)$$

$$y = \frac{-a_{21}x}{a_{22}}. \qquad (9.25b)$$

Both of these functions are represented by straight lines with zero intercept. There

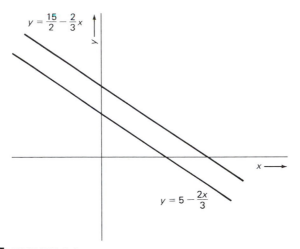

**FIGURE 9.4**    Graphical representation of two inconsistent equations of Example 9.10.

are two possibilities: Either the lines cross at the origin, or they coincide everywhere. In other words, either $x = 0$, $y = 0$ is the solution, or else the equations are linearly dependent. The solution $x = 0$, $y = 0$ is called a *trivial solution*. The two equations must be linearly dependent in order for a nontrivial solution to exist. A nontrivial solution consists of specifying $y$ as a function of $x$, not in finding constant values for both $x$ and $y$. In a later section, we will see an important application of homogeneous linear equations.

### Problem 9.12

Determine whether the set of equations has a nontrivial solution, and find the solution if it exists:

$$7x + 15y = 0$$
$$101x + 195y = 0.$$

## SECTION 9.5. SIMULTANEOUS EQUATIONS WITH MORE THAN TWO UNKNOWNS

Systems of several equations are similar to pairs of equations for two unknowns. For a unique solution, we must have $n$ independent and consistent equations to solve for $n$ unknowns. Sometimes, in practical calculations, you will have more equations than you have unknowns. If all of them are consistent, you can simply pick out $n$ independent equations to solve for $n$ unknowns. If the equations are not all consistent, you have what is called an *overdetermined system of equations*, which has no solution.

If the equations arise from experimental measurements, the source of inconsistency is experimental error. In this case, you can pick various sets of $n$ equations and solve them separately, presumably getting slightly different answers because of experimental error. The variation can be used to get an idea of the effects of the errors.

With sets of several equations, the methods of substitution and elimination can still be used. However, it is convenient to have systematic ways of applying these methods. Two such methods are the application of Cramer's rule and Gauss–Jordan elimination. A third systematic method is matrix inversion.

## Cramer's Rule

This method is equivalent to the method of substitution. To describe the method, we write the set of two linear inhomogeneous equations of Eq. (9.19) in matrix form,

$$\mathbf{AX} = \mathbf{C}, \tag{9.26}$$

where $\mathbf{A}$ is a square matrix, and $\mathbf{X}$ and $\mathbf{C}$ are column vectors (matrices with only one column):

$$\begin{bmatrix} a_{11} & a_{12} \\ a_{21} & a_{22} \end{bmatrix} \begin{bmatrix} x_1 \\ x_2 \end{bmatrix} = \begin{bmatrix} c_1 \\ c_2 \end{bmatrix}. \tag{9.27}$$

---

### Problem 9.13

Use the rules of matrix multiplication to show that Eq. (9.27) is identical with Eq. (9.19).

---

According to Cramer's rule (which we do not prove) the solutions are written as quotients of determinants. The solutions to Eq. (9.19) are given in Eqs. (9.22) and (9.23), and can be rewritten as quotients of determinants:

$$x = \frac{\begin{vmatrix} c_1 & a_{12} \\ c_2 & a_{22} \end{vmatrix}}{\begin{vmatrix} a_{11} & a_{12} \\ a_{21} & a_{22} \end{vmatrix}} \tag{9.28}$$

$$y = \frac{\begin{vmatrix} a_{11} & c_1 \\ a_{21} & c_2 \end{vmatrix}}{\begin{vmatrix} a_{11} & a_{12} \\ a_{21} & a_{22} \end{vmatrix}}. \tag{9.29}$$

Cramer's rule is that the solutions are constructed as follows: Each variable is given as the quotient of two determinants. The denominator in each expression is the determinant of the matrix $\mathbf{A}$, and the numerator is the determinant of this matrix with one of the columns replaced by the column vector $\begin{pmatrix} c_1 \\ c_2 \end{pmatrix}$. In the expression for $x$, the column of coefficients for $x$ is replaced, and in the expression for $y$, the column of coefficients for $y$ is replaced.

**Problem 9.14**

Use Cramer's rule to solve

$$4x + y = 13$$
$$2x - 3y = 3.$$

If there are more than two variables and more than two linear inhomogeneous equations, Cramer's rule gives exactly the same pattern. For the set

$$a_{11}x_1 + a_{12}x_2 + a_{13}x_3 = c_3 \qquad (9.30a)$$
$$a_{21}x_1 + a_{22}x_2 + a_{23}x_3 = c_2 \qquad (9.30b)$$
$$a_{31}x_1 + a_{32}x_2 + a_{33}x_3 = c_3 \qquad (9.30c)$$

the value of $x_1$ is given by

$$x_1 = \frac{\begin{vmatrix} c_1 & a_{12} & a_{13} \\ c_2 & a_{22} & a_{23} \\ c_3 & a_{32} & a_{33} \end{vmatrix}}{\det(\mathbf{A})}, \qquad (9.31)$$

where $\det(\mathbf{A})$ is the determinant of the 3 by 3 matrix of the $a$ coefficients. The values of $x_2$ and $x_3$ are given by similar expressions, with the second column in the determinant in the numerator replaced by the constants $c_1$, $c_2$, and $c_3$ in the expression for $x_2$, and with the third column replaced in the expression for $x_3$. Cramer's rule for more than three equations is completely analogous to this.

Equation (9.30) can be written in matrix notation as

$$\begin{bmatrix} a_{11} & a_{12} & a_{13} \\ a_{21} & a_{22} & a_{23} \\ a_{31} & a_{32} & a_{33} \end{bmatrix} \begin{bmatrix} x_1 \\ x_2 \\ x_3 \end{bmatrix} = \begin{bmatrix} c_1 \\ c_2 \\ c_3 \end{bmatrix} \qquad (9.32)$$

or

$$\mathbf{AX} = \mathbf{C}, \qquad (9.33)$$

where $\mathbf{X}$ and $\mathbf{C}$ are the column vectors shown.

Let $\mathbf{A}_n$ be the matrix which is obtained from $\mathbf{A}$ by replacing the $n$th column by the column vector $\mathbf{C}$. Cramer's rule is now written

$$x_n = \frac{\det(\mathbf{A}_n)}{\det(\mathbf{A})}, \qquad (9.34)$$

**Example 9.12**

Use Cramer's rule to find the value of $x_1$ that satisfies

$$\begin{bmatrix} 2 & 4 & 1 \\ 1 & -1 & 1 \\ 1 & 1 & 1 \end{bmatrix} \begin{bmatrix} x_1 \\ x_2 \\ x_3 \end{bmatrix} = \begin{bmatrix} 21 \\ 4 \\ 10 \end{bmatrix}. \qquad (9.35)$$

**Solution**

$$x_1 = \frac{\begin{vmatrix} 21 & 4 & 1 \\ 4 & -1 & 1 \\ 10 & 1 & 1 \end{vmatrix}}{\begin{vmatrix} 2 & 4 & 1 \\ 1 & -1 & 1 \\ 1 & 1 & 1 \end{vmatrix}} = \frac{21\begin{vmatrix} -1 & 1 \\ 1 & 1 \end{vmatrix} - 4\begin{vmatrix} 4 & 1 \\ 1 & 1 \end{vmatrix} + 10\begin{vmatrix} 4 & 1 \\ -1 & 1 \end{vmatrix}}{2\begin{vmatrix} -1 & 1 \\ 1 & 1 \end{vmatrix} - 1\begin{vmatrix} 4 & 1 \\ 1 & 1 \end{vmatrix} + 1\begin{vmatrix} 4 & 1 \\ -1 & 1 \end{vmatrix}}$$

$$= \frac{21(-1-1) - 4(4-1) + 10(4+1)}{2(-1-1) - (4-1) + (4+1)} = \frac{-4}{-2} = 2.$$

**Problem 9.15**

Find the values of $x_2$ and $x_3$ that satisfy Eq. (9.35).

## Linear Dependence and Inconsistency

We will not discuss completely the questions of consistency and independence for sets of more than two equations, but we will make the following comments, which apply to sets of linear inhomogeneous equations:

1. A set of $n$ equations is said to be *linearly dependent* if a set of constants $b_1, b_2, \ldots, b_n$, not all equal to zero, can be found such that if the first equation is multiplied by $b_1$, the second equation by $b_2$, the third equation by $b_3$, etc., the equations add to zero for all values of the variables. A simple example of linear dependence is for two of the equations to be identical. In this case, we could multiply one of these equations by $+1$ and the other by $-1$ and all of the remaining equations by 0 and have the equations sum to zero. If two equations are identical, one has of course only $n - 1$ usable equations and cannot solve for $n$ variables. More complicated types of linear dependence also occur, and in this case one still has only $n - 1$ usable equations or fewer.

2. In the case of two identical equations, the determinant of the matrix **A** vanishes, from property 2 of determinants, described in Section 8.5. The determinant of **A** will also vanish in more complicated types of linear dependence. If det(**A**) vanishes, either the equations are linearly dependent or they are inconsistent.

3. It is possible for a set of equations to appear to be overdetermined and not actually be overdetermined if some of the set of equations are linearly dependent.

**Problem 9.16**

See if the set of four equations in three unknowns can be solved:

$$x_1 + x_2 + x_3 = 6$$
$$x_1 + x_2 + x_3 = 0$$
$$3x_1 + 3x_2 + x_3 = 12$$
$$2x_1 + x_2 + 4x_3 = 16.$$

### Solution by Matrix Inversion

We have written two sets of linear inhomogeneous equations in matrix notation in Eqs. (9.27) and (9.32), and the same notation can be used for $n$ such equations in $n$ unknowns. We write

$$\boxed{\mathbf{AX} = \mathbf{C}}, \tag{9.36}$$

where $\mathbf{A}$ is now an $n$ by $n$ square matrix, $\mathbf{X}$ is an $n$ by 1 column matrix (column vector), and $\mathbf{C}$ is another $n$ by 1 column matrix.

If we possess the inverse of $\mathbf{A}$, we can multiply both sides of Eq. (9.36) on the left by $\mathbf{A}^{-1}$ to get

$$\boxed{\mathbf{A}^{-1}\mathbf{AX} = \mathbf{X} = \mathbf{A}^{-1}\mathbf{C}}. \tag{9.37}$$

That is, we form the matrix product of $\mathbf{A}^{-1}$ and $\mathbf{C}$. This column matrix is the same as $\mathbf{X}$, so we read off our solution from this column matrix. In order for a matrix to possess an inverse, it must be *nonsingular*, which means that its determinant does not vanish. If the matrix is singular, the system must be linearly independent or inconsistent.

The difficulty with carrying out this procedure by hand is that it is probably more work to invert an $n$ by $n$ matrix than to solve the set of equations by other means. However, with access to TrueBASIC or another computer language that automatically inverts matrices, you can solve such a set of equations very quickly.

---

### Problem 9.17

Solve the following set of simultaneous equations by matrix inversion:

$$2x_1 + x_2 = 1$$
$$x_1 + 2x_2 + x_3 = 2$$
$$x_2 + 2x_3 = 3.$$

The inverse of the matrix has already been obtained in Example 8.10, or you can use the BASIC program in Example 11.13 to obtain the inverse.

---

### Gauss–Jordan Elimination

This is a systematic procedure for carrying out the method of elimination. It is very similar to the Gauss–Jordan method for finding the inverse of a matrix, described in Section 8.4. If the set of equations is written as in Eq. (9.36), we write the augmented matrix consisting of the $\mathbf{A}$ matrix and the $\mathbf{C}$ column vector. For a set of four equations, the augmented matrix is

$$\left[\begin{array}{cccc|c} a_{11} & a_{12} & a_{13} & a_{14} & c_1 \\ a_{21} & a_{22} & a_{23} & a_{24} & c_2 \\ a_{31} & a_{32} & a_{33} & a_{34} & c_3 \\ a_{41} & a_{42} & a_{42} & a_{44} & c_4 \end{array}\right]. \tag{9.38}$$

Row operations are carried out on this augmented matrix: each row can be multiplied by a constant, and one row can be subtracted from or added to another row. These operations will not change the roots to the set of equations, since such operations are equivalent to multiplying one of the equations by a constant or to taking the sum or difference of two equations. In Gauss–Jordan elimination, our aim is to transform the left part of the augmented matrix into the identity matrix, which will transform the right column into the four roots, since the set of equations will then be

$$\mathbf{EX} = \mathbf{C'}. \tag{9.39}$$

The row operations are carried out exactly as in Section 8.4 except for having only one column in the right part of the augmented matrix.

---

### Problem 9.18

Use Gauss–Jordan elimination to solve the set of simultaneous equations in Problem 9.17. The same row operations will be required that were used in Example 8.10.

---

There is a similar procedure known as *Gauss elimination*, in which row operations are carried out until the left part of the augmented matrix is in upper triangular form. The bottom row of the augmented matrix then provides the root for one variable. This is substituted into the equation represented by the next-to-bottom row, and it is solved to give the root for the second variable. The two values are substituted into the next equation, etc.

## Linear Homogeneous Equations

In Section 9.4 we discussed pairs of linear homogeneous equations for two variables. We found that such a pair of equations needed to be linearly dependent in order to have a solution other than the trivial solution $x = 0$, $y = 0$. The same is true of sets with more than two variables.

A set of three homogeneous equations in three unknowns is written

$$a_{11}x_1 + a_{12}x_2 + a_{13}x_3 = 0 \tag{9.40a}$$
$$a_{21}x_1 + a_{22}x_2 + a_{23}x_3 = 0 \tag{9.40b}$$
$$a_{31}x_2 + a_{32}x_2 + a_{33}x_3 = 0. \tag{9.40c}$$

If we attempt to apply Cramer's rule to this set of equations, without asking whether it is legitimate to do so, we find for example that

$$x_1 = \frac{\begin{vmatrix} 0 & a_{12} & a_{13} \\ 0 & a_{22} & a_{23} \\ 0 & a_{32} & a_{33} \end{vmatrix}}{\det(\mathbf{A})}. \tag{9.41}$$

If $\det(\mathbf{A}) \neq 0$, this yields $x_1 = 0$, and similar equations will also give $x_2 = 0$ and $x_3 = 0$. This trivial solution is all that we can have if the determinant of $\mathbf{A}$ is nonzero (i.e., if the three equations are independent). To have a nontrivial solution, the equations must be linearly dependent.

In order to find a possible nontrivial solution, we investigate the condition

$$\det(\mathbf{A}) = 0 \tag{9.42}$$

which in the 3 by 3 case is the same as

$$a_{11}a_{22}a_{33} - a_{11}a_{23}a_{32} - a_{12}a_{21}a_{33} - a_{13}a_{22}a_{31} + a_{12}a_{23}a_{31} + a_{13}a_{21}a_{32} = 0. \tag{9.43}$$

One case in which a set of linear homogeneous equations arises is the *matrix eigenvalue problem*. This is very similar to an eigenvalue equation for an operator, as in Eq. (8.3). The problem is to find a column vector, **X**, and a scalar *eigenvalue b*, such that

$$\mathbf{BX} = b\mathbf{X}, \tag{9.44}$$

where **B** is a given square matrix and **X** is a column vector (the *eigenvector*). Since the right-hand side of Eq. (9.44) is the same as $b\mathbf{EX}$, where **E** is the identity matrix, we can rewrite Eq. (9.44) as

$$(\mathbf{B} - b\mathbf{E})\mathbf{X} = 0 \tag{9.45}$$

which is a set of linear homogeneous equations written in the notation of Eq. (9.36).

---

### Example 9.13

Find the values of $b$ and **X** that satisfy the eigenvalue equation

$$\begin{bmatrix} 1 & 1 & 0 \\ 1 & 1 & 1 \\ 0 & 1 & 1 \end{bmatrix} \begin{bmatrix} x_1 \\ x_2 \\ x_3 \end{bmatrix} = b \begin{bmatrix} x_1 \\ x_2 \\ x_3 \end{bmatrix} \tag{9.46}$$

and also satisfy a "normalization" condition:

$$x_1^2 + x_2^2 + x_3^2 = 1. \tag{9.47}$$

Since the equations must be linearly dependent, this additional equation will provide unique values for the three variables.

---

### Solution

In the form of Eq. (9.45),

$$\begin{bmatrix} 1-b & 1 & 0 \\ 1 & 1-b & 1 \\ 0 & 1 & 1-b \end{bmatrix} \begin{bmatrix} x_1 \\ x_2 \\ x_3 \end{bmatrix} = 0. \tag{9.48}$$

The condition that corresponds to Eq. (9.42) is

$$\begin{vmatrix} y & 1 & 0 \\ 1 & y & 1 \\ 0 & 1 & y \end{vmatrix} = y \begin{vmatrix} y & 1 \\ 1 & y \end{vmatrix} - 1 \begin{vmatrix} 1 & 1 \\ 0 & y \end{vmatrix} = y^3 - 2y = 0, \tag{9.49}$$

where we temporarily let $y = 1 - b$. Equation (9.49) is a cubic equation that can be solved by factoring. It has the three roots

$$y = 0, \qquad y = \sqrt{2}, \qquad y = -\sqrt{2}$$

or

$$b = 1, \qquad b = 1 - \sqrt{2}, \qquad b = 1 + \sqrt{2}. \tag{9.50}$$

The three roots in Eq. (9.50) are three different eigenvalues. It is only when $b$ is equal to one of these three values that Eq. (9.46) has a nontrivial solution. Since we have three values of $b$, we have three different eigenvectors. We find the eigenvectors by substituting each value of $b$ in turn into Eq. (9.48) and solving the set of equations. We begin with $b = 1$ and write

$$0 + x_2 + 0 = 0 \tag{9.51a}$$
$$x_1 + 0 + x_3 = 0 \tag{9.51b}$$
$$0 + x_2 + 0 = 0. \tag{9.51c}$$

It is now obvious that this set of equations is linearly dependent, as required, since the first and third equations are the same. Our solution is now

$$x_2 = 0 \tag{9.52a}$$
$$x_1 = -x_3. \tag{9.52b}$$

We have solved for two of the variables in terms of the third. Since we have only two independent equations, we do not have definite values for $x_1$ and $x_3$ until we apply the normalization condition of Eq. (9.47). Imposing it, we find for our first eigenvector

$$\mathbf{X} = \begin{bmatrix} 1/\sqrt{2} \\ 0 \\ -1/\sqrt{2} \end{bmatrix}. \tag{9.53}$$

The negative of this eigenvector could also have been taken.

We now seek the second eigenvector, for which $y = \sqrt{2}$, or $b = 1 - \sqrt{2}$. Equation (9.48) becomes

$$\sqrt{2}x_1 + x_2 + 0 = 0$$
$$x_1 + \sqrt{2}x^2 + 0x_3 = 0 \tag{9.54}$$
$$0 + x_2 + \sqrt{2}x_3 = 0.$$

With the normalization condition, the solution to this is

$$\mathbf{X} = \begin{bmatrix} 1/\sqrt{2} \\ -1/\sqrt{2} \\ 1/2 \end{bmatrix}. \tag{9.55}$$

---

**Problem 9.19**

    a. Verify Equation (9.55). Show that this is an eigenvector.
    b. Find the third eigenvector for the problem of Example 9.13.

---

In quantum chemistry, orbital functions are often represented as linear combinations of functions from a basis set containing several functions. In this treatment, a set of simultaneous equations very similar to Eq. (9.45) arises that is to be solved for the coefficients in the linear combinations. The condition analogous to Eq. (9.42) is called a *secular equation*, and the eigenvalue $b$ in Eq. (9.45) is replaced by the orbital energy. The simplest theory using this representation for molecular orbitals is the Hückel method,[4] which is known as a semi-empirical method because it relies on experimental data to establish values for certain integrals that occur in the theory while assuming that certain other integrals vanish.

General algorithms exist for finding the eigenvalues of matrices. A popular method is the Jacobi method, which is applied to real symmetric matrices. A discussion is included in the book by Bell, listed at the end of the chapter.

## SUMMARY OF THE CHAPTER

If you have one equation in one unknown, it can be written in the form $f(x) = 0$. Solving the equation means finding those constant values of $x$ such that the equation is correct. In some cases, these values can be obtained by algebraic manipulation such that the relation

$$x = \text{constant}$$

is equivalent to the original equation. For example, the quadratic formula provides two such values of $x$. In other cases, the algebraic manipulation is too difficult to justify the effort or is impossible using known methods. In these cases, graphical or numerical approximation to the root is used. In the graphical method, a graph of $f(x)$ is produced, and the values of $x$ where the curve crosses the $x$ axis are the real roots. The first numerical method discussed was the method of trial and error, along with the method bisection, which is a systematized trial and error method. The other method presented was the method of Newton. In this method, a "trial root" is obtained by guesswork or a rough graphical solution, and values of $f$ and $f'$, the derivative of $f$, are calculated at this value of $x$. These values are used to determine where the tangent line at the trial root crosses the $x$ axis, and this point is used as the next approximation to the root. The method is repeated, or iterated, until a usable approximation to the root is obtained.

To solve for numerical values of two variables, two equations are required, and they must be solved simultaneously, and similarly for more variables. We presented several methods for solving simultaneous equations. First was the method of substitution, which is not limited to linear equations, but which is not practical for more than two or three equations. We then presented several methods which can apply to sets of linear inhomogeneous equations. Cramer's method is a method which uses determinants to obtain the roots. A set of linear equations can be written in matrix form and can be solved by finding the inverse of the matrix of the

---

[4]Donald J. Royer, *Bonding Theory*, pp. 154–163, McGraw–Hill, New York, 1968.

coefficients. The methods of Gauss elimination and Gauss–Jordan elimination were presented.

Finally we examined linear homogeneous equations. With such a set, the equations possess only a trivial solution if the equations are linearly independent. The condition of dependence which must occur in order to have a nontrivial solution is represented by an equation in which the determinant of the matrix of the coefficients is set equal to zero. Matrix eigenvalue equations fall into this category, and we discussed the determination of the eigenvalues and eigenvectors.

## ADDITIONAL READING

W. W. Bell, *Matrices for Scientists and Engineers*, Van Nostrand–Reinhold, New York, 1975. This is a small book that is designed for scientists and engineers rather than for mathematicians and is relatively understandable.

S. D. Conte and Carl de Boor, *Elementary Numerical Analysis*, 2nd ed., McGraw–Hill, New York, 1972. This is a standard beginning textbook in numerical analysis. It includes discussion of Newton's method and of the solution of simultaneous equations.

Lois W. Griffiths, *Introduction to the Theory of Equations*, 2nd ed., Wiley, New York, 1947. This book is an introduction to the theory and practice of solving equations. It includes discussion of single-variable and simultaneous equations, and also of numerical methods for approximations to roots.

Robert W. Hornbeck, *Numerical Methods*, Quantum Publishers, New York, 1975. This is a clearly written book which is available in an inexpensive paperbook version. It has a good discussion on numerical approximations to roots of equations, and another on simultaneous equations and matrix inversion. The discussions include warnings about difficulties that might arise.

Robert G. Mortimer, *Physical Chemistry*, Benjamin–Cummings, Redwood City, CA, 1993. This is a standard physical chemistry textbook. It includes a treatment of the ionization of weak acids in the case that hydrogen ions from the water solvent must be considered.

J. Murphy, D. Ridout, and Bridgid McShane, *Numerical Analysis, Algorithms and Computation*, Wiley, New York, 1988. This book is a good source of algorithms for numerical solutions of differential equations and simultaneous algebraic equations.

J. V. Uspensky, *Theory of Equations*, McGraw–Hill, New York, 1948. This is another book in the same area as Griffiths.

## ADDITIONAL PROBLEMS

### 9.20

Solve the quadratic equations:

a. $x_2 - 3x + 2 = 0$
b. $x^2 - 1 = 0$
c. $x_2 + 2x + 2 = 0$.

**9.21**

Solve the following equations by factorization:

a. $4x^4 - 4x^2 - x - 1 = 0$
b. $x^3 + x^2 - x - 1 = 0$
c. $x^4 - 1 = 0.$

**9.22**

Find the real roots of the following equations by graphing:

a. $x^3 - x^2 + x - 1 = 0$
b. $e^{-x} - 0.5x = 0$
c. $\sin(x)/x - 0.75 = 0.$

**9.23**

Solve the equations of Problem 9.22 numerically, using either Newton's method or the method of bisection.

**9.24**

Repeat the calculation of Example 9.4 without linearizing the exponential function.

a. Use the method of bisection.
b. Use Newton's method.

**9.25**

The van der Waals equation of state is

$$\left(P + \frac{a}{\bar{V}^2}\right)(\bar{V} - b) = RT,$$

where $a$ and $b$ are parameters (not necessarily equal to the parameters with the same symbols in the Dieterici equation of state). For carbon dioxide, $a = 0.3640\,\text{Pa m}^6$ $\text{mol}^{-2}$, $b = 4.267 \times 10^{-5}\,\text{m}^3\,\text{mol}^{-1}$. Write the van der Waals equation of state as a cubic equation in $\bar{V}$. Solve for the value of $\bar{V}$ under the conditions of Example 9.4, using the method of bisection or Newton's method.

**9.26**

Use Newton's method to calculate the pH of a 0.01 molar solution of lactic acid, $C_3H_6O_3$ at $25°C$. The acid dissociation constant, $K_a$, is equal to $1.38 \times 10^{-4}\,\text{mol L}^{-1}$ at this temperature. Use Eqs. (9.10) and (9.12) to calculate the pH and comment on the accuracy of these two approximations.

**9.27**

An approximate equation for the ionization of a weak acid, including consideration of the hydrogen ions from water is[5]

$$[H^+]/c° = \sqrt{K_a c/c° + K_w},$$

---

[5] Henry F. Holtzclaw, Jr., William R. Robinson, and Jerone D. Odom, *General Chemistry*, 9th ed., p. 545, Heath, Lexington, MA, 1991.

where c is the gross acid concentration. This equation is based on the assumption that the concentration of unionized acid is approximately equal to the gross acid concentration. Consider a solution of HCN (hydrocyanic acid) with gross acid concentration equal to $1.00 \times 10^{-5}$ mol L$^{-1}$. $K_a = 4 \times 10^{-10}$ for HCN.

   a. Calculate [H$^+$] using this equation.
   b. Calculate [H$^+$] using Eq. (9.9) and either bisection or Newton's method.
   c. Calculate [H$^+$] using Eq. (9.10).

Comment on your answers.

**9.28**

Write a computer program in BASIC to carry out the method of bisection.

**9.29**

Find the smallest positive root of the equation.

$$\sinh(x) - x^2 - x = 0.$$

**9.30**

In Planck's theory of blackbody radiation, the spectral radiant emittance $\eta$ is given by the formula[6]

$$\eta(\lambda) = \frac{2\pi h c^2}{\lambda^5 (e^{hc/\lambda k_B T} - 1)},$$

where $\lambda$ is the wavelength, $h$ is Planck's constant, $c$ is the speed of light, $k_B$ is Boltzmann's constant, and $T$ is the absolute temperature. Set the first derivative equal to zero and solve the resulting equation numerically to obtain a relation of the form

$$\lambda_{max} = \frac{\text{constant}}{T}.$$

**9.31**

Solve the set of equations, using Cramer's rule,

$$3x_1 + x_2 + x_3 = 19$$
$$x_1 - 2x_2 + 3x_3 = 13$$
$$x_1 + 2x_2 + 2x_3 = 23.$$

**9.32**

Solve the set of equations, using Gauss or Gauss–Jordan elimination.

$$x_1 + x_2 + x_3 = 9$$
$$2x_1 - x_2 - x_3 = 9$$
$$x_1 + 2x_2 - x_3 = 9.$$

---

[6]R. G. Mortimer, *op. cit.*, p. 349.

**9.33**

Determine which, if any, of the following sets of equations are inconsistent or linearly dependent. Draw a graph for each set of equations, showing both equations. Find the solution for any set that has a unique solution.

a.  $x + 3y = 4$         b.  $2x + 4y = 24$         c.  $3x_1 + 4x_2 = 10$
$\quad$ $2x + 6y = 8$         $\quad$ $x + 2y = 8$         $\quad$ $4x_1 - 2x_2 = 6.$

**9.34**

Solve the set of equations

$$x^2 - 2xy + y^2 = 0$$
$$2x + 3y = 5.$$

**9.35**

Solve the sets of equations:

a.  $3x_1 + 4x_2 + 5x_3 = 25$
$\quad$ $4x_1 + 3x_2 - 6x_3 = -7$
$\quad$ $x_1 + x_2 + x_3 = 6$

b.  $\begin{bmatrix} 1 & 1 & 1 & 3 \\ 2 & 1 & 1 & 1 \\ 1 & 2 & 3 & 4 \\ 2 & 0 & 1 & 4 \end{bmatrix} \begin{bmatrix} x_1 \\ x_2 \\ x_3 \\ x_4 \end{bmatrix} = \begin{bmatrix} 6 \\ 5 \\ 10 \\ 7 \end{bmatrix}.$

**9.36**

Find the eigenvalues and eigenvectors of the matrix

$$\begin{bmatrix} 1 & 1 & 1 \\ 1 & 1 & 1 \\ 1 & 1 & 1 \end{bmatrix}.$$

# 10
# THE TREATMENT OF EXPERIMENTAL DATA

## Preview

Sometimes quantities in which we are interested can be measured directly. More often, they must be calculated from other quantities that can be measured. This calculation process is called data reduction. The simplest form of data reduction is the use of a formula into which measured values are substituted. Other forms of data reduction involve a set of data that can be represented by data points on a graph. Construction of such a graph and analysis of features of the graph, such as slopes and intercepts of lines, can provide values of variables. Statistical analysis done numerically can replace graphical analysis, providing better accuracy with less effort. We discuss both of these approaches.

We also discuss the analysis of the accuracy of experimental data and the effects of errors in measured quantities on the calculated values of other variables. In the case that we can directly measure some desired quantity, we need to estimate the accuracy of the measurement. If data reduction must be carried out, we must study the propagation of errors in measurements through the data reduction process. The two principal types of experimental errors, random errors and systematic errors, are discussed separately. Random errors are subject to statistical analysis, and we discuss this analysis.

## Principal Facts and Ideas

1. Every measured quantity is subject to experimental error.
2. When the value of a measured quantity is reported, an estimate of the expected error should be included.

3. Random experimental errors can be analyzed statistically, if the measurement can be repeated a number of times.
4. Systematic errors must usually be estimated by educated guesswork.
5. The mean of a set of repeated measurements is a better estimate of a variable than is a single measurement.
6. The probable error in the mean of a set of repeated measurements can be determined statistically.
7. When a variable is calculated by substitution of measured quantities into a formula, the estimated errors in the measured quantities can be propagated through the calculation.
8. Another type of data reduction involves fitting a set of data to a formula. This can be done graphically or numerically by use of the least-squares (regression) procedure.

## Objectives

After studying the chapter, you should be able to:

1. identify probable sources of error in a physical chemistry experiment and classify the errors as systematic or random;
2. calculate the mean and standard deviation of a sample of numbers;
3. calculate the probable error in the measured value of a directly measured quantity;
4. carry out data reduction using mathematical formulas and do an error propagation calculation to determine the probable error in the final calculated quantity;
5. carry out data reduction using graphical methods and determine the probable error in quantities obtained from the graphs;
6. carry out data reduction numerically using least squares methods and determine probable errors in quantities obtained by these methods.

## SECTION 10.1. EXPERIMENTAL ERRORS IN DIRECTLY MEASURED QUANTITIES

Once we have obtained a value for a directly measured quantity, we should determine how accurate that value is, since experimental error is always present. The best way to begin the analysis of experimental error is to repeat the measurement a number of times. If the repetitions agree well with each other, the set of data is said to have good *precision*. If the set of data agrees well with the correct value, the data set is said to have good *accuracy*. We do not usually know what the correct value is, so we generally can only estimate the accuracy. It is tempting to assume that a data set of high precision also has high accuracy, but this can be a poor assumption.

Experimental errors are classified as *systematic errors* and *random errors*. Systematic errors recur with the same direction and usually the same magnitude on every repetition of the experiment, so they can affect the accuracy without affecting the precision. Random errors do not have the same direction and magnitude every time, so they affect the precision as well as the accuracy. Making several repetitions of a measurement will generally let us know roughly how large the random errors are, but can tell us nothing about systematic errors. One common approach to systematic errors is to make educated guesses about the probable sizes of systematic errors, by thinking about the apparatus and the method used. If no repetitions of the measurement can be made, this method can also be applied to the analysis of random errors.

---

### Example 10.1

The simplest apparatus for measuring the melting temperature of a substance consists of a small bath containing a liquid in which the sample can be suspended in a small capillary tube held next to a thermometer. The bath is slowly heated and the reading of thermometer at the time of melting of the sample is recorded. List some of the possible experimental error sources in this determination. Classify each as systematic or random and guess its relative magnitude.

---

### Solution

1. Faulty thermometer calibration. This is systematic. With an inexpensive thermometer, this error might be as large as several tenths of a degree.
2. Lack of thermal equilibration between the liquid of the bath, the sample, and the thermometer. If the experimental procedure is the same for all repetitions, this will be systematic. If the thermometer is larger than the sample, it will likely be heated more slowly than the sample if the heating is done too rapidly. This error will probably be less than $1°C$.
3. Failure to read the thermometer correctly. This is random. There are two kinds of error here. The first is more of a blunder than an experimental error and amounts to counting the marks on the thermometer incorrectly and, for example, recording $87.5°C$ instead of $88.5°C$, etc. The other kind of error is due to parallax, or looking at the thermometer at some angle other than a right angle. This might produce an error of about two-tenths of a degree Celsius.
4. Presence of impurities in the sample. This is systematic, since impurities that dissolve in the liquid always lower the melting point. If carefully handled samples of purified substances are used, this error should be negligible.
5. Failure to observe the onset of melting. This is systematic, although variable in magnitude. If the heating is done slowly, it should be possible to reduce this error to a few tenths of a degree.

Some of these errors can be minimized by reducing the rate of heating. This suggests a possible procedure: an initial rough determination establishes the approximate value, and a final heating is done with slow and careful heating near the melting point.

---

### Problem 10.1

List as many sources of error as you can for some of the following measurements. Classify as systematic or random and estimate the magnitude of each source of error.

a. The measurement of the diameter of a fine copper wire with a micrometer caliper.
b. The measurement of the length of a piece of glass tubing with a meter stick.
c. The measurement of the mass of a porcelain crucible.

   d. The measurement of the mass of a silver chloride precipitate in a porcelain crucible.

   e. The measurement of the resistance of an electrical heater using a Wheatstone bridge.

   f. The measurement of the time required for an automobile to travel the distance between two highway markers 1 kilometer apart, using a stopwatch.

---

Although it is useful to engage in educated guesswork, it is better to have some kind of objective way to estimate the magnitude of experimental errors. There are two principal ways to gain information about systematic errors. One is to modify the apparatus and repeat the measurement, or to repeat the measurement with a different apparatus. The other is to use the same apparatus to measure a well-known quantity, observing the actual experimental error.

One way to modify an apparatus is to replace a component of the apparatus. For example, in making a voltage measurement with a potentiometer, one compares the voltage with that of a standard cell. One could see if the same result is obtained with a different standard cell. If the result is different, you can assume that the first standard cell is contributing a systematic error about as large as the difference in the values.

The other kind of apparatus modification is to change the design of the apparatus, at least temporarily. For example, the apparatus may include some insulation that minimizes unwanted heat transfer. If the measurement gives a different value when the insulation is temporarily improved, there was probably a systematic error about as big as (or bigger than) the change in the result.

These methods of estimating systematic errors are often not available in physical chemistry laboratory courses. In this event, educated guesswork is nothing to be ashamed of.

## SECTION 10.2. STATISTICAL TREATMENT OF RANDOM ERRORS

Statistics is the study of a large set of people, objects, or numbers, called a *population*. The population is not studied directly, because of its large size or inaccessibility. A subset from the population, called a *sample*, is studied, and the likely properties of the population are inferred from the properties of the sample.

If several repetitions of a measurement can be made, this set of measurements can be considered to be a sample from a population. The population is an imaginary set of infinitely many repetitions of the measurement. Statistical analysis can be used to study the properties of the sample and to infer likely properties of the population, including probable errors in the measurements.

### Properties of a Population

The most important property of a population is its mean value. If there are no systematic errors, the mean of the population of measurements will be the correct value of the measured quantity, since random errors are equally likely in either direction. The infinitely many members in the population will be distributed among all values in some range according to some probability distribution. We have discussed probability distributions in Section 4.8.

A *normalized probability density or normalized probability distribution* $f(x)$ is defined such that

$$(\text{probability of values between } x \text{ and } x + dx) = f(x)\,dx \tag{10.1a}$$

and

$$\int_a^b f(x)\,dx = 1. \tag{10.1b}$$

If the probability distribution $f(x)$ is normalized the population mean $\mu$ is given by Eq. (4.67),

$$\boxed{\mu = \int_a^b x f(x)\,dx}, \tag{10.2}$$

where $a$ is the smallest value of $x$ and $b$ is the largest. In many cases, we will assume that $a = -\infty$ and that $b = +\infty$. This is generally not correct, but if the probability of values of $x$ with large magnitude is small, making this assumption will not introduce significant errors.

The *population standard deviation* is a measure of the spread of the distribution, and is given by Eq. (4.72)

$$\boxed{\sigma = \left[ \int_a^b (x - \mu)^2 f(x)\,dx \right]^{1/2}}. \tag{10.3}$$

---

**Problem 10.2**

Show that the definition of the standard deviation in Eq. (4.72) is the same as that in Eq. (10.3). That is, show that

$$\int_a^b (x - \mu)^2 f(x)\,dx = \int_a^b x^2 f(x)\,dx - \left( \int_a^b x f(x)\,dx \right)^2.$$

---

We will assume that any population of experimental results is governed by the Gaussian probability distribution (also called the normal distribution), discussed in Section 4.8. The important properties of that distribution are as listed in Section 4.8:

1. A total of 68.3% of the members of the population have their values of $x$ lying within one standard deviation of the mean: in the interval $(\mu - \sigma_x, \mu + \sigma_x)$.
2. A total of 95% of the members of the population have their values of $x$ lying within 1.96 standard deviation of the mean: in the interval $(\mu - 1.96\sigma_x, \mu + 1.96\sigma_x)$.
3. The fraction of the population with value of $x$ in the interval $(\mu - x_1, \mu + x_1)$ is $\text{erf}(x_1/\sqrt{2}\sigma)$ where erf stands for the error function, discussed in Appendix 7.

If a random experimental error arises as the sum of many contributions, the *central limit theorem of statistics* gives some justification for assuming that our experimental error will be governed by the Gaussian distribution. The theorem states that if a number of *random variables* (independent variables) $x_1, x_2, \ldots, x_n$ are governed by some probability distributions with finite means and finite standard deviations, then a linear combination (weighted sum)

$$y = \sum_{i=1}^{n} a_i x_i$$

is governed by a probability distribution that approaches a Gaussian distribution as $n$ becomes large.

There are other probability distributions that are used in statistics, including the binomial distribution, the Poisson distribution, and the Lorentzian distribution.[1] We will not discuss these distributions. However, all of them have properties which are qualitatively similar to those of the Gaussian distribution.

## Properties of a Sample

Our sample is a set of repetitions of a measurement, which we think of as being selected randomly from our population. From this sample, we need to get an estimate of the correct value of the measured quantity and the probable error in this estimate. There are three common averages. The *median* is a value such that half of the members of the set are greater than the median and half are smaller than the median. The *mode* is the value that occurs most frequently in the set. The *mean* of a set of numbers is defined by Eq. (4.57)

$$\boxed{\bar{x} = \frac{1}{N} \sum_{i=1}^{N} x_i},\qquad (10.4)$$

where $N$ is the number of members of the set, and $x_1, x_2, \ldots$ are the members of the set. We will use the mean of a set of repetitions as our estimate of the population mean, which is equal to the correct value if systematic errors are absent. The mean of our sample is said by statisticians to be an *unbiased estimate* of the population mean.

In order to estimate the accuracy of our estimate, we need to study the dispersion, or spread, of our set of data. The common quantity used to describe this spread is the standard deviation, which for a population is defined by Eq. (10.3). For a sample of $N$ numbers the *sample standard deviation* is given by a slightly different formula

$$\boxed{s_x = \left[ \frac{1}{N-1} \sum_{i=1}^{N} (x_i - \bar{x})^2 \right]^{1/2}},\qquad (10.5)$$

where $\bar{x}$ is the sample mean given by Eq. (10.4).

---

[1] See Philip R. Bevington, *Data Reduction and Error Analysis for the Physical Sciences*, Chap. 3, McGraw–Hill, New York, 1969, or Hugh D. Young, *Statistical Treatment of Experimental Data*, McGraw–Hill, New York, 1962, for discussions of various distributions.

The square of the standard deviation is called the *variance*. In Eq. (10.5), every term in the sum is positive or zero, since it is the square of a real quantity, so that the variance can vanish only if every member of the set of numbers is equal to the mean, and otherwise must be positive. The larger the differences between members of the set, the larger the standard deviation will be. In most cases, about two-thirds of the members of a sample will have values between $\bar{x} - s$ and $\bar{x} + s$.

The sample standard deviation is not quite the square root of the mean value of the quantity $(x - \bar{x})^2$, while the population standard deviation is the square root of the mean of the quantity $(x - \mu)^2$. It has been shown that the $N - 1$ factor in the denominator of Eq. (10.5) is necessary in order for $s$ to be an unbiased estimate of $\sigma$. This has to do with the fact that in a sample of $N$ members, there are $N - 1$ independent pieces of information in addition to the mean, or $N - 1$ *degrees of freedom* in addition to the mean. We use the sample standard deviation as an estimate of the population standard deviation.

---

**Example 10.2**

Find the mean and the standard deviation of the set of numbers

$$32.41, \quad 33.76, \quad 32.91, \quad 33.04, \quad 32.75, \quad 33.23.$$

---

**Solution**

$$\bar{x} = \frac{1}{6}(32.41 + 33.76 + 32.91 + 33.04 + 32.75 + 33.23)$$
$$= 33.02$$
$$s = \left\{ \frac{1}{5}[(0.39)^2 + (0.74)^2 + (-0.11)^2 + (0.02)^2 + (0.27)^2 + (0.21)^2] \right\}^{1/2}$$
$$= 0.41.$$

One of the six numbers lies below 32.61, and one lies above 33.43, so that two-thirds of them lie between $\bar{x} - s$ and $\bar{x} + s$.

---

**Problem 10.3**

Find the mean, $\bar{x}$, and the standard deviation, $s$, for the following set of numbers:

$$2.876\,\text{m}, \quad 2.881\,\text{m}, \quad 2.864\,\text{m}, \quad 2.879\,\text{m},$$
$$2.872\,\text{m}, \quad 2.889\,\text{m}, \quad 2.869\,\text{m}.$$

Determine how many numbers lie below $\bar{x} - s$ and how many lie above $\bar{x} + s$.

---

## Numerical Estimation of Random Errors

A common practice among scientific workers is to make statements that have a 95% probability of being correct. Such a statement is said to be at the *95% confidence level*. Of course, every statement that we make at this confidence level has a 5% change of being incorrect. For example, we want to make a statement of the form

$$\mu = \text{correct value} = \bar{x} \pm \varepsilon \tag{10.6}$$

that has a 95% chance of being right. That is, we want to state an interval of size $2\varepsilon$ that has a 95% chance of containing the correct value. We call such an interval the *95% confidence interval*, and we call the positive number $\varepsilon$ the *probable error at the 95% confidence level* or (for brevity) the *estimated error*.

If we knew the value of $\sigma$, we could make this statement after a single measurement. Since a single measurement has a 95% change of being within $1.96\sigma$ of $\mu$, we could say with 95% confidence that

$$\mu = x \pm 1.96\sigma, \tag{10.7}$$

where $x$ is the outcome of our single measurement. If we have no opportunity to repeat our measurement, the only thing we can do is to use our educated guess of the expected error as an estimate of $1.96\sigma$ and make this statement.

However, let us assume that we can make $N$ measurements of the same quantity. Think of our set of $N$ measurements as only one possible set of $N$ measurements from the same population. If we could take infinitely many sets of $N$ measurements from our population, the means of these sets would themselves form a new population. If the original population is governed by a Gaussian distribution, the population of sample means will also be governed by a Gaussian distribution (although we do not prove this fact). The mean of this new population is the same as the mean of the original population, and $\sigma_m$, the standard deviation of the new population, is given by

$$\sigma_m = \frac{\sigma}{\sqrt{N}}, \tag{10.8}$$

where $\sigma$ is the standard deviation of the original population. The new population of sample means has a smaller spread than the original population, and its spread becomes smaller as $N$ becomes larger. That is, the means of the sets of measurements cluster more closely about the population mean than do individual measurements. Thus, the mean of a set of measurements is more likely to lie close to the correct value than a single measurement.

If we knew the standard deviation of the original population, we would now be able to write at the 95% confidence level

$$\varepsilon = \frac{1.96\sigma}{\sqrt{N}}. \tag{10.9}$$

In general, we do not know the population standard deviation. However, since we use the sample standard deviation as our estimate of the population standard deviation, we could write as an approximation

$$\varepsilon = \frac{1.96s}{\sqrt{N}}, \tag{10.10}$$

where $s$ is the sample standard deviation. However, there is a statistically precise formula, derived by Gossett,[2] which enables us to write an exact version of Eq. (10.10). Gossett defined the *Student t factor*

$$t = \frac{(\bar{x} - \mu)N^{1/2}}{s},$$  (10.11)

where $\mu$ is the population mean, $\bar{x}$ is the sample mean, and $s$ is the sample standard deviation. There is a different value of $t$ for every sample. Although $\mu$ is not known, Gossett derived the probability distribution that $t$ obeys.[3] From this distribution, which is called *Student's t distribution*, the maximum value of $t$ corresponding to a given confidence level can be calculated for any value of $N$. The notation used is $t(\nu, 0.05)$. The quantity $\nu$ is the number of degrees of freedom, equal to $N-1$, and 0.05 represents the confidence level of 95%. Table 10.1 gives these values for various values of $N$ and for four different confidence levels. Notice that as $N$ approaches infinity, the maximum Student $t$ value for 95% confidence (also called 0.05 significance) approaches 1.96, the factor in Eq. (10.9). This is because $s$ approaches $\sigma$ and the Student $t$ distribution approaches a Gaussian distribution as $N$ becomes large. Unfortunately, some authors use a different notation for the critical value of $t$, such as $t_\nu(0.025)$ to represent the Student $t$ factor for $\nu + 1$ data points at the 95% confidence level.[4]

Using a value from Table 10.1, we can write a formula for the expected error in the mean at the 95% confidence level

$$\varepsilon = \frac{t(\nu, 0.05)s}{\sqrt{N}}$$  (10.12)

for a sample of $\nu + 1$ members ($N$ members).

## Example 10.3

Assume that the melting temperature of calcium nitrate tetrahydrate, $Ca(NO_3)_2$ $4H_2O$, has been measured 10 times, and that the results are 42.70, 42.60, 42.78, 42.83, 42.58, 42.68, 42.65, 42.76, 42.73, and 42.71°C. Ignoring systematic errors, determine the 95% confidence interval for the set of measurements.

## Solution

Our estimate of the correct melting temperature is the sample mean:

$$\bar{t} = \frac{1}{10}(42.70°C + 42.60°C + 42.78°C + 42.83°C + 42.58°C + 42.68°C$$
$$+ 42.65°C + 42.76°C + 42.73°C + 42.71°C)$$
$$= 42.70°C.$$

---

[2] William Sealy Gossett, 1876–1937, English chemist and statistician who published under the pseudonym "Student" in order to keep the competitors of his employer, a brewery, from knowing what statistical methods he was applying to quality control.

[3] See Walter Clark Hamilton, *Statistics in Physical Science*, pp. 78ff, The Ronald Press Company, New York, 1964.

[4] John A. Rice, *Mathematical Statistics and Data Analysis*, Wadsworth & Brooks/Cole, Pacific Grove, CA, 1988.

**TABLE 10.1**    **Some Values of Student's t Factor**[*]

| Number of degrees of freedom | Maximum value of student's t factor for the significance levels indicated | | | |
|---|---|---|---|---|
| $\nu$ | $t(\nu, 060)$ | $t(\nu, 0.10)$ | $t(\nu, 0.05)$ | $t(\nu, 0.01)$ |
| 1 | 1.376 | 6.314 | 12.706 | 63.657 |
| 2 | 1.061 | 2.920 | 4.303 | 9.925 |
| 3 | 0.978 | 2.353 | 3.182 | 5.841 |
| 4 | 0.941 | 2.132 | 2.776 | 4.604 |
| 5 | 0.920 | 2.015 | 2.571 | 4.032 |
| 6 | 0.906 | 1.943 | 2.447 | 3.707 |
| 7 | 0.896 | 1.895 | 2.365 | 3.499 |
| 8 | 0.889 | 1.860 | 2.306 | 3.355 |
| 9 | 0.883 | 1.833 | 2.262 | 3.250 |
| 10 | 0.879 | 1.812 | 2.228 | 3.169 |
| 11 | 0.876 | 1.796 | 2.201 | 3.106 |
| 12 | 0.873 | 1.782 | 2.179 | 3.055 |
| 13 | 0.870 | 1.771 | 2.160 | 3.012 |
| 14 | 0.868 | 1.761 | 2.145 | 2.977 |
| 15 | 0.866 | 1.753 | 2.131 | 2.947 |
| 16 | 0.865 | 1.746 | 2.120 | 2.921 |
| 17 | 0.863 | 1.740 | 2.110 | 2.898 |
| 18 | 0.862 | 1.734 | 2.101 | 2.878 |
| 19 | 0.861 | 1.729 | 2.093 | 2.861 |
| 20 | 0.860 | 1.725 | 2.086 | 2.845 |
| 21 | 0.859 | 1.721 | 2.080 | 2.831 |
| 22 | 0.858 | 1.717 | 2.074 | 2.819 |
| 23 | 0.858 | 1.714 | 2.069 | 2.807 |
| 24 | 0.857 | 1.711 | 2.064 | 2.797 |
| 25 | 0.856 | 1.708 | 2.060 | 2.787 |
| 26 | 0.856 | 1.706 | 2.056 | 2.479 |
| 27 | 0.855 | 1.703 | 2.052 | 2.771 |
| 28 | 0.855 | 1.701 | 2.048 | 2.763 |
| 29 | 0.854 | 1.699 | 2.045 | 2.756 |
| 30 | 0.854 | 1.697 | 2.042 | 2.750 |
| 40 | 0.851 | 1.684 | 2.021 | 2.704 |
| 60 | 0.848 | 1.671 | 2.000 | 2.660 |
| $\infty$ | 0.842 | 1.645 | 1.960 | 2.576 |

[*]John A. Rice, *Mathematical Statistics and Data Analysis*, p. 560, Wadsworth & Brooks/Cole, Pacific Grove, CA, 1988.

The sample standard deviation is

$$s = \left\{ \frac{1}{9}[(0.00°C)^2 + (0.10°C)^2 + (0.08°C)^2 + (0.13°C)^2 + (0.12°C)^2 + (0.02°C)^2 \right.$$
$$\left. + (0.05°C)^2 + (0.16°C)^2 + (0.13°C)^2 + (0.01°C)^2] \right\}^{1/2}$$
$$= 0.08°C.$$

The value of $t(9, 0.05)$ is found from Table 10.1 to equal 2.26, so that

$$\varepsilon = \frac{(2.26)(0.08°C)}{\sqrt{10}} = 0.06°C.$$

Therefore, at the 95% confidence level,

$$t = 42.70°C \pm 0.06°C.$$

## Problem 10.4

Assume that the H–O–H bond angles in various crystalline hydrates have been measured to be $108°$, $109°$, $110°$, $103°$, $111°$, and $107°$. If these measurements all come from the same population, give your estimate of the population mean and its 95% confidence interval.

## Rejection of Discordant Data

Sometimes a repetition of a measurement yields a value that differs greatly from the other members of the sample (a *discordant value*). For example, say that we repeated the measurement of the melting temperature of $Ca(NO_3)_2 \cdot 4H_2O$ in Example 10.3 one more time and obtained a value of $39.75°C$. If we include this eleventh data point, we get a sample mean of $42.43°C$ and a sample standard deviation of $0.89°C$. Using the table of Student's $t$ value in Table 10.1, we obtain a value for $\varepsilon$ of $0.60°C$ at the 95% confidence level. Some people think that the only honest thing to do is to report the melting temperature as $42.43 \pm 0.60°C$.

If we assume that our sample standard deviation of $0.89°C$ is a good estimate of the population standard deviation, our data point of $39.75°C$ is 3.01 standard deviations away from the mean. From the table of the error function in Appendix 7, the probability of a randomly chosen member of a population differing from the mean by this much or more is 0.003, or 0.3%. There is considerable justification for asserting that such an improbable event was due not to random experimental errors but to some kind of a mistake such as misreading a thermometer, etc. If you assume this, you discard the suspect data point and recompute the mean and standard deviation just as though the discordant data point had not existed. Do not discard more than one data point from a set of data. If two or more apparently discordant data points occur in a set, you should regard it as a signal that the data are simply imprecise.

There are several rules for deciding when to discard a data point. Some people discard data points that are more than 2.7 standard deviations from the mean (this means 2.7 standard deviations calculated with the suspect point left in). This discards points that have less than a 1% chance of having arisen through normal experimental error. Pugh and Winslow[5] suggest that for a sample of $N$ data points, a data point should be discarded if there is less than one chance in $2N$ that the point came from the

[5]Emerson M. Pugh and George H. Winslow, *The Analysis of Physical Measurements*, Addison–Wesley, Reading, MA, 1966.

**TABLE 10.2  Critical Q Values at the 95% Confidence Level**

| $N$ | 3 | 4 | 5 | 6 | 7 | 8 | 9 | 10 | 20 | 30 |
|---|---|---|---|---|---|---|---|---|---|---|
| $Q$ | 0.97 | 0.84 | 0.72 | 0.63 | 0.58 | 0.53 | 0.50 | 0.47 | 0.36 | 0.30 |

same population. This rule discards more points than the first rule for a sample of 10 measurements, since it would discard a point lying 1.96 standard deviations from the mean in a sample of 10 measurements. Since you would expect such a point to occur once in 20 times, the probability of occurring in a sample of 10 by random chance is fairly large.

There is also a test called the $Q$ *test*, in which $Q$ is defined as the difference between an "outlying" data point and its nearest neighbor divided by the difference between the highest and the lowest values:

$$Q = \frac{\text{(outlying value)} - \text{(value nearest the outlying value)}}{\text{(highest value)} - \text{(lowest value)}}. \tag{10.13}$$

An outlying data point is discarded if its value of $Q$ of exceeds a certain critical value, which depends on the number of members in the sample. Table 10.2 contains the critical value of $Q$ at the 95% confidence level for samples of $N$ members.[6]

**Problem 10.5**

Apply the $Q$ test to the 39.75°C data point appended to the data set of Example 10.3.

## SECTION 10.3. DATA REDUCTION AND THE PROPAGATION OF ERRORS

Many values that are obtained by measurement in a physical chemistry laboratory are used along with other values to calculate some quantity that is not directly measured. An experimental error in a measured quantity will affect the accuracy of any quantity that is calculated from it.

### The Combination of Errors

Assume that we have measured two quantities, $a$ and $b$, and have established a probable value and a 95% confidence interval for each,

$$a = \bar{a} \pm \varepsilon_a \tag{10.14a}$$
$$b = \bar{b} \pm \varepsilon_b, \tag{10.14b}$$

[6]W. J. Dixon, *Ann. Math. Statist.* **22**, 68 (1951); and R. B. Dean and W. J. Dixon, *Anal. Chem.* **23**, 636 (1951).

where $\bar{a}$ is our probable value of $a$, perhaps computed as an average of several measurements, and $\varepsilon_a$ is our expected error in $a$, perhaps computed from Eq. (10.12), and analogous quantities are defined for $b$.

Assume that we want to obtain a probable value and a 95% confidence interval for $c$, the sum of $a$ and $b$. The probable value of $c$ is the sum of $\bar{a}$ and $\bar{b}$:

$$\bar{c} = \bar{a} + \bar{b}. \tag{10.15}$$

A simple estimate of the probable error in $c$ is the sum of $\varepsilon_a$ and $\varepsilon_b$, which corresponds to the assumption that the errors in the two quantities always add together:

$$\varepsilon_c = \varepsilon_a + \varepsilon_b \qquad \text{(simple preliminary estimate).} \tag{10.16}$$

However, Eq. (10.16) provides an overestimate, because there is some chance that errors in $a$ and in $b$ will cancel.

If $a$ and $b$ are both governed by Gaussian distributions, $c$ is also governed by a Gaussian distribution, and the probable error in $c$ is given by

$$\boxed{\varepsilon_c = \left(\varepsilon_a^2 + \varepsilon_b^2\right)^{1/2}}. \tag{10.17}$$

This formula allows for the statistically correct probability of cancellation of errors.

---

### Example 10.4

Two lengths have been measured as $24.8 \pm 0.4$ m and $13.6 \pm 0.3$ m. Find the probable value of their sum and its probable error.

---

### Solution

The probable value is $24.8 + 13.6 = 38.4$ m, and the probable error is

$$\varepsilon = (0.4^2 + 0.3^2)^{1/2} = 0.5\,\text{m}.$$

Therefore,

$$\text{sum} = 38.4 \pm 0.5\,\text{m}.$$

---

### Problem 10.6

Two time intervals have been clocked as $56.57 \pm 0.13$ s and $75.12 \pm 0.17$ s. Find the value of their sum and its probable error.

---

## The Combination of Random and Systematic Errors

In Section 10.2, we discussed the use of statistics to determine the probable error due to random errors in the case that the measurements could be repeated a number of times. The probable error due to systematic errors can be estimated by apparatus modification or by guesswork. These errors combine in the same way as the errors in Eq. (10.17). If $\varepsilon_r$ is the probable error due to random errors and $\varepsilon_s$ is the probable error due to systematic errors, the total probable error is given by

$$\varepsilon_t = \left(\varepsilon_s^2 + \varepsilon_r^2\right)^{1/2}.$$

(10.18)

In using this formula, you must try to make your estimate of the systematic error conform to the same level of confidence as your random error. If you use the 95% confidence level for the random errors, do not estimate the systematic errors at the 50% or 70% confidence level, which is what most people instinctively tend to do when asked what they think a probable error is. You might make a first guess at your systematic errors and then double it to avoid this underestimation.

## Example 10.5

Assume that you estimate the total systematic error in the melting temperature measurement of Example 10.3 as 0.20°C, at the 95% confidence level. Find the total expected error.

### Solution

$$\lambda_t = [(0.06°C)^2 + (0.20°C)^2]^{1/2} = 0.21°C.$$

## Error Propagation in Data Reduction Using Mathematical Formulas

In the Dumas method for determining the molar mass of a volatile liquid,[7] one uses the formula

$$M = \frac{wRT}{PV},$$

(10.19)

where $M$ is the molar mass, $w$ is the mass of the sample of vapor of the substance contained in volume $V$ at pressure $P$ and temperature $T$, and $R$ is the ideal gas constant. We think of Eq. (10.19) as being an example of a general formula,

$$y = y(x_1, x_2, x_3, \ldots, x_n).$$

(10.20)

Let us assume that we have a 95% confidence interval for each of the independent variables, such that

$$x_i = \bar{x}_i \pm \varepsilon_i \qquad \text{(one equation for each value of } i\text{).}$$

Our problem is to take the uncertainties in $x_1$, $x_2$, etc., and calculate the uncertainty in $y$, the dependent variable. This is called the *propagation of errors*.

If the errors are not too large, we can take an approach based on the differential calculus of several independent variables. The fundamental equation of differential

[7]Lawrence J. Sacks, *Experimental Chemistry*, pp. 26–29, Macmillan Co., New York, 1971.

calculus is Eq. (5.11),

$$dy = \left(\frac{\partial y}{\partial x_1}\right)dx_1 + \left(\frac{\partial y}{\partial x_2}\right)dx_2 + \left(\frac{\partial y}{\partial x_3}\right)dx_3 + \cdots + \left(\frac{\partial y}{\partial x_n}\right)dx_n. \quad (10.21)$$

This equation gives an infinitesimal change in $y$ due to arbitrary infinitesimal changes in the independent variables.

If finite changes are made in the independent variables, we could write as an approximation

$$\Delta y \approx \left(\frac{\partial y}{\partial x_1}\right)\Delta x_1 + \left(\frac{\partial y}{\partial x_2}\right)\Delta x_2 + \cdots + \left(\frac{\partial y}{\partial x_n}\right)\Delta x_n. \quad (10.22)$$

If we had some known errors in $x_1$, $x_2$, etc., we could use Eq. (10.22) to calculate a known error in $y$. Since all we have is probable errors in the independent variables and do not know whether these are positive or negative, one cautious way to proceed would be to assume that the worst might happen and that all the errors would add:

$$\varepsilon_y \approx \left|\left(\frac{\partial y}{\partial x_1}\right)\varepsilon_1\right| + \left|\left(\frac{\partial y}{\partial x_2}\right)\varepsilon_2\right| + \left|\left(\frac{\partial y}{\partial x_3}\right)\varepsilon_3\right|. \quad (10.23)$$

Equation (10.23) overestimates the errors because there is always some statistical probability that the errors in the $x$ values will cancel instead of adding. An equation that is analogous to Eq. (10.17) is

$$\varepsilon_y \approx \left[\left(\frac{\partial y}{\partial x_1}\right)^2\varepsilon_1^2 + \left(\frac{\partial y}{\partial x_2}\right)^2\varepsilon_2^2 + \left(\frac{\partial y}{\partial x_3}\right)^2\varepsilon_3^2\right]^{1/2}. \quad (10.24)$$

This equation includes the correct probability of cancellation if the independent variables are governed by Gaussian distributions. It will be our working equation for the propagation of errors through formulas. Since it is based on a differential formula, it becomes more nearly exact if the errors are small.

---

**Example 10.6**

Find the expression for the propagation of errors for the Dumas molar mass determination. Apply this to the following set of data for $n$-hexane:

$$T = 373.15 \pm 0.25 \, \text{K}$$
$$V = 206.34 \pm 0.15 \, \text{mL}$$
$$P = 760 \pm 0.2 \, \text{torr}$$
$$w = 0.585 \pm 0.005 \, \text{g}.$$

---

**Solution**

$$M = \frac{(0.585 \, \text{g})(0.082057 \, \text{L atm K}^{-1}\text{mol}^{-1})(373.15 \, \text{K})}{(1.000 \, \text{atm})(0.20634 \, \text{liters})}$$
$$= 86.81 \, \text{g mol}^{-1}.$$

The analog of Eq. (10.24) for our equation is

$$\varepsilon_M \approx \left[ \left( \frac{RT}{PV} \right)^2 \varepsilon_w^2 + \left( \frac{wR}{PV} \right)^2 \varepsilon_T^2 + \left( \frac{wRT}{P^2V} \right)^2 \varepsilon_P^2 + \left( \frac{wRT}{PV^2} \right)^2 \varepsilon_V^2 \right]^{1/2}.$$

(10.25)

Substituting the numerical values into Eq. (10.25), we obtain

$$\varepsilon_M = 0.747 \, \text{g mol}^{-1}$$
$$M = 86.8 \pm 0.7 \, \text{g mol}^{-1}.$$

There is no point in reporting insignificant digits in the calculated quantity and the expected error, so one significant digit suffices in an expected error. The digit 8 after the decimal point is not completely significant, but since the error is smaller than $1.0 \, \text{g mol}^{-1}$, it provides a little information, and we include it. The accepted value is $86.17 \, \text{g mol}^{-1}$, so that our expected error is larger than our actual error, as it should be about 95% of the time.

---

**Problem 10.7**

In the cryoscopic determination of molar mass,[8] the molar mass in $\text{g mol}^{-1}$ is given by

$$M = \frac{(1000 \, \text{g}/1 \, \text{kg})wK_f}{W \Delta T_f}(1 - k_f \Delta T_f),$$

where $W$ is the mass of the solvent, $w$ is the mass of the unknown solute, $\Delta T_f$ is the amount by which the freezing point of the solution is less than that of the pure solvent, and $K_f$ and $k_f$ are constants characteristic of the solvent. Assume that in a given experiment, a sample of an unknown substance was dissolved in benzene, for which $K_f = 5.12 \, \text{K kg mol}^{-1}$ and $k_f = 0.011 \, \text{K}^{-1}$. For the following data, calculate $M$ and its probable error:

$$W = 13.185 \pm 0.003 \, \text{g}$$
$$w = 0.423 \pm 0.002 \, \text{g}$$
$$\Delta T_f = 1.263 \pm 0.020 \, \text{K}.$$

---

## SECTION 10.4. GRAPHICAL DATA REDUCTION PROCEDURES

When digital computers did not exist, physical chemists frequently drew graphs by hand and used them to perform quantitative data reduction. We now have access to computers with sophisticated graphical software and can construct graphs to represent

---

[8]David P. Shoemaker, Carl W. Garland, and Joseph W. Nibler, *Experiments in Physical Chemistry*, 6th ed., pp. 179ff, McGraw–Hill, New York, 1996.

**TABLE 10.3  Experimental Vapor Pressures of Pure
Ethanol at Various Temperature**

| t/°C | T/K | Vapor pressure/torr | Expected error/torr |
|------|------|---------------------|---------------------|
| 25.00 | 298.15 | 55.9 | 3.0 |
| 30.00 | 303.15 | 70.0 | 3.0 |
| 35.00 | 308.15 | 93.8 | 4.2 |
| 40.00 | 313.15 | 117.5 | 5.5 |
| 45.00 | 318.15 | 154.1 | 6.0 |
| 50.00 | 323.15 | 190.7 | 7.6 |
| 55.00 | 328.15 | 241.9 | 8.0 |
| 60.00 | 333.15 | 304.15 | 8.8 |
| 65.00 | 338.15 | 377.9 | 9.5 |

experimental data quickly and easily. We can easily perform various kinds of numerical data reduction based on graphs without actually making the graphs. However, we discuss manual graphical procedures for historical perspective and to show what underlies the numerical procedures. We discuss a particular example, the treatment of a set of measurements of the vapor pressure of a pure liquid as a function of the temperature. Since the temperature is controlled and the pressure is measured, we write

$$P = P(T). \tag{10.26}$$

This functional form is assumed to represent a continuous function, so the data points should lie somewhere near a smooth curve representing the correct values. Table 10.3 contains a set of student data for the vapor pressure of pure ethanol. Figure 10.1 is a graph on which the data have been plotted. A fairly smooth curve has been drawn near the points, passing through or nearly through the confidence intervals given in the table.

The curve in Fig. 10.1 gives us a means to interpolate between the data points, and it gives us "smoothed" values at the data points. When the curve misses a measured value, the value at the curve is probably a better estimate of the correct value than the measured value.

The Clapeyron equation for any phase transition is

$$\frac{dP}{dT} = \frac{\Delta H}{T \Delta V}, \tag{10.27}$$

where $P$ is the pressure, $\Delta H$ is the enthalpy change of the phase transition, $T$ is the absolute temperature, and $\Delta V$ is the volume change of the phase transition. If $\Delta V$ is known and the value of the derivative $dP/dT$ can be determined, then the enthalpy change of vaporization can be calculated.

The derivative $dP/dT$ is the slope of the tangent line at the point being considered (see Chapter 3). After a smooth curve has been drawn as in Fig. 10.1, a tangent line can be constructed. Two line segments can be drawn parallel to the coordinate axes, forming a right triangle with the tangent line. The slope is equal to the height of the triangle divided by its base ("rise" over "run").

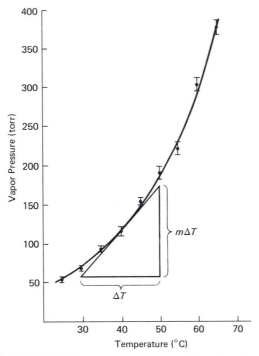

**FIGURE 10.1**   The vapor pressure of ethanol as a function of temperature.

### Example 10.7

Find the enthalpy change of vaporization of ethanol at 40°C by graphically determining the slope of the tangent line at 40°C in the graph of Fig. 10.1.

### Solution

The tangent line and a suitable right triangle have been drawn in Fig. 10.1. The height of the triangle is 115 torr and the base is 20.0 K,

$$\frac{dP}{dT} = \frac{115 \, \text{torr}}{20.0 \, \text{K}} = 5.75 \, \text{torr K}^{-1}.$$

To obtain the volume change of the vaporization, we assume that the vapor behaves as an ideal gas and that the volume of the liquid is negligible compared with the volume of the vapor,

$$
\begin{aligned}
\Delta H &= \left(\frac{dP}{dT}\right) T \Delta V \approx \frac{RT^2}{P} \frac{dP}{dT} \\
&\approx \frac{(8.3145 \, \text{JK}^{-1}\text{mol}^{-1})(313.15 \, \text{K})^2(5.75 \, \text{torr K}^{-1})}{117.5 \, \text{torr}} \\
&\approx 40 \times 10^3 \, \text{J mol}^{-1} = 40 \, \text{kJ mol}^{-1}.
\end{aligned}
\tag{10.28}
$$

## Problem 10.8

The rate of a first-order chemical reaction obeys the equation

$$-\frac{dc}{dt} = kc,\qquad(10.29)$$

where $c$ is the concentration of the reactant and $k$ is a function of temperature called the rate constant. The following is a set of data for the reaction at $25°C$.[9]

$$(CH_3)_3CBr + H_2O \rightarrow (CH_3)_3COH + HBr$$

| Time/hr | $[(CH_3)_3CBr]/mol\ L^{-1}$ |
|---------|------------------------------|
| 0       | 0.1039 |
| 3.15    | 0.0896 |
| 6.20    | 0.0776 |
| 10.0    | 0.0639 |
| 18.3    | 0.0353 |
| 30.8    | 0.0207 |
| 43.8    | 0.0101 |

Plot the data as given and draw a smooth curve nearly through the data points. Draw a tangent to the curve at some point and evaluate the rate constant.

## Linearization

The procedure used in Example 10.7 can be simplified if we linearize our graph. This means finding new variables such that the curve in the graph is expected to be a line. In the vapor pressure example, these variables are found by manipulation of the Clapeyron equation, which is a differential equation that can be solved by separation of variables. We assume that $\Delta H$ is a constant, that the volume of the liquid is negligible compared to that of the gas, and that the gas is ideal. After separation of variables and integration, we obtain the Clausius–Clapeyron equation

$$\ln(P) = -\frac{\Delta H}{RT} + K,\qquad(10.30)$$

where $K$ is a constant of integration.

## Problem 10.9

Solve Eq. (10.28) to obtain Eq. (10.30), using the stated assumptions.

Equation (10.30) represents a linear function if we use $1/T$ as the independent variable and $\ln(P)$ as the dependent variable. Figure 10.2 is a graph of the same data as Fig. 10.1, using these variables. A straight line passing nearly through the points

[9]L. C. Bateman, E. D. Hughes, and C. K. Ingold, "Mechanism of Substitution at a Saturated Carbon Atom. Part XIX. A Kinetic Demonstration of the Unimolecular Solvolysis of Alkyl Halides," *J. Chem. Soc.* 960 (1940).

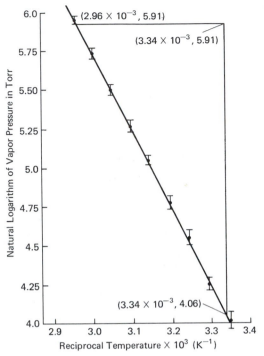

**FIGURE 10.2**    The natural logarithm of the vapor pressure of ethanol as a function of the reciprocal of the absolute temperature.

has been drawn in the figure. The expected errors in $\ln(P)$ were obtained by use of Eq. (10.24), with $y = \ln(P)$.

$$\varepsilon_y = \left(\frac{dy}{dP}\right)\varepsilon_P = \frac{1}{P}\varepsilon_P. \tag{10.31}$$

---

### Problem 10.10

Calculate the expected error in $\ln(P)$ for a few data points in Table 10.3, using Eq. (10.31).

---

### Example 10.8

Find the enthalpy change of vaporization of ethanol from the graph in Fig. 10.2.

---

### Solution

The necessary right triangle has been drawn in Fig. 10.2 and the coordinates of the vertices are given in the figure. If $m$ is the slope, then

$$\Delta H = -mR = -(-4.87 \times 10^3 \text{ K})(8.3145 \text{ J K}^{-1}\text{mol}^{-1})$$
$$= 40.5 \times 10^3 \text{ J mol}^{-1} = 40.5 \text{ kJ mol}^{-1}.$$

---

**Problem 10.11**

Linearize the graph of the data in Problem 10.8. Do this by solving Eq. (10.29) to obtain

$$\ln(c) = -kt + K. \tag{10.32}$$

Find the value of the rate constant $k$.

---

### Error Propagation in Graphical Data Reduction

Consider the determination of the probable error in the enthalpy change of vaporization of ethanol from the vapor pressure data given. In Fig. 10.2, the error bars give the expected errors in the values of the dependent variable. We assume that the errors in the independent variable are much smaller and can be neglected.

One way to proceed is to draw two additional lines on the graph, one that has the largest slope consistent with the data and one that has the smallest slope consistent with the data. Figure 10.3 is another graph of the values of $\ln(P)$ versus $1/T$, with these lines drawn in.

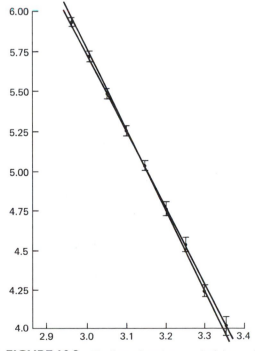

**FIGURE 10.3** The lines of maximum and minimum slope in the graph of the natural logarithm of the vapor pressure of ethanol as a function of the reciprocal of the temperature.

**Example 10.9**

Find the slopes of the two extreme lines in Fig. 10.3 and from them deduce the probable error in the enthalpy change of vaporization determined in Example 10.8.

**Solution**

Each slope is determined as in Example 10.8. The right triangles are not shown in Fig. 10.3, but the results are

$$\Delta H(\text{minimum}) = 39.4 \times 10^3 \text{ J mol}^{-1}$$
$$\Delta H(\text{maximum}) = 42.2 \times 10^3 \text{ J mol}^{-1}.$$

The difference between the minimum value of the enthalpy change of vaporization and the value of Example 10.8 is 1.1 kJ mol$^{-1}$, while the difference between the maximum value and the value of Example 10.8 is 1.7 kJ mol$^{-1}$. We specify a probable error of 1.7 kJ mol$^{-1}$.

The method can be modified to include uncertainties in the independent variable. In this case, instead of drawing an error bar at each point, you draw an *error box*, whose width is equal to the size of the confidence interval for the independent variable and whose height is equal to the size of the confidence interval for the dependent variable. The lines of maximum and minimum slope are then drawn to pass through all or nearly all of the boxes. In the graphical method, it is somewhat difficult to determine the confidence level of our error interval. If the error bars or error boxes are at the 95% confidence level, the error interval in the slope is probably somewhat larger than a 95% confidence interval.

**Problem 10.12**

Assume that each concentration of tert-butyl bromide in Problem 10.8 is uncertain by 1.5% of its value. Find the expected error in the rate constant, $k$, by graphical means.

If you do not have the probable errors in the values of the dependent variable, you can still draw a line of maximum slope and one of minimum slope. One way to do this is to draw the line through the points and then to measure the vertical distances from the points to the line. These distances are called residuals. If the line were the correct function, these *residuals* would be the experimental errors.

Although the line is not necessarily the correct function, we assume that the residuals collectively have about the same sizes as the experimental errors. Pick a distance that is longer than about 95% of the residuals, and draw an error bar that extends this far above and below each point. Then proceed to draw the line of maximum slope and the line of minimum slope as before.

## Other Graphical Data Reduction Procedures

We have discussed graphical procedures involving the plotting of a curve and the determination of the derivative of the function represented by the curve. It is also possible to use graphical methods of integration to reduce data. For example, the area under an absorption peak in an NMR spectrum is proportional to the number of nuclei producing the absorption, and the area under a peak in a gas or liquid chromatogram is proportional to the amount of the substance producing the peak.

There are several ways to carry out graphical integration. One tedious method is the counting of squares on the graph paper. Another method that has been used by chemists is to cut out the area and weigh it. The mass of the piece of paper will nearly be proportional to its area. It is also possible to use a mechanical device called a planimeter, which registers the area around which the stylus of the device is passed. However, numerical integration techniques are so much easier than graphical techniques that graphical integration is now almost unknown.

Error estimates can be obtained in graphical integration much as in graphical differentiation. If expected errors are available for individual data points, you can find the maximum error by incrementing each value by its expected error and determining the area again.

## SECTION 10.5. NUMERICAL DATA REDUCTION PROCEDURES

Graphical techniques have lost favor because of the availability of computers and software packages that make numerical procedures almost painless. Furthermore, numerical procedures are less subjective and are usually more accurate than graphical procedures.

### Numerical Differentiation

Sometimes we need a value for the derivative of a function that is represented approximately by a set of data points. One procedure is based on choosing polynomial functions that provide "smoothed" values of the function. If we have the set of data points $(x_1, y_1)$, $(x_2, y_2)$, $(x_3, y_3)$, etc., a smoothed value for the dependent variable to replace $y_i$ is given by[10]

$$y_i = \frac{1}{35}[17y_i + 12(y_{i+1} + y_{i-1}) - 3(y_{i+2} + y_{i-2})].$$
(10.33)

This equation corresponds to the value of the parabolic function that most nearly fits the five data points included in the formula and is valid only for equally spaced values of $x$. There are also similar formulas that involve a larger number of points.

We now define a set of differences, which are used to calculate numerical approximations to derivatives. The *first difference* for the $i$th point is defined by

$$\Delta y_i = y_{i+1} - y_i.$$
(10.34a)

The *second difference* for the $i$th point is defined by

$$\Delta^2 y_i = \Delta y_{i+1} - \Delta y_i = y_{i+2} - 2y_{i+1} + y_i.$$
(10.34b)

The *third difference* for the $i$th point is defined by

$$\Delta^3 y_i = \Delta^2 y_{i+1} - \Delta^2 y_i.$$
(10.34c)

---

[10] A. G. Worthing and I. Geffner, *Treatment of Experimental Data*, p. 7, Wiley, New York, 1943.

Higher-order differences are defined in a similar way. These definitions apply only to data sets for which the values of $x$ are equally spaced. For data sets of ordinary accuracy, the values of $y$ in Eqs. (10.34) should be the smoothed values given by Eq. (10.33). This set of differences for points number $i$ involves only points with subscripts greater than or equal to $i$. Other schemes can be defined that use points on both sides of the $i$th data point.

A value for the derivative $dy/dx$ at the $i$th point is given by [11]

$$\left.\frac{dy}{dx}\right|_{x=x_i} = \frac{1}{w}\left(y_i - \frac{1}{2}\Delta^2 y_i + \frac{1}{3}\Delta^3 y_i - \frac{1}{4}\Delta^4 y_i + \cdots\right), \tag{10.35}$$

where $w$ is the spacing between values of $x$,

$$w = x_{i+1} - x_i. \tag{10.36}$$

Equation (10.35) is based on the Gregory–Newton interpolation formula.[12]

The second derivative is given by

$$\left.\frac{d^2 y}{dx^2}\right|_{x=x_i} = \frac{1}{w^2}\left(\Delta^2 y_i - \Delta^3 y_i + \frac{11}{12}\Delta^4 y_i - \frac{10}{12}\Delta^5 y_i + \cdots\right). \tag{10.37}$$

If the data of Table 10.3 are smoothed and the derivative at 40°C is found by Eq. (10.35), a value for the enthalpy change of vaporization equal to 38 kJ mol$^{-1}$ is found using terms in the series up to the third difference. This value is probably less reliable than the value obtained graphically.

## Numerical Curve Fitting: The Method of Least Squares (Regression)

This method is a numerical procedure for finding a continuous function to represent a set of data points. This goal was accomplished graphically in Section 10.4 by drawing a straight line by inspection nearly through the data points in Fig. 10.2. Our data points are $(x_1, y_1)$, $(x_2, y_2)$, $(x_3, y_3)$, etc. We assume that there is some correct function

$$y = y(x) \tag{10.38}$$

governing the behavior of $y$ as a function of $x$, and that we know a family of functions to which the correct function belongs,

$$y = f(x, a_1, a_2, \ldots, a_p), \tag{10.39}$$

where the $a$'s are parameters. We want to find the values of the parameters to specify the member of the assumed family that most nearly fits the data points.

We have defined the residual for the $i$th data point as the difference between the measured value and the value of the function at that point:

$$r_i = y_i - f(x_i, a_1, a_2, \ldots, a_p). \tag{10.40}$$

When the function fits the points well, these residuals will collectively be small. Under certain conditions, it has been shown by statisticians that the best fit is obtained when the sum of the squares of the residuals is minimized. The reason for naming the method the method of least squares is that we minimize the sum of the squares of the residuals.

[11] Shoemaker, Garland, and Steinfeld, *op. cit.*, p. 757ff.

[12] *Ibid.*

The method is also called *regression*. It was first applied by Sir Francis Galton (1822–1911), a famous geneticist who studied the sizes of seeds and their offspring, and also the heights of fathers and sons.[13] He found in both these cases that there was a correlation between the trait in the second generation and in the earlier generation. However, he also found that the offspring tended to be closer to the means than the earlier generation. He called this tendency "regression toward mediocrity" and it has also been called "regression toward the mean." The name "regression" has stuck to the method.

We seek the minimum of the function $R$, which is the sum of the squares of the residuals,

$$R = \sum_{i=1}^{N} [y_i - f(x_i, a_1, a_2, \ldots, a_p)]^2, \tag{10.41}$$

where $N$ is the number of data points.

This minimum occurs where all of the partial derivatives of $R$ with respect to $a_1, a_2, \ldots, a_p$ vanish:

$$\frac{\partial R}{\partial a_i} = 0 \qquad (i = 1, 2, \ldots, p). \tag{10.42}$$

This is a set of simultaneous equations, one for each parameter. For some families of functions, these simultaneous equations are nonlinear equations and must be solved by successive approximations.[14] For linear functions or polynomial functions, the equations are linear equations, and we can solve them by the methods of Chapter 9.

In the method of *linear least squares* or *linear regression*, we assume that the family of functions in Eq. (10.39) is the family of linear functions, and then find the linear function that best fits our points. In many cases, we have a theory that predicts a certain dependence of $y$ on $x$, and we can often obtain a linear dependence (linearize) by changing variables. In the vapor pressure example, we linearized by choosing $\ln(P)$ as our dependent variable and $1/T$ as our independent variable, according to the Clausius–Clapeyron equation.

The family of linear functions is given by

$$y = mx + b. \tag{10.43}$$

We seek that value of the slope $m$ and that value of the intercept $b$ that give us the best fit. For the linear function of Eq. (10.43), the sum of the squares of the residuals is

$$R = \sum_{i=1}^{N} (y_i - m x_i - b)^2. \tag{10.44}$$

The simultaneous equations are

$$\frac{\partial R}{\partial m} = 2 \sum_{i=1}^{N} (y_i - mx_i - b)(-x_i) = 0 \tag{10.45a}$$

$$\frac{\partial R}{\partial b} = 2 \sum_{i=1}^{N} (y_i - mx_i - b)(-1) = 0. \tag{10.45b}$$

---

[13] John A. Rice, *op. cit.*, p. 473.

[14] Shoemaker, Garland, and Steinfeld, *op. cit.*, pp. 710ff.

This is a set of linear inhomogeneous simultaneous equations in $m$ and $b$. We write them in the form of Eq. (9.19),

$$S_{x2}m + S_x b = S_{xy} \tag{10.46a}$$
$$S_x m + Nb = S_y, \tag{10.46b}$$

where

$$S_x = \sum_{i=1}^{N} x_i \tag{10.47}$$

$$S_y = \sum_{i=1}^{N} y_i \tag{10.48}$$

$$S_{xy} = \sum_{i=1}^{N} x_i y_i \tag{10.49}$$

$$S_{x2} = \sum_{i=1}^{N} x_i^2. \tag{10.50}$$

These equations can be solved by any of the techniques of Chapter 9. For two equations in two variables, Cramer's rule is the easiest method. From Eqs. (9.28) and (9.29),

$$\boxed{m = \frac{1}{D}(N S_{xy} - S_x S_y)} \tag{10.51a}$$

$$\boxed{b = \frac{1}{D}(S_{x2} S_y - S_x S_{xy})}, \tag{10.51b}$$

where

$$\boxed{D = N S_{x2} - S_x^2}. \tag{10.52}$$

These are our working equations.

---

### Example 10.10

Calculate the slope $m$ and the intercept $b$ for the least-squares line for the data in Table 10.3, using $\ln(P)$ as the dependent variable and $1/T$ as the independent variable. Calculate the enthalpy change of vaporization from the slope.

---

### Solution

When the numerical work is done, the results are

$$m = -4854\,\text{K}$$
$$b = 20.28$$
$$\Delta H = -mR = -(-4854\,\text{K})(8.3145\,\text{J}\,\text{K}^{-1}\,\text{mol}^{-1}) = 40.36 \times 10^3\,\text{J}\,\text{mol}^{-1}.$$

This compares well with the accepted value of $40.3 \times 10^3\,\text{J}\,\text{mol}^{-1}$.

## Problem 10.13

The following data give the vapor pressure of water at various temperatures.[15] Find the least-squares line for the data, using $\ln(P)$ for the dependent variable and $1/T$ for the independent variable. Find the enthalpy change of vaporization.

| Temperature/°C | Vapor pressure/torr |
|---|---|
| 0 | 4.579 |
| 5 | 6.543 |
| 10 | 9.209 |
| 15 | 12.788 |
| 20 | 17.535 |
| 25 | 23.756 |

In some problems, it is not certain in advance what variables should be used for a linear least-squares fit. In the vapor pressure case, we had the Clausius–Clapeyron equation, Eq. (10.30), which indicated that $\ln(P)$ and $1/T$ were the variables that should produce a linear relationship. In the analysis of chemical rate data, it may be necessary to try two or more hypotheses to determine which gives the best fit.

In a reaction involving one reactant, the concentration $c$ of the reactant is given by Eq. (10.32) if there is no back reaction and if the reaction is a first-order reaction. If there is no back reaction and the reaction is a second-order reaction, the concentration of the reactant is given by

$$\frac{1}{c} = kt + K, \tag{10.53}$$

where $k$ is the rate constant and $K$ is a constant of integration. If there is no back reaction and the reaction is third order, the concentration $c$ of the reactant is given by

$$\frac{1}{2c^2} = kt + K. \tag{10.54}$$

If the order of a reaction is not known, it is possible to determine the order by trying different linear least-squares fits and finding which one most nearly fits the data. One way to see whether a given hypothesis produces a linear fit is to examine the residuals. Once the least-squares line has been found, the residuals can be calculated from Eq. (10.40), which for a linear function becomes

$$r_i = y_i - mx_i - b. \tag{10.55}$$

If a given dependent variable and independent variable produce a linear fit, the points will deviate from the line only because of experimental error, and the residuals will be either positive or negative without any pattern.

However, if there is a general curvature to the data points in a graph, the residuals will have the same sign near the ends of the graph and the other sign in the middle. This indicates that a different pair of variables should be tried or a nonlinear least-squares fit attempted.

[15]R. Weast, Ed., *Handbook of Chemistry and Physics*, 51st ed., p. D-143, CRC Press, Boca Raton, FL, 1971–1972.

**Example 10.11**

The following is a fictitious set of data for the concentration of the reactant in a chemical reaction with only one reactant. Determine whether the reaction is first, second, or third order. Find the rate constant and the initial concentration.

| Point number | Time/min | Concentration/mol L$^{-1}$ |
|---|---|---|
| 1 | 5.0 | 0.715 |
| 2 | 10.0 | 0.602 |
| 3 | 15.0 | 0.501 |
| 4 | 20.0 | 0.419 |
| 5 | 25.0 | 0.360 |
| 6 | 30.0 | 0.300 |
| 7 | 35.0 | 0.249 |
| 8 | 40.0 | 0.214 |
| 9 | 45.0 | 0.173 |

**Solution**

a. We first attempt a linear fit using $\ln(c)$ as the dependent variable and $t$ as the independent variable. The result is

$$m = -0.03504 \, \text{min}^{-1} = -k$$
$$b = -0.1592 = \ln[c(0)]$$
$$c(0) = 0.853 \, \text{mol liter}^{-1}.$$

The following set of residuals was obtained:

$$
\begin{array}{ll}
r_1 = -0.00109 & r_6 = 0.00634 \\
r_2 = 0.00207 & r_7 = -0.00480 \\
r_3 = -0.00639 & r_8 = 0.01891 \\
r_4 = -0.00994 & r_9 = -0.01859. \\
r_5 = 0.01348 &
\end{array}
$$

This is a good fit, with no pattern of general curvature shown in the residuals.

b. We test the hypothesis that the reaction is second order by attempting a linear fit using $1/c$ as the dependent variable and $t$ as the independent variable. The result is

$$m = 0.1052 \, \text{L mol}^{-1} \, \text{min}^{-1} = k$$
$$b = 0.4846 \, \text{L mol}^{-1} = \frac{1}{c(0)}$$
$$c(0) = 2.064 \, \text{mol liter}^{-1}.$$

The following set of residuals was obtained:

$$
\begin{array}{ll}
r_1 = 0.3882 & r_6 = -0.3062 \\
r_2 = 0.1249 & r_7 = -0.1492 \\
r_3 = -0.0660 & r_8 = -0.0182 \\
r_4 = -0.2012 & r_9 = 0.5634. \\
r_5 = -0.3359 &
\end{array}
$$

This is not such a satisfactory fit as in part a of the solution, since the residuals show a general curvature, beginning with positive values, becoming negative, and then becoming positive again.

c. We now test the hypothesis that the reaction is third order by attempting a linear fit using $1/(2c^2)$ as the dependent variable and $t$ as the independent variable. The results are

$$m = 0.3546 \text{ liter}^2 \text{ mol}^{-2} \text{ min}^{-1} = k$$
$$b = -3.054 \text{ liter}^2 \text{ mol}^{-2}.$$

This is obviously a bad fit, since the intercept $b$ should not be negative. The residuals are

$$
\begin{array}{ll}
r_1 = 2.2589 & r_6 = -2.0285 \\
r_2 = 0.8876 & r_5 = -1.2927 \\
r_3 = -0.2631 & r_8 = -0.2121 \\
r_4 = -1.1901 & r_9 = 3.8031. \\
r_5 = -1.9531 &
\end{array}
$$

Again, there is considerable curvature. The reaction is apparently first order, with the rate constant and initial concentration given in part a of the solution.

---

In addition to inspecting the residuals, we can calculate a quantity called the *correlation coefficient*, which gives information about the closeness of a least-squares fit. For linear least squares, the correlation coefficient is defined by

$$r = \frac{N S_{xy} - S_x S_y}{\left[\left(N S_{x2} - S_x^2\right)\left(N S_{y2} - S_y^2\right)\right]^{1/2}}, \tag{10.56}$$

where $S_x$, $S_y$, $S_{xy}$, and $S_{x2}$ are defined in Eqs. (10.47)–(10.50) and where

$$S_{y2} = \sum_{i=1}^{N} y_i^2. \tag{10.57}$$

If the data points lie exactly on the least-squares line, the correlation coefficient will be equal to 1 if the slope is positive or to $-1$ if the slope is negative. If the data points are scattered randomly about the graph so that no least-squares line can be found, the correlation coefficient will vanish. The magnitude of the correlation coefficient will be larger for a close fit than for a poor fit. In a fairly close fit, its magnitude might equal 0.99. Most software packages give the square of the correlation coefficient rather than the correlation coefficient itself.

---

### Example 10.12

Calculate the correlation coefficients for the three linear fits in Example 10.11.

---

### Solution

Use of Eq. (10.56) gives the results:

a. For the first-order fit, $r = -0.9997$.
b. For the second-order fit, $r = 0.9779$.
c. For the third-order fit, $r = 0.9257$.

Again, the first-order fit is the best.

---

**Problem 10.14**

Do three linear least-squares fits on the data of Problem 10.8. Calculate the correlation coefficients for the three fits and show that the reaction is first order. If you wish, you can use a software package such as CricketGraph or Excel, which will plot the data and carry out the least-squares fit for you. These programs will also carry out several different nonlinear least-square fits.

---

The correlation coefficient is related to a quantity called the *covariance*, defined by[16]

$$s_{xy} = \frac{1}{N-1} \sum_{i-1}^{N} (x_i - \bar{x})(y_i - \bar{y}),$$

where $\bar{x}$ is the average of the $x$'s,

$$\bar{x} = \frac{1}{N} S_x,$$

and where $\bar{y}$ is the average of the $y$'s,

$$\bar{y} = \frac{1}{N} S_y.$$

The covariance has the same general behavior as the correlation coefficient. If large values of $x$ tend to occur with small values of $y$, the covariance will be negative, and if large values of $x$ tend to occur with large values of $y$, the covariance will be positive. If there is no relationship between $x$ and $y$, the covariance will equal zero.

## Error Propagation in Linear Least Squares

We discuss two cases: (1) the expected error in each value of the dependent variable is known, and (2) the expected error in these values is not known. In both cases, we assume that the errors in the values of the independent variable $x$ are negligible.

**Case 1.** Let the expected error in the value of $y_i$ be denoted by $\varepsilon_i$. Equations (10.51a) and (10.51b) are formulas for the slope and the intercept of the least-squares line in which the $y$'s can be considered to be independent variables, so we can apply Eq. (10.24). The expected error in the slope is given by

$$\varepsilon_m = \left[ \sum_{i=1}^{N} \left( \frac{\partial m}{\partial y_i} \right)^2 \varepsilon_i^2 \right]^{1/2} = \left[ \frac{1}{D^2} \sum_{i=1}^{N} (Nx_i - S_x)^2 \varepsilon_i^2 \right]^{1/2}, \qquad (10.58)$$

where $D$ and $S_x$ are given by Eqs. (10.52) and (10.47). In the case that all of the expected errors in the $y$'s are equal, Eq. (10.58) becomes

$$\boxed{\varepsilon_m = \left( \frac{N}{D} \right)^{1/2} \varepsilon_y}, \qquad (10.59)$$

where $\varepsilon_y$ is the value of all the $\varepsilon_i$'s.

---

[16]John E. Freund, *Modern Elementary Statistics*, 7th ed., p. 459, Prentice–Hall, Englewood Cliffs, NJ, 1988.

The expected error in the intercept is

$$
\varepsilon_b = \left[ \sum_{i=1}^{N} \left( \frac{\partial b}{\partial y_i} \right)^2 \varepsilon_i^2 \right]^{1/2} = \left[ \frac{1}{D^2} \sum_{i=1}^{N} (S_{x2} - S_x x_i)^2 \varepsilon_i^2 \right]^{1/2}. \tag{10.60}
$$

For the case that all of the $\varepsilon_i$'s are assumed to be equal,

$$
\boxed{\varepsilon_b = \left( \frac{S_{x2}}{D} \right)^{1/2} \varepsilon_y}, \tag{10.61}
$$

where $S_{x2}$ is given by Eq. (10.50).

---

## Problem 10.15

Verify Eqs. (10.59) and (10.61).

---

## Example 10.13

Assume instead of the given expected errors that the expected error in the logarithm of each vapor pressure in Table 10.3 is 0.040. Find the expected error in the least-squares slope and in the enthalpy change of vaporization.

---

### Solution

From the data,

$$
D = 1.327 \times 10^{-6} \, \mathrm{K}^{-2}
$$

so that

$$
\varepsilon_m = \left( \frac{9}{1.327 \times 10^{-6} \, \mathrm{K}^{-2}} \right)^{1/2} (0.040) = 104 \, \mathrm{K}
$$

$$
\varepsilon_{\Delta H} = R \varepsilon_m = 8.7 \times 10^2 \, \mathrm{J \, mol}^{-1}.
$$

---

## Problem 10.16

Assume that the expected error in the logarithm of each concentration in Example 10.11 is 0.010. Find the expected error in the rate constant, assuming the reaction to be first order.

---

**Case 2.** Since we do not have information about the expected errors in the dependent variable, we assume that the residuals are a sample from the population of actual experimental errors. This is a reasonable assumption if systematic errors can be ignored.

The variance of the $N$ residuals is given by

$$
s_r^2 = \frac{1}{N-2} \sum_{i=1}^{N} r_i^2 \tag{10.62}
$$

and the standard deviation of the residuals is the square root of the variance:

$$s_r = \left( \frac{1}{N-2} \sum_{i=1}^{N} r_i^2 \right)^{1/2}. \tag{10.63}$$

This differs from Eq. (10.5) in that a factor of $N-2$ occurs in the denominator instead of $N-1$. The number of degrees of freedom is $N-2$ because we have calculated two quantities, a least-squares slope and a least-squares intercept from the set of numbers, "consuming" two of the degrees of freedom. The mean of the residuals does not enter in the formula, because the mean of the residuals in a least-squares fit always vanishes.

---

### Problem 10.17

Sum the residuals in Example 10.11 and show that this sum vanishes in each of the three least-square fits.

---

Equation (10.63) provides an estimate of the standard deviation of the population of experimental errors. The errors in $y$ are at least approximately distributed according to the Student $t$ distribution, so the expected error in $y$ at the 95% confidence level is given by

$$\varepsilon_y = t(v, 0.05)s_r, \tag{10.64}$$

where $t(v, 0.05)$ is the Student $t$ factor for $v = N - 2$; the number of degrees of freedom for $N$ data points after the slope and the intercept have been calculated. If there were a very large number of data points, the Student $t$ distribution would approach the Gaussian distribution, and this factor would approach 1.96.

We can now write expressions similar to Eqs. (10.59) and (10.60) for the expected errors at the 95% confidence level:

$$\varepsilon_m = \left( \frac{N}{D} \right)^{1/2} t(v, 0.05)s_r \tag{10.65}$$

and

$$\varepsilon_m = \left( \frac{1}{D} \sum_{i=1}^{N} x_i^2 \right)^{1/2} t(v, 0.05)s_r. \tag{10.66}$$

The standard deviations of the slope and intercept are given by similar formulas, without the Student $t$ factor:

$$s_m = \left( \frac{1}{D} \sum_{i=1}^{N} x_i^2 \right)^{1/2} s_r \tag{10.67}$$

$$s_m = \left( \frac{N}{D} \right)^{1/2} s_r. \tag{10.68}$$

The slope and the intercept of a least-squares line are not independent of each other, since they are derived from the same set of data, and their *covariance* is given by[17]

$$\text{Cov}(m, b) = s_{m,b} = \frac{-s_r^2 S_x}{D}. \tag{10.69}$$

### Example 10.14

Calculate the residuals for the linear least-squares fit of Example 10.10. Find their standard deviation and the probable error in the slope and in the enthalpy change of vaporization, using the standard deviation of the residuals.

### Solution

Numbering the data points from the 25°C point (number 1) to the 65°C point (number 9), we find the residuals:

$$
\begin{array}{ll}
r_1 = 0.0208 & r_6 = -0.0116 \\
r_2 = -0.0228 & r_7 = -0.0027 \\
r_3 = 0.0100 & r_8 = 0.0054 \\
r_4 = -0.0162 & r_9 = 0.0059. \\
r_5 = 0.0113 &
\end{array}
$$

The standard deviation of the residuals is found to be

$$s_r = 0.0154.$$

Using the value of $D$ from Example 10.13, the uncertainty in the slope is

$$\varepsilon_m = \left( \frac{9}{1.327 \times 10^{-6}\,\text{K}^{-2}} \right)^{1/2} (2.365)(0.0154) = 94.9\,\text{K},$$

where we have used the value of the Student's $t$ factor for seven degrees of freedom from Table 10.1. The uncertainty in the enthalpy change of vaporization is

$$\varepsilon_{\Delta H} = R\varepsilon_m = (8.3145\,\text{J}\,\text{K}^{-1}\,\text{mol}^{-1})(96.9\,\text{K}) = 789\,\text{J}\,\text{mol}^{-1}.$$

### Problem 10.18

Assuming that the reaction in Problem 10.14 is first order, find the expected error in the rate constant, using the residuals.

## Expected Errors in the Dependent Variable

If the linear least-squares method is used to find the line that best represents a set of data, this line can be used to predict a value for the dependent variable corresponding to any given value of the independent variable. We now consider the probable error

[17] John A. Rice, *op. cit.*, p. 460.

in such a prediction.[18] Since the dependent variable $y$ is a function of the slope $m$ and the intercept $b$, we might try to apply Eq. (10.24):

$$\varepsilon_y = \left( \left( \frac{\partial y}{\partial m} \right)^2 \varepsilon_m^2 + \left( \frac{\partial y}{\partial b} \right)^2 \varepsilon_b^2 \right)^{1/2} \qquad \text{(NOT APPLICABLE).} \qquad (10.70)$$

However, this equation is incorrect, because $m$ and $b$ have been derived from the same set of data and are not independent of each other, as was assumed in obtaining Eq. (10.24).

The correct equation is obtained by including the covariance of $m$ and $b$. We consider the standard deviation of $y$,

$$s_y^2 = \left( \frac{\partial y}{\partial m} \right)^2 s_m^2 + \left( \frac{\partial y}{\partial b} \right)^2 s_b^2 + 2 \left( \frac{\partial y}{\partial m} \right) \left( \frac{\partial y}{\partial b} \right) s_{m,b} \qquad (10.71)$$

$$= x^2 s_m^2 + s_b^2 + 2x s_{m,b}, \qquad (10.72)$$

where $s_m$ is the standard deviation of $m$, $s_b$ is the standard deviation of $b$, and $s_{m,b}$ is the covariance of $m$ and $b$. In this equation, $x$ stands for the value of $x$ for which we want the value of $y$. The expected error in $y$ at the 95% confidence level is

$$\varepsilon_y = t(N - 2, 0.05)s_y. \qquad (10.73)$$

If you want to determine a value of $x$ for a given value of $y$, a similar analysis can be carried out, considering that for a given value of $y$, $x$ is a function of $m$ and $b$.

---

**Problem 10.19**

a. From the least-squares fit of Example 10.10, find the predicted value of the vapor pressure of ethanol at 70.0°C. Find the expected error in the natural logarithm of the vapor pressure. Compare the values of the three terms in Eq. (10.68). Find the expected error in the predicted value of the vapor pressure.

b. From the least-squares fit of Example 10.10, find the predicted temperature at which the vapor pressure of ethanol is equal to 350.0 torr. Find the expected error in the reciprocal of the absolute temperature. Compare the values of the three terms analogous to those in Eq. (10.68).

---

## A Computer Program for Linear Least Squares

The following is a computer program in BASIC that will carry out a linear least-squares fit. This program runs using the TrueBASIC processor (see Chapter 11). It was written and run on a Macintosh computer, but the TrueBASIC processor on a Windows machine accepts the same programs. The program calculates the expected error in the slope and the intercept from the residuals, as well as the square of the

---

[18]Edwin F. Meyer, *J. Chem. Educ.* **74**, 1339 (1997). The author of this article refers to a software package by Ramette, called FLEXFIT, which is available on JCE Online at http://jchemed.chem.wisc.edu/. This package can be manipulated to give the covariance of m and b.

correlation coefficient. A similar program was used to work out the examples of this section.

```
10 REM PROGRAM SIMPLESQUARE
20 REM THIS PROGRAM CARRIES OUT A LEAST SQUARES FIT TO
330 REM TO 30 DATA POINTS
35 OPTION NOLET
40 DIM T(30)
50 DIM X(30)
60 DIM Y(30)
70 DIM R(30)
75 DEF FNM = (N*XY - X1*Y1)/(N*X2 - X1*X1)
85 DEF FNB = (X2*Y1 - X1*XY)/(N*X2 - X1*X1)
95 DEF FNR2 = ((N*XY - X1*Y1)^2)/((N*X2 - X1*X1)*
(N*Y2 - Y1*Y1))
100 FOR I = 1 TO 28
110 READ T(I)
120 NEXT I
190 GOTO 3010
2140 X1=0
2150 X2=0
2160 Y1=0
2170 Y2=0
2180 XY=0
2190 FOR I = 1 TO N
2200 X1=X1+X(I)
2202 X2=X2+X(I)*X(I)
2204 Y1=Y1+Y(I)
2206 Y2=Y2+Y(I)*Y(I)
2208 XY=XY+X(I)*Y(I)
2210 NEXT I
2220 M = FNM
2222 B = FNB
2224 R2 = FNR2
2230 PRINT "SLOPE =";M, "INTERCEPT =";B
2250 IF B$ = "YES" THEN 2252 ELSE 2256
2252 PRINT "RESIDUALS"
2256 FOR I = 1 TO N
2260 R(I) = Y(I) - M*X(I) - B
2270 IF B$ = "YES" THEN 2280 ELSE 2290
2280 PRINT "RESIDUAL R(";I;") = ";R(I)
2290 R1 = R1 + R(I)*R(I)
2300 NEXT I
2310 S = SQR(R1/(N -2))
2320 PRINT "STANDARD DEVIATION OF THE RESIDUALS = "; S
```

```
2330 D = N*X2 - X1*X1
2340 E1 = SQR(N/D)*T(N-2)*S
2350 PRINT "UNCERTAINTY IN THE SLOPE (95%
CONFIDENCE) = "; E1
2360 E2 = SQR(X2/D)*T(N-2)*S
2370 PRINT "UNCERTAINTY IN THE INTERCEPT (95%
CONFIDENCE) = ";E2
2380 R2 = FNR2
2390 PRINT "CORRELATION COEFFICIENT SQUARED = "; R2
2400 GOTO 4000
3010 PRINT "DO YOU WANT A LIST OF RESIDUALS?"
3012 INPUT B$
3014 PRINT "TYPE IN THE NUMBER OF DATA POINTS."
3016 INPUT N
3020 PRINT "INDEPENDENT VARIABLE MUST BE ENTERED
FIRST."
3030 FOR I = 1 TO N
3040 PRINT "TYPE IN X(";I;"), Y(";I;")"
3050 INPUT X(I), Y(I)
3060 NEXT I
3070 GOTO 2140
3081 IF B >= 0 THEN 3082 ELSE 3086
3082 PRINT "Y = ";M; "TIMES X + ";B
3084 GOTO 2250
3086 PRINT "Y = ";M; "TIMES X ";B
3090 GOTO 2250
3900 DATA 12.706, 4.303, 3.182, 2.776, 2.571, 2.477, 2.365
3910 DATA 2.306, 2.262, 2.228, 2.201, 2.179, 2.160, 2.145
3920 DATA 2.131, 2.120, 2.110, 2.101, 2.093, 2.086, 2.080
3930 DATA 2.074, 2.039, 2.064, 2.060, 2.056, 2.052, 2.048
4000 END
```

## Some Warnings about Linear Least-Squares Procedures

It is a poor idea to rely blindly on a numerical method. You should always determine whether your results are reasonable. It is possible to spoil your results by entering one number incorrectly or by failing to recognize a bad data point. Remember the *first maxim of computing: Garbage in, garbage out.*

You should always look at your correlation coefficient. A low magnitude usually indicates a problem. Another way to make sure that a linear least-squares procedure has given you a good result is to make a plot of the data points and the least-squares line on the same graph. Software packages such as CricketGraph, KaleidaGraph, and Excel will do this for you automatically. If some error such as an incorrectly entered data point has produced the wrong line, you will probably be able to tell by looking at the graph. If the data points show a general curvature, you will probably be able to tell that as well from the graph.

You should examine your residuals. There are calculators that have linear least-squares programs built into them, but these do not generally provide you with the values of the residuals. As explained earlier, examination of the residuals can reveal a general curvature in your data points and can also reveal the presence of a bad data point.

If you use CricketGraph or some similar graphical software to carry out a least-squares analysis, you will get the graph automatically, as well as the correlation coefficient. Unfortunately, you will not get the residuals. A spreadsheet such as Excel will provide you with residuals and expected errors, as well as the expected errors in the slope and intercept.

A final warning is that in making a change in variables in order to fit a set of data to a straight line rather than to some other function, one is actually changing the relative importance, or weight, of the various data points.[19] Because of this, fitting $\ln(c)$ to a straight line

$$\ln(c) = -kt + K \tag{10.74}$$

will not necessarily give the same value of $k$ as will fitting $c$ to the function

$$c = e^K e^{-kt}. \tag{10.75}$$

We now discuss a way to compensate for this and also to compensate for errors of different sizes in different data points.

## Weighting Factors in Linear Least Squares

Say that we have a set of data in which the probable errors in the values of the independent variable are negligible, and in which the probable errors in the values of the dependent variable are not all of the same size. In this case, instead of minimizing the sum of the squares of the residuals, it has been shown that one should minimize the sum of the squares of the residuals divided by the square of the probable error for each data point. If $\sigma_i$ is the standard deviation of the distribution from which $r_i$ is drawn, we should minimize[20]

$$R' = \sum_{i=1}^{N} \frac{r_i^2}{\sigma_i^2} = \sum_{i=1}^{N} \frac{1}{\sigma_i^2}(y_i - mx_i - b)^2. \tag{10.76}$$

The factors $1/\sigma_i^2$ in the sum are called *weighting factors*. The effect of this weighting is to give a greater importance, or greater weight, to those points that have smaller expected errors. You must decide whether it is advisable to use a weighted least-squares procedure.[21]

---

[19]Donald E. Sands, "Weighting Factors in Least Squares," *J. Chem. Educ.* **51**, 473 (1974).

[20]P. R. Bevington and D. K. Robinson, *Data Reduction and Error Analysis for the Physical Sciences*, 2nd ed., McGraw–Hill, New York, 1992.

[21]R. deLevie, *J. Chem. Educ.* **63**, 10 (1986).

We can now proceed to minimize $R'$. The equations are very similar to Eqs. (10.45)–(10.52), except that each sum includes the weighting factors. The results for the slope and intercept are

$$m = \frac{1}{D'}(S_1' S_{xy}' - S_x' S_y') \tag{10.77a}$$

$$b = \frac{1}{D'}(S_{x2}' S_y' - S_x' S_{xy}'), \tag{10.77b}$$

where

$$D' = S_1' S_{x2}' - S_x'^2, \tag{10.78}$$

and where

$$S_1' = \sum_{i=1}^{N} \frac{1}{\sigma_i^2} \tag{10.79a}$$

$$S_x' = \sum_{i=1}^{N} \frac{x_i}{\sigma_i^2} \tag{10.79b}$$

$$S_y' = \sum_{i=1}^{N} \frac{y_i}{\sigma_i^2} \tag{10.79c}$$

$$S_{xy}' = \sum_{i=1}^{N} \frac{x_i y_i}{\sigma_i^2} \tag{10.79d}$$

$$S_{x2}' = \sum_{i=1}^{N} \frac{x_i^2}{\sigma_i^2}. \tag{10.79e}$$

---

**Problem 10.20**

Verify Eqs. (10.77)–(10.79).

---

There are two ways in which we will use Eq. (10.77). The first is in a case when the probable errors in the data are known and not equal, and the second is in a case when the probable errors are assumed to be equal and a change in variables is required for a linear fit.

---

**Example 10.15**

Find the least-squares line for the data of Table 10.3, assuming that the weighting factors are inversely proportional to the squares of the expected errors in the logarithms.

---

**Solution**

The expected errors in the logarithm of $P$ were calculated by Eq. (10.31) from the expected errors in the pressures given in the table. These were substituted into

Eqs. (10.69)–(10.79) in place of the $\sigma_i$'s. The results were

$$m = -4872 \text{ K}$$
$$b = 20.34$$

These figures differ slightly from those of Example 10.10 and are probably more nearly accurate. The slope gives a value of the enthalpy change of vaporization of 40.51 kJ mol$^{-1}$.

In the following example, we see what an inaccurate point can do if the unweighted least-squares procedure is used.

### Example 10.16

Change the data set of Table 10.3 by adding a value of the vapor pressure at 70°C of 421±40 torr. Find the least-squares line using both the unweighted and weighted procedures.

--------

### Solution

After the point was added, the unweighted procedure was carried out as in Example 10.10, and the weighted procedure was carried out as in Example 10.15. The results were:

unweighted procedure,

$$m = \text{slope} = -4752 \text{ K}$$
$$b = \text{intercept} = 19.95;$$

weighted procedure,

$$m = \text{slope} = -4855 \text{ K}$$
$$b = \text{intercept} = 20.28.$$

The data point of low accuracy has done more damage in the unweighted procedure than in the procedure with weighting factors.

--------

If the values of the original dependent variable have equal expected errors, an unweighted least-squares fit is appropriate if we use that variable in our procedure. However, if we take a function of the original variable in order to use a linear fit, then the original expected errors, which are all equal, will not generally produce equal errors in the new variable, and the weighted least-squares procedure is preferred.

--------

### Example 10.17

Assume that each concentration in the data set of Example 10.11 is uncertain by 0.0070 mol L$^{-1}$. Calculate the probable error in each value of $\ln(c)$ and carry out a linear least-squares fit using $\ln(c)$ as the dependent variable and the square of the reciprocal of the probable error as the weighting factor.

**Solution**

For each point, we use the following, where $y = \ln(c)$:

$$\varepsilon_y = \frac{dy}{dc}\varepsilon_c = \frac{1}{c}\varepsilon_c.$$

We have

| Point number | $\varepsilon_y$ |
|---|---|
| 1 | 0.010 |
| 2 | 0.012 |
| 3 | 0.014 |
| 4 | 0.017 |
| 5 | 0.019 |
| 6 | 0.023 |
| 7 | 0.028 |
| 8 | 0.033 |
| 9 | 0.040 |

Notice the differences in the accuracy of the logarithms.

The result of carrying out the procedure with weighting factors equal to the reciprocals of the squares of the expected error is

$$m = \text{slope} = -0.03494 \text{ min}^{-1} = -k$$
$$b = \text{intercept} = -0.1612.$$

If the errors are as stated, the present results are more reliable than those of Example 10.12.

---

This discussion suggests a possible procedure to use if you carry out a least-squares fit and a few points lie a long way from the line: Carry out the fit a second time using weighting factors, utilizing the residuals from the first fit in place of the $\sigma_i$'s of Eqs. (10.69)–(10.79). This procedure should give a better fit than use of the unweighted procedure alone.

## Error Propagation in Weighted Least Squares

We can derive equations analogous to Eqs. (10.59) and (10.61) for the weighted least-squares procedure. We continue to assume that the errors in the values of the independent variable, $x$, are negligible, and use the symbol $\varepsilon_i$ for the probable error in $y_i$. The standard deviations $\sigma_1$, $\sigma_2$, etc., in Eq. (10.79) are generally unknown, but will be assumed to be proportional to $\varepsilon_1$, $\varepsilon_2$, etc., so we replace the $\varepsilon_i$'s by the $\varepsilon_i$'s. This will not affect the slope and the intercept. We now proceed as in the unweighted case, writing as in Eq. (10.58)

$$\varepsilon_m = \left[\sum_{i=1}^{N}\left(\frac{\partial m}{\partial y_i}\right)^2 \varepsilon_i^2\right]^{1/2} = \left[\sum_{i=1}^{N}\frac{1}{D'^2}\left(\frac{S_1'}{\varepsilon_i^2} - \frac{S_x'}{\varepsilon_i^2}\right)^2 \varepsilon_i^2\right]^{1/2}$$
$$= \frac{1}{D'}\left(S_1'^2 S_{x2}' - S_1' S_x'^2\right)^{1/2}, \tag{10.80}$$

where the sums are those of Eq. (10.79) except that the $\sigma_i$'s have been replaced by the $\varepsilon_i$'s.

---

### Problem 10.21

Verify Eq. (10.80). A similar procedure gives

$$\varepsilon_b = \frac{1}{D'}\left(S_1' S_{x2}'^2 - S_{x2}' S_x'^2\right)^{1/2}. \tag{10.81}$$

---

### Problem 10.22

Verify Eq. (10.81).

---

### Example 10.18

a. Find the expected error in the slope and intercept for the least-squares line found in Example 10.15.
b. Find the expected error in the slope and the intercept for both of the least-squares fits of Example 10.16.

---

### Solution

a. The results are

$$\varepsilon_m = 97 \text{ K}$$
$$\varepsilon_b = 0.30.$$

b. For the unweighted procedure, we find from the residuals,

$$\varepsilon_m = 137 \text{ K}$$
$$\varepsilon_b = 0.43.$$

For the weighted procedure, we find

$$\varepsilon_m = 96 \text{ K}$$
$$\varepsilon_b = 0.30.$$

In the unweighted procedure, the introduction of the bad data point has raised the uncertainty in the slope from 78 K to 137 K, but in the weighted procedure, there was no effect. If a bad data point has a large known error, its effect in the weighted least-squares procedure is minimal.

---

### Problem 10.23

Modify the BASIC program given above to carry out a weighted linear least-squares fit. Carry out the weighted procedure on the data of Problem 10.13, assuming that each vapor pressure is uncertain by 0.050 torr, so that the logarithms have different uncertainties.

---

## Linear Least Squares with Fixed Slope or Intercept

At times it is necessary to do a least-squares fit with the constraint that the slope or the intercept have a fixed value. For example, the Bouguer–Beer law states that the absorbance of a solution is proportional to the concentration of the colored substance. In fitting the absorbance of several solutions to their concentrations, one would specify that the intercept of the least-squares line had to be zero. In other cases, a particular slope might be required.

In the minimization of the sum of the squares of the residuals, one minimizes only with respect to the slope $m$ if the intercept $b$ is fixed. This is the same as Eq. (10.45a):

$$\frac{dR}{dm} = 2 \sum_{i=1}^{N} (y_i - mx_i - b)(-x_i) = 0. \tag{10.82}$$

The solution to this is

$$m = \frac{S_{xy} - bS_x}{S_{x2}}. \tag{10.83}$$

If the slope $m$ is required to have a fixed value, we have only one equation, which is the same as Eq. (10.45b),

$$\frac{dR}{db} = 2 \sum_{i=1}^{N} (y_i - mx_i - b)(-1) = 0. \tag{10.84}$$

The solution to this is

$$b = \frac{S_y - mS_x}{N}. \tag{10.85}$$

If the required slope is equal to zero, the resulting intercept is equal to the mean of the $y$ values:

$$b = \frac{S_y}{N} = \bar{y}. \tag{10.86}$$

## SUMMARY OF THE CHAPTER

We discussed several related techniques in this chapter. The first is the estimation of probable errors in directly measured quantities. We assumed the existence of a population of infinitely many repetitions of the measurement. If several repetitions of the measurement can be made, we considered this set of measurements to be a sample from the population. We took the sample standard deviation to be an unbiased estimate of the population standard deviation and the sample mean to be an estimate of the population mean (which is the correct value if systematic error is absent). The probable error in the mean was determined by a formula of Student.

If a formula is used to calculate values of some variable from measured values of other variables, it is necessary to propagate the errors in the measured quantities through the calculation. We provided a scheme to calculate the expected error in the dependent variable, based on the total differential of the dependent variable.

We also discussed graphical and numerical data reduction procedures. The most important numerical data reduction procedure is the least squares, or regression, method, which finds the best member of a family of functions to represent a set of data. We discussed the propagation of errors through this procedure and presented a version of the procedure in which different data points are given different weights, or importances, in the procedure.

## ADDITIONAL READING

P. R. Bevington and D. K. Robinson, *Data Reduction and Error Analysis for the Physical Sciences*, 2nd ed., McGraw–Hill, New York, 1992. This is a very nice book, which includes a lot of useful things, including a discussion of different probability distributions, including the Gaussian distribution, and a discussion of weighted least-squares procedures.

R. A. Day, Jr., and A. L. Underwood, *Quantitative Analysis*, 3rd ed., Prentice–Hall, Englewood Cliffs, NJ, 1974. This is a typical quantitative analysis textbook and contains a discussion of experimental errors and some statistical techniques involving indirectly measured quantities.

Walter Clark Hamilton, *Statistics in Physical Science*, The Ronald Press Company, New York, 1964. This is a useful small book that contains just about everything that a chemist might need to know about the use of statistics.

Eugene Jahnke and Fritz Emde, *Tables of Functions*, Dover, New York, 1945. This is a very useful paperback book that contains a lot of tables and formulas, including the error function and related quantities.

Emerson M. Pugh and George H. Winslow, *The Analysis of Physical Measurements*, Addison–Wesley, Reading, MA, 1966. This small book is available in paperback version and contains quite a lot of information.

John A. Rice, *Mathematical Statistics and Data Analysis*, Wadsworth & Brooks/ Cole, Pacific Grove, CA, 1988. This is a standard textbook for mathematical statistics. It includes numerous examples from experimental chemistry and is a good reference for chemists.

David P. Shoemaker, Carl W. Garland, and Joseph W. Nibler, *Experiments in Physical Chemistry*, 6th ed., McGraw–Hill, New York, 1996. This is a standard physical chemistry laboratory textbook and contains a good section on the treatment of experimental errors as well as most of the experiments commonly done in physical chemistry courses.

Archie G. Worthing and Joseph Geffner, *Treatment of Experimental Data*, Wiley, New York, 1943. This is a useful source of formulas and facts about the application of statistics to experimental results.

Hugh D. Young, *Statistical Treatment of Experimental Data*, McGraw–Hill, New York, 1962. This is a small paperback book, written at an elementary level, and explains on a physical basis what is behind some of the things discussed in this chapter.

## ADDITIONAL PROBLEMS

### 10.24

A sample of 10 sheets of paper has been selected randomly from a ream (500 sheets) of paper. Regard the ream as a population, even though it has only a finite number of members. The width and length of each sheet of the sample were measured, with the following results:

| Sheet number | Width/inch | Length/inch |
|---|---|---|
| 1 | 8.50 | 11.03 |
| 2 | 8.48 | 10.99 |
| 3 | 8.51 | 10.98 |
| 4 | 8.49 | 11.00 |
| 5 | 8.50 | 11.01 |
| 6 | 8.48 | 11.02 |
| 7 | 8.52 | 10.98 |
| 8 | 8.47 | 11.04 |
| 9 | 8.53 | 10.97 |
| 10 | 8.51 | 11.00 |

a. Calculate the sample mean length and its sample standard deviation, and the sample mean width and its sample standard deviation.
b. Give the expected ream mean length and width, and the expected error in each at the 95% confidence level.
c. Calculate the expected ream mean area from the width and length, and give the 95% confidence interval for the area.
d. Calculate the area of each sheet in the sample. Calculate from these areas the sample mean area and the standard deviation in the area.
e. Give the expected ream mean area and its 95% confidence interval from the results of part d.
f. Compare the results of parts c and e. Would you expect the two results to be identical? Why (or why not)?

**10.25**

The following is a set of student data on the vapor pressure of pure liquid ammonia, obtained in a physical chemistry laboratory course. Find the indicated enthalpy change of vaporization, using the graphical procedure once and the least-squares procedure once.

| Temperature/°C | Pressure/torr |
|---|---|
| −76.0 | 51.15 |
| −74.0 | 59.40 |
| −72.0 | 60.00 |
| −70.0 | 75.10 |
| −68.0 | 91.70 |
| −64.0 | 112.75 |
| −62.0 | 134.80 |
| −60.0 | 154.30 |
| −58.0 | 176.45 |
| −56.0 | 192.90 |

**10.26**

a. Ignoring the systematic errors, find the 95% confidence interval for the enthalpy change of vaporization, using the residuals, for the data in Problem 10.25.
b. Assuming that the apparatus used to obtain the data in Problem 10.25 was about like that found in most undergraduate physical chemistry laboratories, make a reasonable estimate of the systematic errors and find the 95%

confidence interval for the enthalpy change of vaporization, including both systematic and random errors.

### 10.27

The vibrational contribution to the molar heat capacity of a gas of nonlinear molecules is given in statistical mechanics by the formula

$$\bar{C}(\text{vib}) = R \sum_{i=1}^{3n-6} \frac{u_i^2 e^{-u_i}}{(1 - e^{-u_i})^2},$$

where $u_i = h\nu_i/k_B T$. Here $\nu_i$ is the frequency of the $i$th normal mode of vibration, of which there are $3n - 6$ if $n$ is the number of nuclei in the molecule (assumed nonlinear), $h$ is Planck's constant, $k_B$ is Boltzmann's constant, $R$ is the gas constant, and $T$ is the absolute temperature. The $H_2O$ molecule has three normal modes. If their frequencies are given by

$$\nu_1 = 4.78 \times 10^{13} \pm 0.002 \times 10^{13} \text{ s}^{-1}$$
$$\nu_2 = 1.095 \times 10^{14} \pm 0.004 \times 10^{14} \text{ s}^{-1}$$
$$\nu_3 = 1.126 \times 10^{14} \pm 0.004 \times 10^{14} \text{ s}^{-1}$$

calculate the vibrational contribution to the heat capacity of $H_2O$ vapor at 500 K and find the 95% confidence interval.

### 10.28

The *nth moment of a probability distribution* is defined by

$$M_n = \int (x - \mu)^n f(x)\, dx.$$

The second moment is the variance, or square of the standard deviation. Show that for the Gaussian distribution, $M_3 = 0$, and find the value of $M_4$. For this distribution, the limits of integration are $-\infty$ and $+\infty$.

### 10.29

Vaughan[22] obtained the following data for the dimerization of butadiene at 326°C.

| Time (min) | Partial pressure of butadiene |
|---|---|
| 0 | to be deduced |
| 3.25 | 0.7961 |
| 8.02 | 0.7457 |
| 12.18 | 0.7057 |
| 17.30 | 0.6657 |
| 24.55 | 0.6073 |
| 33.00 | 0.5573 |
| 42.50 | 0.5087 |
| 55.08 | 0.4585 |
| 68.05 | 0.4173 |
| 90.05 | 0.3613 |
| 119.00 | 0.3073 |
| 259.50 | 0.1711 |
| 373.00 | 0.1081 |

---

[22]W. E. Vaughan, "The Homogeneous Thermal Polymerization of 1,3-Butadiene," *J. Am. Chem. Soc.* **54**, 3863 (1932).

Determine whether the reaction is first, second, or third order, using the least-squares method. Find the rate constant and its 95% confidence interval, ignoring systematic errors. Find the initial pressure of butadiene.

### 10.30

Make a graph of the partial pressure of butadiene as a function of time, using the data in Problem 10.29. Find the slope of the tangent line at 33.00 min and deduce the rate constant from it. Compare with the result from Problem 10.29.

### 10.31

Use Eq. (10.33) to "smooth" the data given in Example 10.11. Using the smoothed data and Eq. (10.35) find the derivative $dc/dt$ at $t = 25$ min. Find the rate constant.

### 10.32

Assuming that the ideal gas law holds, find the number of moles of nitrogen gas in a container if

$$P = 0.856 \pm 0.003 \text{ atm}$$
$$V = 17.85 \pm 0.08 \text{ liter}$$
$$T = 297.3 \pm 0.08 \text{ K}.$$

Find the expected error in the number of moles.

### 10.33

The Bouguer–Beer law (sometimes called the Lambert–Beer law) states

$$A = abc,$$

where $A$ is the absorbance of a solution, defined as $\log_{10}(I_0/I)$ where $I_0$ is the incident intensity of light at the appropriate wavelength and $I$ is the transmitted intensity; $b$ is the length of the cell through which the light passes; and $c$ is the concentration of the absorbing substance. The coefficient $a$ is called the *molar absorptivity* if the concentration is in moles per liter. The following is a set of data for the absorbance of a set of solutions of disodium fumarate at a wavelength of 250 nm. Using Eq. (10.83), find the least-squares value of $a$ if $b = 1.000$ cm.

| $A$ | 0.1425 | 0.2865 | 0.4280 | 0.5725 | 0.7160 | 0.8575 |
|---|---|---|---|---|---|---|
| $c$(mol liter$^{-1}$) | $1.00 \times 10^{-4}$ | $2.00 \times 10^{-4}$ | $3.00 \times 10^{-4}$ | $4.00 \times 10^{-4}$ | $5.00 \times 10^{-4}$ | $6.00 \times 10^{-4}$ |

# ██
# ██ USING COMPUTERS IN
# PHYSICAL CHEMISTRY

## Preview

In this chapter, we introduce several software packages, which are computer programs that can accept instructions from the user and translate them into a form which can actually operate the computer. The introductions are sufficient only to allow a beginner to get started using the software. We also provide a brief introduction to programming in the BASIC language.

## Principal Facts and Ideas

1. A computer is ordinarily operated by using a program that accepts easily understood instructions and translates them into machine language, which can operate the computer. In addition to the computer operating system, the principal programs used by physical chemistry students are word processors, spreadsheets, graphics packages, high-level programming languages such as BASIC, and complete mathematics packages such as Mathematica and MathCad.

2. A word processor can produce documents of a quality that previously required a sophisticated printing establishment.

3. A spreadsheet can carry out calculations on sets of numbers and display results in tabular and graphical form.

4. Graphics software can produce good quality graphs with a minimum of effort.

5. A high-level computer programming language such as BASIC can accept instructions written in an easily learned form and translate them into machine language.
6. A complete mathematics package such as Mathematica can carry out symbolic as well as numerical mathematics.

## Objectives

After studying this chapter, you should be able to:

1. compose a document containing equations and tables using a word processor;
2. carry out simple calculations using a spreadsheet such as Excel;
3. construct graphs using a graphics program such as CricketGraph;
4. write and run a simple program in the BASIC language;
5. carry out elementary mathematical operations using a complete mathematics pack such as Mathematica.

## SECTION 11.1. THE OPERATION OF A COMPUTER

Several decades ago, computers were used only by persons who either wrote their own computer programs or acquired copies of programs from other users, usually in the form of decks of punched cards. Computers were large and were too expensive for individuals to own. They resided in university and corporation computer centers and could be used by only one person at a time. In many centers, individuals were not allowed to enter their own programs, but had to submit them to be run by an employee of the computer center. The situation is now vastly different. There are many computer users, who usually use personal computers, and there are many commercially available computer programs ("software") that carry out various tasks, including word processing, graphics generation, and a variety of numerical and symbolic calculations.

All of the operations carried out by a computer are done in the *central processing unit* (CPU). A CPU contains an oscillator which produces a signal of a fixed frequency, and the CPU carries out one operation per cycle of the oscillator. Personal computers now operate at speeds up to 350 million cycles per second (350 megahertz) or even higher. The actual calculations are carried out using binary numbers (numbers to the base 2, expressed with digits that equal either 0 or to 1). Numbers used in calculations are stored in *core memory*, which consists of sets of devices that store binary numbers. A single binary digit is called a *bit*, and 8 bits constitute a *byte*. The size of a core memory is specified in terms of kilobytes or megabytes. A *kilobyte* (kB) consists of 1024 bytes, a *megabyte* (mB) consists of 1024 kilobytes, and a *gigabyte* (gB) consists of 1024 megabytes. A personal computer might have up to 40 megabytes or more of core memory. A portion of the memory is used to store a program, which is a set of codes that tell the CPU to carry out operations and tell it where the numbers are stored on which to operate. The CPU moves automatically through these instructions to accomplish the goal of the program.

Computers also have disk storage devices on which binary numbers are stored. A personal computer might have a hard disk drive with several gigabytes of storage space and a diskette drive in which a diskette capable of storing 1.4 megabytes can be inserted. Data of all sorts is stored on the hard drive, as well as software that is not being used at the moment. Although everything is stored as binary numbers, there are standard codes for translating letters into numbers, so that verbal information can also be stored. There are also codes for translating pictorial information into digital form. Every small area of a video display is designated as a picture element, or "pixel," and a location in memory or on the hard drive can be specified to store numerically the pictorial information to be displayed in that pixel.

## SECTION 11.2. COMPUTER SOFTWARE

Computer users are likely to use several kinds of prepared software. The first type is the *operating system*, which is a program that enables the user to load software from a disk drive, to store files in an organized way on a disk, to adjust the way in which images are displayed on the video screen, to send information to a printer, and a lot of other things. This software is automatically loaded from the disk drive when the computer is turned on ("booted up"). It is in memory and running even when another program is running. Personal computers now have operating systems with a "graphical user interface" (GUI) that displays output and lists of commands ("menus") on a video display screen and enables the user to select commands from menus by moving a "mouse" while watching a cursor move among various little pictures ("icons") on the video screen. The various Windows operating systems and the various Macintosh operating systems are of this type.

In addition to the operating system, the student of physical chemistry is likely to use several kinds of software, known as "*applications*": (1) a word processor; (2) a spreadsheet; (3) a graphics package; (4) a computer programming language processor; and (5) a "complete" mathematics package capable of carrying out symbolic mathematical operations, making all kinds of graphs, etc. In this chapter, we present a brief introduction to the use of these kinds of software. We make no attempt to be general and provide only enough introduction to get a beginner started. We confine our discussion to software used on the Macintosh computer. Similar procedures apply to the "Windows" operating system.

The Macintosh and Windows operating systems use the mouse to enter commands. When the mouse is moved on a flat surface, a *cursor* (a small picture of an arrow, etc.) moves on the screen. Software and documents (*files*) are stored on a disk in a way that corresponds to their being contained in "file folders" which can be contained in other file folders, etc. Windows are displayed on the screen which contain icons that represent programs, folders, or files.

All Macintosh and Windows applications are written so that operations can be chosen in a standardized way. To open a folder and view its contents, move the mouse so that the cursor on the screen is superimposed on the icon for that folder and then press twice on the mouse button (this is called *double-clicking*). To use an application, move the mouse so that the cursor on the screen is superimposed on the application's icon and double-click. This action starts the application program and opens a blank file. To open an existing file, double-click on the icon for that file, which starts the application program that created the file if it is not already in memory and opens that file. When a file is open, it is loaded in memory and the applications program has access to it. Part or all of the file is shown on the video screen. For example, a file created by a word processor is a document containing text, and a file created by a graphing program is a pictorial graph. A file or folder can be "closed" by moving the mouse so that the cursor is in a small square at the upper left of the screen (the *close box*) and then pressing the mouse button once (*single-clicking*).

When an application is running there is a list of menu headings displayed across the top of the video screen. Different applications have different lists of headings, but two headings that nearly always occur are the "File" heading and the "Edit" heading. A command in a menu is chosen by moving the mouse so that the cursor is on the menu heading. The mouse button is then depressed and held. The appropriate menu appears on the screen and any command on the menu can be chosen by moving the cursor to that command while holding the mouse button down ("dragging" the cursor), and then releasing the mouse button. For example, every Macintosh application has the "Quit" command in the "File" menu. This command closes the file and then shuts down the application, returning control to the operating system. There are keyboard equivalents for some commands, which can be used instead of the mouse/menu method. Most of these keyboard equivalents consist of holding down the "Command" key next to the space bar and typing a letter, or holding down both the shift key and the "Command" key and typing a letter. Directions for the keyboard equivalents are displayed next to the command names when the menu is displayed.

A convenient part of the operating system of the Macintosh computer is the "Clipboard." This is a portion of memory devoted to short-term storage of a file or part of a file. For example, if you have a document produced by a word processor or spreadsheet and if want to move a part of it to a different location in the document, you "select" that portion of the text by dragging the cursor over it while holding down the mouse button. You then choose the "Cut" command from the "Edit" menu, and the selected portion disappears from the document and is stored in the Clipboard. You can place the cursor in the desired new location of the deleted text, click on that location, and choose the "Paste" command in the "Edit" menu. The selected text reappears in that location. You can also use the "Copy" command to put a copy of something into the Clipboard without deleting it from the document. The Clipboard belongs to the operating system, not to a particular application. For example, if you have a graph created by a graphing program, you can copy it into the Clipboard, open a verbal document created by a word processor, and paste the graph into the verbal document.

Typical operating systems also contain a variety of typographical fonts to be used in displaying and printing verbal information. In addition to characters obtained on the keyboard with and without pressing the "shift" key, there are characters available by pressing the "option" key or by pressing both the "option" and "shift" keys while typing a letter. Most fonts contain a few Greek letters and mathematical symbols, but there is a font called the "Symbol" font that contains all of the Greek letters and a good selection of mathematical symbols.

## SECTION 11.3. WORD PROCESSORS

Two commonly used word processors are Corel WordPerfect and Microsoft Word. There is also a word processor contained in Microsoft Works. We provide some elementary comments that might help you in writing a laboratory report or a similar document, using Microsoft Word for the Macintosh. Its usage is very similar on a Windows machine. At the time of this writing, the latest version of Microsoft Word for the Macintosh is called Word 98. Previous versions were Word 6.0, 5.0, etc. Unfortunately, some of the features of the earlier versions have been deleted from Word 98. If you use Word 98 to open a document created with an earlier version, some parts might be garbled.

To begin typing a document using Microsoft Word, you double-click on the icon for the application in the usual way, and a blank document appears on your video display screen. In addition to the usual row of menu headings at the top of the screen, there is a row of symbols called the "toolbar" and a double row called the "ribbon." Below that there is a ruler on which the margins and tab stops are shown. If you do not want to see the ruler, you can select "Ruler" in the "View" menu, and the item will disappear. You can recover it with the same command. When an item is in view, there is a check-mark next to its name in the "View" menu. The first item in the ribbon is the name of the font that the application is using. You can change the font for any part of your document by selecting that part by dragging the cursor over it, and then clicking on a small triangle to the right of the font name. A list of available fonts appears, and you can drag the cursor to the desired font name and release the mouse button. Not only do the various fonts look different, but they have somewhat different selections of characters available. There are two sets of characters available as with a typewriter, entered by pressing a key or by pressing the "shift" key while pressing the key. Another set is available by holding the "option" key while pressing a key and a fourth set is available by holding both the "option" key and the "shift" key and pressing a key. Most fonts contain a few Greek letters, and all of the Greek letters are found in the Symbol font. When Microsoft Word is running, the complete set of characters can be seen in a window by choosing the "Symbol" command in the "Insert" menu. Single-clicking on any character in the list inserts it into the document at the location of the cursor.

Microsoft Word provides means for placing footnotes, tables, and equations in a document. To insert a footnote, you choose the "Footnote" command from the "Insert" menu, and a window will open in which you can choose whether to use consecutive numbers or other symbols for the footnotes, and can choose to use endnotes instead of footnotes. You can click on "OK" and a footnote symbol will be inserted at the location of your cursor. A small window will open at the bottom of the screen, and you can type the text of the footnote in this window and then close it by clicking on "Close." The footnote will disappear from view. At any time, you can view the footnotes by choosing "Footnotes" from the "View" menu.

You can also view the "header" (an area at the top of the page) and "footer" (an area at the bottom of the page) by selecting "Header and Footer" in the "View" menu. An area for the header and an area for the footer will appear, as well as a small toolbar. The view will also change from "Normal" to "Page Layout," which shows an image of a page as it will be printed, including header, footer, and footnotes, if any. You can insert the date and the time into the header or footer by placing the cursor in the header or footer area and clicking on icons in this toolbar. You can move the cursor to the middle or the right of the header or footer with the "Tab" key. You can also automatically number your pages in the header or footer by clicking on an icon with a large number symbol (#). There is another icon with a # symbol and a picture of a hand. You can click on this icon to specify a number other than 1 for your first page number. You can also put text in the header or footer by typing it in while viewing the header or footer. When you close the header–footer toolbar, the view reverts to "Normal." You can also return to "Page Layout" by clicking on this item in the "View" menu. If the break between pages is not where you want it, you can place the cursor where you want the page break and choose "Page Break" from the "Insert" menu.

Tables are entered by clicking on the "Table" icon in the main toolbar. This icon looks like a window with 12 panes. A small window opens up with an array of squares. Drag the cursor from upper left toward the lower right until you have selected the number of rows and columns you want and release the mouse button. When you close this window, the blank table is inserted into the document, and you can type the desired entries in cells that appear.

If you wish to construct an equation, choose "Object" from the "Insert" menu, and then choose "Microsoft Equation 3.0" in the window which opens up. You will see a window with a set of symbols above it. Included are integral signs, summation signs, etc., as well as templates for fractions, subscripts, superscripts, etc. Put the cursor on the appropriate symbol, and a "palette" of symbols and templates opens up. Drag the cursor to the desired symbol or template while holding down the mouse button, and release. The desired symbol or template will be placed in the working area. For example, if you choose a definite integral symbol, the integral sign will be placed in the working area, along with small areas in which you can type the limits and the integrand function. Greek letters and other symbols can be inserted by selecting them from a palette, and ordinary letters and digits can be typed into the equation. When you have finished constructing the equation, click on the "close" box, and the equation will be inserted into the document at the location which the cursor had prior to opening the equation editor. Versions of Microsoft Word earlier than Word 98 also had a means of typing equations using typesetting commands, but this option has been removed from Word 98.

If you need boldface, italic, or underlined type, you can select the desired portion of your text, and then click on the appropriate icon in the ribbon. If you need subscripts or superscripts in the body of your document, Microsoft Word 98 provides the following way to do this. Type the subscript or superscript as a regular character and then select it by dragging the cursor over it (or hold down the "shift" key while moving the cursor over it with the arrow keys). You then click on one of two small triangles in the ribbon just to the right of the icon for underlined type. Earlier versions of Word provided another way: select the subscript or superscript character and then type a plus sign ($+$) or a minus sign ($-$) while holding down both the "shift" key and the "command" key.

You can copy your document on the hard drive or on a diskette by selecting the "Save" or the "Save As" command in the "File" menu. The "Save As" command allows you to specify a folder on the hard drive in which to save the document and to specify a name for the file. It also allows you to specify a diskette in the diskette drive and any folder in that diskette. Use the "Save As" command the first time you save the document, and the "Save" command thereafter, unless you want to save it in a different place. You should save modifications to your document frequently as you make them so that an unexpected power outage won't cause you to lose a lot of your work.

We do not discuss further editing techniques included in Microsoft Word. It is a powerful program that can do a lot of things. You can learn a lot just by trying out different commands in the various menus. If you have done something that you want to undo, choose the "Undo" command in the "Edit" menu. You can use it repeatedly to undo a number of operations in reverse order. There is a "Help" item in the "Window" menu which can be used, but a serious user must consult the manuals provided by the software manufacturer. The Corel WordPerfect word processor is quite similar to Word and contains a similar equation editor, but we do not discuss it.

## SECTION 11.4. SPREADSHEETS

A spreadsheet is a program that can perform various mathematical and other operations on sets of items that are entered by the user and displayed in the form of a table. Two common spreadsheets are Lotus 1-2-3 and Microsoft Excel. Microsoft Works contains a spreadsheet that is a simplified version of Excel, and Claris Works contains a similar spreadsheet program. At the time of this writing, the latest version of Excel for the Macintosh is called Excel 98. Previous versions were called Excel 4.0, 3.0, etc.

### Creating and Editing a Workbook in Excel

To use Excel on the Macintosh, one first opens the folder containing the software by double-clicking on the icon for that folder and then opens Excel by double-clicking on the Excel icon. A window is automatically displayed on the screen with a number of rectangular areas called *cells* arranged in rows and columns. This window is called a *workbook*. The rows are labeled by numbers and the columns are labeled by letters. Any cell can be specified by giving its column and its row (its address). For example, the address of the cell in the third row of the second column is B3. A list of menu headings appears across the top of the screen, as with Microsoft Word, and a double strip of small icons called a "toolbar" appears under the menu headings.

Any cell can be selected by using the arrow buttons on the keyboard or by moving the mouse until its cursor is in the desired cell and clicking the mouse button. One can then type one of three kinds of information into the cell: a number, some text, or a formula. For example, one might want to use the top cell in each column for a title. One would first select the cell and then type the title. As the title is typed, it appears in a line above the cells. It is then entered into the cell by moving the cursor away from that cell. A number is entered into a cell in the same way. To enter a number but treat it as text, precede the number with a single quotation mark (').

The use of formulas in cells is one of the most useful properties of spreadsheets. A formula is entered by typing an equal sign followed by the formula, using the symbol * (asterisk) for multiplication, / (slash) for division, + (plus) for addition, and − (minus) for subtraction. Since the formula must be typed on a single line, parentheses are used as necessary to make sure that the operations are carried out correctly, using the rule that all operations inside a pair of parentheses are carried out before being combined with anything else. Other operations are carried out from left to right, with multiplications and divisions carried out before additions and subtractions. If a number stored in another cell is needed in a formula, one types the address of that cell into the formula in place of the number. The spreadsheet provides for evaluation of common functions, whose abbreviations can also be typed into formulas. The argument of a function is enclosed in parentheses, as indicated:

| Abbreviation | Function |
| --- | --- |
| SIN($\cdots$) | Sine |
| COS($\cdots$) | Cosine |
| EXP($\cdots$) | Exponential ($e$ raised to the argument) |
| LOG($\cdots$) | Common logarithm (base 10) |
| LN($\cdots$) | Natural logarithm (base $e$) |

Lower case letters can also be used. The argument of the sine and cosine must be expressed in radians. An arithmetic expression can be used as the argument and will automatically be evaluated when the function is evaluated. After the formula is typed one enters it into a cell by pressing the "return" key. When a formula is entered into a cell, the computer will automatically calculate the appropriate number from whatever constants and cell contents and will display the numerical result in the cell. If the value of the number in a cell is changed, any formulas in other cells containing the first cell's address will automatically recalculate the numbers in those cells.

---

### Example 11.1

Enter a formula into cell C1 to compute the sum of the number in cell A1 and the number in cell B2, divide by 2, and take the common logarithm of the result.

---

### Solution

With the cursor in cell C1 we type the following:

$$= \text{LOG}((\text{A1} + \text{B2})/2)$$

We then press the "return" key to enter the formula into that cell.

---

### Problem 11.1

Enter a formula into cell D2 that will compute the mean of the numbers in cells A2, B2, and C2.

---

If you move a formula from one cell to another, any addresses entered as in the above example will change. Such addresses are called *relative addresses* or *relative references*. For example, say that the address A1 and the address B2 were typed into a formula placed in cell C1. If this formula is copied and placed into another cell, the address A1 is replaced by the address of whatever cell is two columns to the left of the new location of the formula. The address B2 is replaced by the address of whatever cell is one column to the left and one row below the new location of the formula. This feature is very useful, but you must get used to it. If you want to move a formula to a new location but still want to refer to the contents of a particular cell, put a dollar sign ($) in front of the column letter and another dollar sign in front of the row number. For example, $A$1 would refer to cell A1 no matter what cell the formula is placed in. Such an address is an *absolute address* or an *absolute reference*.

There is a convenient way to put the same formula or the same number in an entire column or an entire row of a table. Type a formula with the cursor in the topmost cell of a given portion of a column, and then press the "return" key to enter the formula in the first cell. Then select a portion of the column by dragging the cursor down the column (while holding down the mouse button) from the cell containing the formula as far as desired. Then choose the "Fill" command in the "Edit" menu and choose "Down" from a small window that appears. When you do this, the cells are all "filled" with the formula. The formulas in different cells will refer to different cells according to the relative addressing explained above. Selecting a given cell will show the formula for that cell, with the addresses that will actually be used. A similar procedure is used to fill a portion of a row by entering the formula in the left-most cell of a portion of a row, selecting the portion of the row, and using the "Fill" command in the "Edit" menu and choosing "Right" in the next window. The same procedures can be used to fill a column or a row with the same number in every cell.

A block of cells can be selected by moving the cursor to the upper left cell of the block and then moving it to the opposite corner of the block while holding down the mouse button ("dragging" the cursor). The contents of the cell or block of cells can then be cut or copied into the clipboard, using the "Cut" command or the "Copy" command in the "Edit" menu. The contents of the clipboard can be pasted into a new location. One selects the upper left cell of the new block of cells and then uses the "Paste" command in the "Edit" menu to paste the clipboard contents into the workbook. If you put something into a set of cells and want to change it, you can select the cells and then choose "Clear" in the "Edit" menu. To clear the cells completely, choose "All" in the window that appears.

We illustrate the above procedures by describing the construction of a workbook containing the ethanol vapor pressure data in Table 10.2. We open a new workbook by double-clicking on the "Excel" icon. We choose column A for the Celsius temperatures. We type the title t/°C in the first row of this column, and enter the nine values 25 through 65 in rows 2 through 10 of column A. Numerical values can be entered either with the keypad or the number in the regular keyboard. We type the title T/K and enter it in the first row of Column B, and enter the following formula into cell B2:

$$= A2 + 273.15$$

We drag the cursor from the second row to the tenth row in Column B, and choose the command "Fill" and "Down" from the "Edit" menu. The computer uses the formulas to enter the appropriate Kelvin temperatures in Column B.

We now enter the vapor pressure values. We type the title P/torr and enter it in cell C1, and enter the appropriate vapor pressure values in rows 2 through 10, typing and entering each value individually.

Since the Clausius–Clapeyron equation implies that ln(P/torr) should be a linear function of 1/T, we choose column D and E for these functions. We type the title 1/(T/K) and enter it in cell D1. We type the following formula into cell D2:

$$= 1/B2$$

We enter it by pressing the "return" key. We drag the cursor over rows 2 through 10 of Column D, and choose the command "Fill Down" from the "Edit" menu. The computer fills Column D with the appropriate reciprocal temperature values. We type the title ln(P/torr) and enter it into cell E1. We then type the following formula

| t/°C | T/K | P/torr | 1/(T/K) | ln(P/torr) |
|------|--------|--------|------------|------------|
| 25 | 298.15 | 55.9 | 0.00335402 | 4.02356438 |
| 30 | 303.15 | 70.0 | 0.00329870 | 4.24849524 |
| 35 | 308.15 | 93.8 | 0.00324517 | 4.54116486 |
| 40 | 313.15 | 117.5 | 0.00319336 | 4.76643833 |
| 45 | 318.15 | 154.1 | 0.00314317 | 5.03760174 |
| 50 | 323.15 | 190.7 | 0.00309454 | 5.25070151 |
| 55 | 328.15 | 241.9 | 0.00304739 | 5.48852442 |
| 60 | 333.15 | 304.15 | 0.00300165 | 5.71752100 |
| 65 | 338.15 | 377.9 | 0.00295727 | 5.93462961 |

**FIGURE II.I**   Excel worksheet for the vapor pressure of ethanol.

into cell E2 and enter it by pressing the "return" key after typing it:

$$= \text{LN(C2)}$$

We drag the cursor over rows 2 through 10 of Column E and choose the command "Fill Down" from the "Edit" menu. The computer fills Column E with the appropriate logarithm values.

We now save the workbook. We choose the "Save As" command in the "File" menu and choose the name "Ethanol Vapor Pressure." If we wish, we can now print the workbook, choosing the "Print" command in the "File" menu. The result is shown in Fig. 11.1. This workbook should be satisfactory for a physical chemistry laboratory report, but would be improved by adding more columns containing the expected errors or the maximum and minimum logarithm values corresponding to adding or subtracting the expected errors. If a workbook contains more than six columns, Excel prints it on more than one page, six columns to a page. You can see how things will fit on the pages by choosing "Print Preview" from the "File" menu.

---

### Problem II.2

Using Excel, reproduce the workbook of Fig. 11.1 and generate columns with the error values, the incremented and decremented vapor pressure values, and also with the incremented and decremented logarithm values.

---

### Creating Graphs with Excel

Once columns of a workbook have been filled with numbers, either by typing them in or by the use of formulas, the spreadsheet can be used to produce graphs of various kinds (Excel refers to graphs as "charts"). We present a procedure to construct a two-dimensional graph from one column containing values of an independent variable and another column containing the corresponding values of a dependent variable.

1. Save the spreadsheet, and if you want a printed copy of it without the graph, print it now by using the "Print" command in the "File" menu.
2. Select the two columns by dragging the mouse cursor over them. If the two columns are not adjacent, drag the cursor over the first column, and then hold down the command key (next to the space-bar) while dragging the cursor over the second column. The values of the independent variable must be in the column to the left of

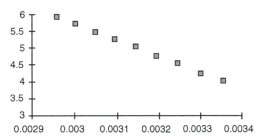

**FIGURE 11.2** The ethanol vapor pressure graph created with excel.

the other column. Copy and paste a column of values if necessary. There are various procedures that can be used from this point on. In Excel 98, one can use a part of the spreadsheet called the "Chart Wizard," as follows.

3. Click on the icon for the Chart Wizard in the toolbar. It looks like a small bar graph.

4. A window appears showing 10 different types of graphs to choose from. Most of the graph types plot categories, not values of a variable, on the horizontal axis. That is, the row number is used as the variable on this axis. To put values of a variable on the horizontal axis, choose the type of graph called "XY(Scatter)" by putting the cursor on this icon and clicking. Put the cursor on the "Next" button and click. This is called step 1.

5. Another window appears. Since our data are in columns, you put the cursor on a little circle corresponding to this fact and click the mouse button if that circle is not already dark. You can verify or change the range of cells from which the graph will be made. Click on the "Next" button. This is called step 2.

6. A window appears into which you can type a title for the graph (called a "legend") and labels for the axes. There are several areas that look like tabs on file folders. You can click on them and choose such things as grid lines, whether a labeled symbol appears to the right of the graph (the "legend"), and whether data values are printed next to the data points in the graph. Click on the "Next" button. This is step 3.

7. A window appears that allows you to choose whether to place the graph in your workbook or on a separate sheet. After you do this, click on the "Finish" button.

Printing a graph in Excel is done by using the "Print" command in the "File" menu in the usual way. You can use the "Print Preview" command in the "File" menu to see on the screen how the printed version will be arranged. You can change this by choosing "Page Setup." Figure 11.2 shows the graph that we produced.

## Curve Fitting with Excel

In order to carry out a linear least squares fit using Excel, one enters the values of the independent variable into one column and the corresponding values of the dependent variable in another column. The first row of each column can be used for a column title or for the first data point. For example, in the workbook which we constructed for the ethanol vapor pressure data, the values of the reciprocal of the Kelvin temperature were in Column D and the values of the logarithm of the vapor pressure were in Column E.

To carry out the least squares fit, we move the cursor to the "Tools" menu heading and drag the cursor down to the " Data Analysis" command. Another menu appears, and we choose the "Regression" command by clicking on it. We then click on the "OK" button.

A window appears in which all of the instructions for the curve fitting can be entered. For the y range, we enter $F$2:$F$10. We must use the dollar signs to indicate absolute addresses, and must use a colon (:) between the first and the last address. We enter $E$2:$E$10 for the x range. We now enter the location on the workbook where we want the output to be printed, by entering the absolute address of the upper left cell of the output area in the area called "Output Range." We type in $A$12. The output requires 7 columns and about 18 lines if we choose to print 10 residuals, and this choice will put the output under our data in the workbook. Standard errors in the slope and intercept will be printed (one standard deviation, or 68% confidence level), along with the correlation coefficient and the upper and lower values of the slope and intercept at the 95% confidence level. The expected error at the 95% confidence level is obtained by subtracting the value of the slope from the upper value. If you want expected errors at some confidence level other than 95%, you can enter that choice. In order to obtain a list of the residuals, we click on a small box next to "Residuals." We now click on the "OK" button and the computer carries out the fit and places the results on the workbook. The workbook, including the output from the least-squares fits, can be printed in the usual way.

A spreadsheet such as Excel is a large and powerful program, and can perform many different tasks, most of which are not needed by a physical chemistry student. You can consult the manual provided by the software manufacturer to learn more about Excel. There are also two textbooks listed at the end of the chapter that are more easily used than the manufacturer's manual.


## SECTION 11.5. GRAPHICS SOFTWARE

There exists a wide variety of graphics software, capable of drawing structural formulas, perspective views of molecular models, atomic and molecular orbital regions, etc., and also capable of constructing graphs of many sorts, including perspective views of three-dimensional graphs. A Macintosh program which can be used to draw structural formulas for ring-containing organic substances is Chemintosh.[1] There is also a version called ChemWindow, which runs on IBM-compatible personal computers using the Windows operating system. A similar program is called CHEMEDIT.[2] CHEMEDIT is distributed without cost. That is, anyone is permitted to copy the program from someone else's disk. These programs are relatively easy to use, but we do not discuss them.

---

[1] Sold by Softshell International, Ltd., 715 Horizon Drive, Grand Junction, CO 81506.

[2] Written by N. A. B. Gray and W. D. Wibowo, University of Wollongong, Wollongong, N. S. W. 2500, Australia.

## Drawing Graphs with CricketGraph

CricketGraph[3] is a program that can be used to produce good-quality two-dimensional graphs. A similar program is called KaleidaGraph, and various other graphing programs are available. We give only a brief introduction to CricketGraph III, version 1.5 for the Macintosh. Once the program has been installed on the hard disk of a Macintosh computer, you can open the folder in which it resides, and then open the CricketGraph application by double-clicking on its icon. A data window appears on the screen, along with a list of menu headings. This window contains rows and columns that are similar to a spreadsheet, along with a list of menu headings across the top of the screen. The columns have numerical labels instead of letters as in Excel, and there is an additional row of cells above row number 1 for column labels. When the window first appears, these cells contain the labels "Column 1," "Column 2," etc. You can enter a title of your choice by selecting any one of these cells and typing the title. Data values can be typed into cells, moving the cursor from cell to cell with the arrow buttons on the keyboard or with the mouse.

A workbook from Excel or Microsoft Works can be "imported" into Cricket-Graph using the clipboard. The numerical values will be imported, but not any formulas. Select the desired block of cells in the workbook, excluding the top row if it contains labels. Copy this block of cells into the clipboard using the "Copy" command in the "Edit" menu. Close the workbook by clicking on the "close" box at the upper left of the window, or by choosing "Quit" from the "File" menu (which also closes the Excel program). Open CricketGraph and place the cursor in the upper left cell of the block of cells where you want to place the contents of the clipboard. Paste these contents into the window with the "Paste" command in the "Edit" menu.

CricketGraph does not have the same capabilities as a spreadsheet, and no formulas can be typed in. However, it is possible to carry out arithmetic operations and function evaluations on columns of data, as illustrated in the following example:

---

### Example 11.2

Make a graph of the ethanol vapor pressure data equivalent to that produced using Excel in the previous section.

---

### Solution

1. We open the CricketGraph application by double-clicking on the CricketGraph icon. A blank data table appears in the window, and a list of menu headings appears across the top of the window. We type the nine Celsius temperatures in the first nine cells of the first column and type the column title "$t/^\circ C$" in place of "Column 1" in the column heading. Numerical data can be typed in using either the keypad or the numbers in the regular keyboard.

2. We fill the second column with the number 273.15 in each cell. We then choose the command "Simple Math" from the "Data" menu. A window appears with two lists of the columns, separated by symbols for addition, subtraction, multiplication, and division. We select the first column (now called "$t/^\circ C$") in the left list, Column 2 in the

---

[3] Sold by Cricket Software, 40 Valley Stream Parkway, Malvern, PA 19355.

right list, and the addition operator in the list of operations, by clicking on each item in turn. We see that the software has chosen Column 3 as the destination column. This choice is acceptable, so we click on the "OK" button, and the computer fills column 3 with the Kelvin temperatures. We type the title "T/K" into the heading for Column 3. If we had wished to do so, we could have computed the Kelvin temperatures by hand and typed them into a column.

3. We now put the reciprocal of the Kelvin temperature into Column 4. We choose "Transform" from the "Data" menu. A window appears. The default window is "Transcendental Functions." We choose "Other Functions" and another window appears. We choose the $x^n$ option, and specify that $n = -1$. We select Column 3 and click on the "OK" button, and the computer fills column 4 with the reciprocals of the Kelvin temperatures. We type the title "1/(T/K)" into the heading for Column 4.

4. We type in the vapor pressure values into Column 5 by placing the cursor into each cell in turn and typing the appropriate value. We type in the title "P/torr" into the heading for Column 5.

5. We choose the command "Transform" from the "Data" menu. The appropriate window appears with a list of columns on the left and a list of operations on the right. The "default" list contains transcendental operations (sine, cosine, exponential, etc.). You can also choose other lists, including the "Other functions" list used to get the reciprocal of the temperature. We choose Column 5 (now called "P/torr") by clicking on it in the list, and choose the operation "ln(x)" by clicking on the small circle next to it. The computer chooses Column 6 for the destination column. This is acceptable, so we click on the "OK" button, and the computer fills Column with the logarithms of the values in column 5. We type in the title "ln(P/torr)" in the heading for column 6. Our Data Table is now complete.

6. We place the cursor on the "Graph" menu, and icons for a number of different types of graphs are displayed. We choose the "Scatter" item by dragging the cursor to its icon and releasing the mouse button. (This option will plot the data points without curves or lines.) A new window appears with two lists of columns. The left list is for the horizontal (x) axis and the right list is for the vertical (y) axis. Only one column can be chosen for the x axis, but several can be chosen for the y axis by clicking on the first and holding down the shift key while clicking on the others. A graph with several curves will result if you do that. We choose the "1/(T/K)" entry for the x axis by clicking on that entry in the left list, and click on "ln(P/torr)" in the right list. We then click on the "New Plot" button and the graph appears on the video screen, along with a new set of menu headings. An area appears at the upper left that represents various "tools" that can be used to modify the graph.

Any part of the graph can be selected by placing the cursor on the desired feature and double-clicking. You can then edit that part of the graph. For example, if you want a different scale or different minimum and maximum values on an axis, you place the arrowhead of the cursor anywhere on the axis and double-click. A window appears in which you can make changes by placing the cursor on the appropriate quantity, clicking, and then typing in a new value. The axis choices which the software chose for our graph are acceptable. A symbol at the right of the graph represents the symbol placed at each data point. We can double-click on this symbol and change the symbol by choosing another symbol from the menu by placing the cursor on this symbol and single-clicking the mouse button. The letter "N" stands for omission of any symbol. We click on the "ABC" area in the tool area at the upper left, and then click on the title.

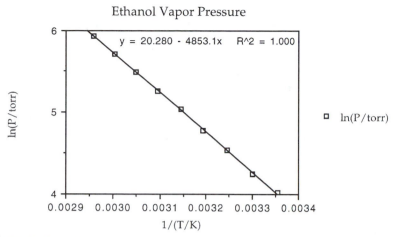

**FIGURE 11.3** Graph of ethanol vapor pressure created with CricketGraph.

A small window appears in which we type the new title. We can change the font for any part of the title by selecting it (dragging the cursor over it), then choosing "Font" from the "Type" menu, and finally choosing the desired font. We can also change the size of the font by choosing "Size" from the "Type" menu. We can selected boldface or italic type, or make subscripts and superscripts by choosing "Style" from the "Type" menu. We can also change each axis label by clicking on the "ABC" tool and then double-clicking in the axis label area. When all changes have been made, we click on the "OK" button, and the corrected graph appears on the screen.

CricketGraph has the capability of fitting curves to a set of data points by the least squares procedure. It can perform a linear fit, polynomial fits up to fifth degree, a log-arithmic fit, and an exponential fit, or can interpolate. Since the Clausius–Clapeyron equation indicates that ln(P/torr) should be a linear function of 1/(T/K), we do a linear fit. We select "Curve Fit" under the "Options" menu. A small window appears, and we select "Linear" from the "Methods" menu. We choose to exhibit the square of the correlation coefficient by selecting it in the "Coefficients Display" menu. The line is placed on the graph, and the equation for the line and the correlation coefficient squared are also placed on the graph. No expected errors in the slope and intercept are provided.[4]

The software originally placed the equation so that it covered part of the curve, so we place the cursor in the area of the equation, depress the mouse button, and drag the equation slightly to the right. The graph is printed in the usual way, with the "Print" command in the "File" menu. Figure 11.3 shows the resulting graph.

There are several additional useful things that can be done with CricketGraph, and some of these are found in the "Options" menu. For example, we can add error bars in the horizontal and vertical directions. Since we have left the data table without entering error values, we must close the graph and return to the data table. We enter

[4]There is a "shareware" program called "MacCurve Fit" that functions much like CricketGraph, which provides these expected errors.

the errors in the vapor pressure values from the rightmost column of Table 10.2 into an empty column, and then use the "Simple Math" command in the "Data" menu to add these values to the vapor pressure values. We then use the "Data Transformation" item to take the natural logarithms of the incremented vapor pressure values. We then use the "Simple Math" item to subtract the logarithms of the original values from the logarithms of the incremented values and place these error values in still another column.

Since we closed the original graph that was based on data that did not include the error values, we must construct a new graph. After producing a new graph just like that in Fig. 11.3, we then move the cursor to the "Options" menu heading and drag it to the "Y Error Bars" item. There are four choices, and we click on the fourth choice, corresponding to errors stored in the data table. We specify the column in which the errors are found, and click on the "OK" button. The error bars appear on the graph.

The legend at the right of the graph, which is needed to distinguish between several curves, is unnecessary since we have a single curve. We move the cursor to the "Options" menu heading and choose the "Show Graph Items" command. A window appears in which a number of items are listed with boxes. Those boxes containing "X" correspond to things that are displayed. We click on the "Legend" box and the "X" disappears. We close the small window, and we have the graph shown in Fig. 11.4, without the legend at the right. The legend can be restored if desired.

We can add text, arrows, and boxes around rectangular areas within a graph. If the "tools" are not displayed, drag the cursor to the "Windows" item in the "View" menu, and select "Show Tools" in the window that appears. A set of icons for the tools appears at the left of the graph on the screen. If you want text in a graph, put the cursor on the item "ABC" and click. Then move the cursor to the center of the region in the graph where you want the text and click the mouse button. Any text that you then type will be inserted at that point. You can draw curves or line segments by clicking on the appropriate icon and then clicking in your graph where you want to place the line segment or curve. You can also draw a rectangle, an ellipse, etc., in the same way.

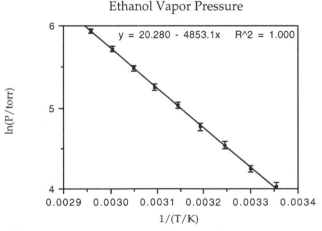

**FIGURE 11.4** Graph of ethanol vapor pressure with error bars.

### Problem 11.3

Obtain an estimated error in the slope of the least-squares line in the graph of Fig. 11.4. One way to proceed is to graph a new line by taking the logarithm of the first vapor pressure value and incrementing it by its expected error and taking the logarithm of the last vapor pressure value and decrementing it by its expected error. The line fitted to these two points should have a slope differing from that of the least-squares line roughly by the expected error.

Once a graph has been created, you can print it with the "Print" command in the "File" menu. You can copy the graph into the clipboard by using the "Copy" command in the "Edit" menu. When CricketGraph is closed, the graph remains in the clipboard, and it can be imported into a Microsoft Word document which you subsequently open. You can read the CricketGraph manual for further information.

## SECTION 11.6. PROGRAMMING IN BASIC

BASIC is one of several "high-level" computer programming languages. The name is an acronym for Beginners All-Purpose Symbolic Instruction Code. Most introductory college courses in computer programming now use another language such as the PASCAL language, but BASIC is still a good choice for elementary applications in chemistry. It is a simplified version of the FORTRAN language (an abbreviation for FORmula TRANslation), which remains the high-level language most commonly used in computational chemistry research.

A computer program as written in the BASIC language is called a *source program*. It consists of a set of statements that must be translated into machine language to be executed by the computer. The software which does this translation is called a *processor*. In order to be translated correctly, each statement in a source program must exactly conform to the rules built into the BASIC processor. All processors can translate the program one statement at a time as it is run, but some processors can also produce a permanently translated machine-language program, which is called *compiling* the program. One BASIC processor that is available is TrueBASIC[5], which can be used as a compiler. It is a versatile and powerful processor with built-in graphic capabilities. It also carries out matrix operations automatically. It functions in exactly the same way for Macintosh and Windows computers.

### Variables and Operations in BASIC

A BASIC source program consists mostly of *statements*. BASIC statements resemble ordinary formulas. Quantities that are represented in formulas by letters are called *variables* and are given names consisting of letters and digits. In the original BASIC, only names consisting of single capital letters and single capital letters followed by a

[5]Written by J. G. Kemeny and T. E. Kurtz, and sold by True Basic, Inc., 12 Commerce Avenue, West Lebanon, NH 03784.

single digit were allowed, but most processors will now allow longer variable names consisting of any combinations of capital letters, lower case letters, and digits. A variable name must start with a letter, not a digit. To the BASIC processor, a variable name refers to a location in memory. If a variable called X1 is introduced in a program, the computer chooses a memory location and retrieves whatever value is stored there when X1 is called for.

In addition to variables, statements can contain constants. In BASIC, constants can be integers, decimal fractions ("floating-point numbers"), or numbers expressed in scientific notation. Subscripts and superscripts are not used, so a modified version of scientific notation is used. For example, $3.2435 \times 10^{-3}$ is written as 3.2435E-3, 3.2435E-03, or 0.32435E-02, etc.

The BASIC language uses the same symbols as Excel for the four elementary arithmetic operations:

|                 |              |
|-----------------|--------------|
| addition        | +            |
| subtraction     | −            |
| multiplication  | * (asterisk) |
| division        | / (slash).   |

The caret symbol ($^\wedge$) is used for exponentiation.[6] For example, X^2 means $X^2$.

A combination of constants, variables, and symbols for operations is called an *expression*. One or more expressions are combined in a statement. If you are a beginning programmer, put only one statement on a line. Older BASIC processors required an integer statement number followed by a space in front of every BASIC statement. The computer stored the statements in numerical order, even if they were not typed in that order. Some modern processors, like TrueBASIC, do not require statement numbers, but store the statements in the order they are typed in. However, if you number any statements, you must number them all.

The most common type of statement is the *arithmetic assignment statement*, which has the form

$$\text{LET variable = expression} \tag{11.1}$$

An example of an arithmetic assignment statement is

$$\text{LET X1 = Y/Z2} \tag{11.2}$$

When the program is executed, this statement would cause the computer to divide the current value of Y by the current value of Z2 and copy the result into the storage unit called X1. If X1 has not yet been introduced into the program, the computer will at this point choose a memory storage unit to be called X1. If X1 has already occurred in the program, whatever value previously assigned to the variable X1 will be erased and replaced. You should not put the symbol for a variable on the right-hand side of the equal sign if that variable has not yet been given a value. If you do this, some processors will proceed to choose a memory unit and assign the value zero to the new variable, while some processors will stop the execution of the program and print an error message. Others might assign a memory unit and use whatever value happened to be stored there previously.

---

[6]A positive quantity can be raised to any power, but a negative quantity can be raised only to an integer power.

With some processors, the word LET is optional, so that Eq. (11.2) can be written

$$X1 = Y/Z2 \tag{11.3}$$

TrueBASIC requires that you include a statement **OPTION NOLET** at the first of your program if you wish to omit the word LET in your statements.

Equation (11.3) looks just like a mathematical equality, but it is not. The old value of X1 might or might not be to Y divided by Z2, but the new value of X1 is caused to be equal to Y divided by Z2. In fact, it is reasonable to write

$$X1 = X1 + Y/Z2 \tag{11.4}$$

This statement will cause the old value of X1 to be replaced by the old value of X1 plus Y divided by Z2. If you include the word LET, there is no confusion about the role of the equal sign. The equal sign in the arithmetic assignment statement is sometimes called a *replacement operator*. A BASIC statement can contain any number of arithmetic operations, but only one equal sign.

BASIC processors rank arithmetic operations (decide which operations to do first), according to the rules:

1. Exponentiations are done first.
2. Multiplications and divisions are done next.
3. Additions, subtractions, and negations are done next. Negation is a unary operation consisting of changing the sign of a quantity. This is treated as a subtraction from zero, not as a multiplication by $-1$.
4. Operations of the same rank are performed from left to right.

Multiplication and division are equal in rank, so that a division will be performed before a multiplication if it is to its left in the expression, and after it if it is to its right. Similarly, a subtraction will be performed before an addition if it is to its left, and after it if it is to its right. If there are some unperformed operations left in the expression when the expression has been scanned from left to right, the computer will then perform the operations in reverse order, from right to left.

You can modify the order in which operations are performed by using parentheses. All operations inside a pair of parentheses will be performed before the resulting quantity is combined with anything else. It is possible to nest parentheses inside parentheses to just about any extent that might be necessary. You must include a right parenthesis for every left parenthesis. If you are not sure whether you need parentheses, put them in.

---

### Example 11.3

Write BASIC statements to evaluate the following formulas:

a. $Y = [(A^2 + B^2)/2C]^N$
b. $Z = [AC + BC - DC(A + E)]/[(X + 1) - (2D)^2]$.

---

### Solution

```
a. LET Y = ((A^2 + B^2)/(2*C))^N
b. LET Z = C*(A + B + D*(A + E))/(X + 1 - (2*D)^2).
```

### Problem 11.4

Write BASIC statements to evaluate the following formula:

a. $R = \left[\dfrac{V}{4/3\pi}\right]^{1/3}$

b. $R = (X^2 + Y^2)^{1/2}$.

Many BASIC processors recognize the name PI for $\pi = 3.14159\ldots$ and supply its value when this name is included on the right-hand side of the equal sign. Others require you to specify its value in the program.

### Example 11.4

Write BASIC statements to find the values of the roots of the quadratic equation

$$ax^2 + bx + c = 0. \qquad (11.5)$$

Treat only the case that both roots are real.

### Solution

The roots are given by the formula

$$x = \frac{-b \pm \sqrt{b^2 - 4ac}}{2a}. \qquad (11.6)$$

There are two roots, which we name X1 and X2. We use A, B, and C for the quantities called $a$, $b$, and $c$ in the formula,

```
LET D = (B^2 - 4*A*C)^.5
LET X1 = (-B + D)/(2*A)
LET X2 = (-B - D)/(2*A).
```

### Problem 11.5

a. Write BASIC statements to evaluate the mean of six numbers, called N1, N2, N3, N4, N5, N6.

b. Write BASIC statements to evaluate the square root of the mean of the squares of the six numbers.

c. Write BASIC statements to calculate the standard deviation of the six numbers.

A variable can be designated to take on only integer values. The advantage is that the computer uses less storage space to store an integer and operates on it more quickly. An integer variable is denoted by a variable name ending in a percent symbol, as, for example, B%. The computer will not round a calculated quantity to the nearest integer when assigning it to an integer variable. It will discard the fractional part, which is called *truncating* the number. For example, the statement

```
LET B% = 15/6
```

will assign a value of 2 to the variable B%.

## Library Functions

BASIC processors have programs written into them to evaluate functions such as sines, cosines, logarithms, etc. These functions are called *library functions*, or *predefined functions*. For example, if you need the natural logarithm of the variable X3 to be assigned to the variable Y, you can write the statement

$$LET \; Y \; = \; LOG(X3) \tag{11.7}$$

The LOG function provides the natural logarithm, not the common logarithm. The LOG10 function provides the common logarithm.

Library functions can be included in statements containing arithmetic operations, as in the statement

$$LET \; Z3 \; = \; (3*LOG(X) \; + \; (A*B)^2) \tag{11.8}$$

Table 11.1 lists the library functions that are available in all BASIC processors. Some processors might have additional functions, such as the inverse sine and inverse cosine.

**TABLE 11.1  Predefined Mathematical Functions in BASIC**

| Name | Function | Result | Example |
|------|----------|--------|---------|
| ABS | Absolute value or magnitude | Positive constant | Y = ABS(X) |
| ATN | Arc tangent (in radians) | Constant, $-\frac{\pi}{2} < c < \frac{\pi}{2}$ | Y = ATN(X) |
| COS | Cosine of an angle in radians | Constant, $-1 < c < 1$ | Y = COS(X) |
| EXP | Exponential function, $e^x$ | Positive constant | Y = EXP(X) |
| INT | Truncation (drops decimal part of the argument)* | Integer constant | Y = INT(X) |
| LOG | Natural logarithm (base $e$) of positive argument | Constant | Y = LOG(X) |
| LGT or CLG or LOG10 | Common logarithm (base 10) of positive argument | Constant | Y = LGT(X) |
| RND | Random number generation | Constant, $0 < c < 1$ | Y = RND or Y = RND(X)** |
| SGN | Sign of argument ($-1$ if argument is negative, 0 if argument is 0, $+1$ if argument is positive) | 1, 0, 1 | Y = SGN(X) |
| SIN | Sine of an angle in radians | Constant, $-1 < c < 1$ | Y = SIN(X) |
| SQR | Square root | Positive constant | Y = SQR(X) |
| TAN | Tangent of angle in radians | Constant | Y = TAN(X) |

*Some processors have a FIX function that truncates either a positive or negative number. If so INT will do the same for a positive number. Check to see what INT does to a negative number with your processor.

**Argument optional, and makes no diffrence.

**Example 11.5**

Write statements to determine the angle $A$ from its cosine, stored as the variable X.

**Solution**

We use the identities

$$\tan(A) = \frac{\sin(A)}{\cos(A)}$$

$$\sin^2(A) + \cos^2(A) = 1.$$

We write the statements

```
LET S = SQR(1 - X^2)
LET A = ATN(S/X).
```

It is not necessary for the argument of a function to be a single variable. All of the BASIC functions will accept an expression as the argument, as long as it is written according to the BASIC rules. However, a function is not allowed to have the same function in its argument:

$$\text{LET Y = SQR(SQR(A))} \qquad \textit{not permitted.} \qquad (11.9)$$

It is permissible to put other functions in the argument of a function. For example, you can write the statement

$$\text{LET Z = LOG(EXP(X) - A)} \qquad \textit{permitted.} \qquad (11.10)$$

**Example 11.6**

Write BASIC statements to evaluate the side of a triangle $A_1$, opposite the angle $A$, given the other two sides, $B_1$ and $C_1$ and the angle $A$, using the formula

$$A_1^2 = B_1^2 + C_1^2 - 2B_1C_1 \cos(A). \qquad (11.11)$$

**Solution**

$$\text{LET A1 = SQR(B1^2 + C1^2 - 2*B1*C1*COS(A)).} \qquad (11.12)$$

**Problem 11.6**

Write BASIC statements to evaluate the three angles of a triangle if the three sides are given. You will need Eq. (11.11), and if your BASIC processor does not have the inverse cosine function, you will need the statements in the solution to Example 11.5.

## Programmer-Defined Functions

In addition to predefined functions you can define your own functions. The form of the defining statement is

$$110 \ \text{DEF FNA(X)} \ = \ \text{expression.} \tag{11.13}$$

The letter A in the function name can be replaced by any of the other letters of the alphabet, so that in a single program it is possible to have 26 different functions. The defining statement in Eq. (11.13) can be placed anywhere in the program.

The function is used by including the name of the function in a BASIC statement in much the same way as a predefined function, as, for example,

$$190 \ \text{LET Y} \ = \ 5*\text{FNA(Y1)}$$

or

$$190 \ \text{LET Y} \ = \ 5*\text{FNA(Z2)}$$

It is not necessary for the argument of the function to be the same variable as was used in defining the function. The BASIC processor will evaluate the function using the current value of Y1, Z2, or any other variable.

## Subroutines

If you have a function that cannot be expressed by one BASIC statement, you can use an *internal subroutine*, which is a self-contained program. Using such a subroutine is referred to as "calling" the subroutine and is done with a *GOSUB statement* like

$$110 \ \text{GOSUB s} \tag{11.14}$$

where s represents the first statement number in the subroutine. During execution of the main program, when the computer reaches the statement in Eq. (11.14), it branches to the first statement of the subroutine. That is, it leaves the part of the program it has been executing and jumps to the statement with the number represented by s. It executes the statements of the subroutine and then returns to the next statement following the GOSUB statement. The last statement in the subroutine must be the RETURN statement.

---

### Example 11.7

Write a subroutine to evaluate $n! = n$ factorial $= n(n-1)(n-2)\cdots(2)(1)$.

---

### Solution

We use the variable F% for the factorial, and N% for n:

```
710 F% = 1
720 IF N% = 1, THEN 760
730 IF N% = 0, THEN 760
730 FOR J% = 2 TO N% STEP 1
740 F% = F%*J%
750 NEXT J%
760 RETURN
```

We have used the IF–THEN statement, discussed below. This subroutine could be called in any part of the program where N has been given a value. The following might be part of a program that uses this subroutine and calculates $2^N/N!$.

```
350 PRINT "VALUE OF N%?"
360 INPUT N%
370 GOSUB 710
380 Y = 2^N%/F%
390 more statements
```

## Transfer Statements and Loops

The computer will execute the statements of a program in the sequence in which they are stored, unless instructed to do otherwise. *Transfer statements* cause the computer to *branch*, or to go to a different point in the program, and *loops* cause the computer to go through a part of a program repeatedly.

We describe two kinds of transfer statements: the GOTO statement and the IF–THEN statement.

## The GOTO Statement

This statement does what its name implies. The statement has the form

$$\text{GOTO s} \tag{11.15}$$

where s stands for the number of the statement that you want the computer to execute next. For example, if you want the computer to branch to statement number 170, the GOTO statement is, if its own number is 120,

```
120 GOTO 170
```

If you want a portion of a program to be repeated an indefinite number of times, one way to do this is to put a GOTO statement at the end of this portion of the program. If the first executable statement to be repeated is statement number 100, such a statement would be

```
970 GOTO 100
```

However, the computer will continue this repetition forever unless you make provision to stop it.

## The IF–THEN Statement

There are three principal types of IF–THEN statements. The simplest IF–THEN statement has the form

$$\text{IF exp1 rel exp2 THEN s} \tag{11.16}$$

where exp1 is an arithmetic expression, exp2 is a second arithmetic expression, and rel stands for a *relational operator*, or *logical operator*. The combination exp1 rel exp2 is called a *logical expression*, and can have either of two values: true or false.

**TABLE 11.2   Relational Operators in BASIC**

| Relation | Algebraic symbol | BASIC symbol |
|---|---|---|
| Is equal to | $=$ | $=$ |
| Is greater than | $>$ | $>$ |
| Is less than | $<$ | $<$ |
| Is greater than or equal to | $\geq$ | $>=$ |
| Is less than or equal to | $\leq$ | $<=$ |
| Is not equal to | $\neq$ | $<>$ |
| Is approximately equal to | $\approx$ | $==$ |

If the logical expression is true, the computer will branch to the statement whose number is s, and if it is false, the computer will continue to the next statement after the IF–THEN statement.

The relational operators are listed in Table 11.2. The approximate equality does not exist in all BASIC compilers. Its purpose is to prevent round-off error from spoiling a comparison between two expressions that are supposed to be equal.

An example of an IF–THEN statement is

$$250 \quad \text{IF} \quad X \quad >= \quad 0 \quad \text{THEN} \quad 490 \tag{11.17}$$

which will cause the computer to branch to statement number 490 if X is either zero or positive, or will let it go on to whatever statement follows statement number 250 if X is negative.

Another version of the IF–THEN statement includes an executable statement within the IF–THEN statement itself. It has the form

$$\text{IF} \quad \text{exp1} \quad \text{rel} \quad \text{exp2} \quad \text{THEN:} \quad S \tag{11.18}$$

where S represents an executable statement. For example, if you want to replace a negative value of X by zero, but want to leave X alone if it has a positive or zero value, you can use the statement

$$390 \quad \text{IF} \quad X \quad < \quad 0 \quad \text{THEN:} \quad X \quad = \quad 0 \tag{11.19}$$

A third version of the IF–THEN statement has the form

$$\text{IF} \quad \text{exp1} \quad \text{rel} \quad \text{exp2} \quad \text{THEN} \quad s_1 \quad \text{ELSE} \quad s_2 \tag{11.20}$$

If the logical expression exp1 rel exp2 is true, this will cause branching to statement number $s_1$, and if it is false, it will cause branching to statement number $s_2$. For example,

$$760 \quad \text{IF} \quad B \quad == \quad 0 \quad \text{THEN} \quad 590 \quad \text{ELSE} \quad 980$$

will cause branching to statement 590 if B is approximately equal to zero, and to statement 980 otherwise.

---

**Example 11.8**

Write BASIC statements that will find the roots of the quadratic equation

$$ax^2 + bx + c = 0$$

if the roots are equal (case 1), if the roots are real and unequal (case 2), or if the roots are complex (case 3).

### Solution

```
100 OPTION NOLET
200 D = B^2 - 4*A*C
210 IF D = 0 THEN 400
220 IF D > 0 THEN 500
230 D = -D
240 R = -B/(2*A)
250 I1 = SQR(D)/(2*A)
260 I2 = -SQR(D)/(2*A)
270 GOTO 600
400 X1 = -B/(2*A)
410 GOTO 600
500 X1 = (-B + SQR(D))/(2*A)
510 X2 = (-B - SQR(D))/(2*A)
520 GOTO 600
600 (further statements)
```

### Loops

The BASIC statements that cause the computer to execute a portion of a program repeatedly are called a *FOR–NEXT loop*. The general form of the statements is

$$
\begin{aligned}
&\texttt{FOR I = exp1 TO exp2 STEP exp3}\\
&\text{executable statements}\\
&\texttt{NEXT I}
\end{aligned}
\tag{11.21}
$$

The quantity I is the *loop variable*, which has as its initial value the value of the expression exp1. When the computer reaches the statement NEXT I, the loop variable is incremented by the value of the expression exp3 and the computer branches back to the first executable statement after the FOR statement. However, if after the increment I has a value greater than the value of exp2, the computer leaves the loop and goes on to the next statement after the NEXT I statement. Usually exp1, exp2, and exp3 are chosen to be integer constants, but this is not necessary.

### Problem 11.7

Write a FOR–NEXT loop to sum up the squares of all the integers from 1 to 20 and to find the mean value of these squares.

It is possible to branch out of a loop before all the repetitions have been executed, by inserting an IF–THEN statement, as in the following example:

**Example 11.9**

Write a FOR–NEXT loop to sum up the series

$$1 - \frac{1}{2} + \frac{1}{3} - \frac{1}{4} + \cdots$$

to four significant digits.

---

**Solution**

This is a convergent alternating series, so the difference between any partial sum and the complete sum is smaller in magnitude than the next term after the partial sum. Our error must be smaller in magnitude than $5 \times 10^{-5}$, so we demand that the first neglected term be smaller than $5 \times 10^{-5}$ in magnitude. Our loop is

```
90 OPTION NOLET
100 S = 1
110 FOR N = 2 TO 100000 STEP 2
120 IF 1/N < 5E-5 THEN 150
130 S = S - 1/N + 1/(N + 1)
140 NEXT N
150 further statements
```

We have added two terms at a time to accommodate the alternating signs.

---

**Problem 11.8**

Write a FOR–NEXT loop to sum up the series of Eq. (3) of Appendix 3. Stop summing when a term divided by the partial sum is less than $1 \times 10^{-4}$.

---

It is possible to "nest" one FOR–NEXT loop inside another, so that the computer will run through the inner loop a number of times for each time that it goes through the outer loop.

---

**Example 11.10**

Write BASIC statements to sum up all the products $nm$, where $n$ and $m$ both range from 1 to 100.

---

**Solution**

```
90 S = 0
100 FOR N = 1 TO 100 STEP 1
110 FOR M = 1 TO 100 STEP 1
120 S = S + N*M
130 NEXT M
140 NEXT N
```

The entire inner loop (the M loop) must lie completely inside the outer loop (the N loop). It is possible to branch from the inner loop to the outer loop, but not from the outer loop into the inner loop, unless the inner loop has been executed at least once. Unless you are absolutely sure what you are doing, do not branch into a loop.

---

### Problem 11.9

Modify the loops of Example 11.10 so that all terms with $m > n$ are omitted.

---

Here are a few comments about the use of loops:

1. After the computer has left a loop, whether by branching or by completing the specified repetitions, the loop variable will have the value that it had upon leaving the loop. With some processors, if the loop has been executed the maximum number of times, the loop variable will have the value given by exp2 in Eq. (11.21). In others, it will have been incremented one time beyond this.

2. If the increment in the loop variable is equal to 1, the STEP 1 part of the loop statement can be omitted.

3. A negative increment is permissible, but if exp3 in Eq. (11.21) is negative, exp2 must then be smaller than exp1.

## Arrays, Vectors, and Matrices

BASIC processors provide an automatic way to handle lists of numbers, such as components of vectors, elements of matrices, etc. Such lists are called *arrays*. A vector is a *one-dimensional array*, and a matrix is a *two-dimensional array*. Most processors also permit three-dimensional arrays. The indices are enclosed in parentheses, but are still often called subscripts. The components of the vector **A** are denoted by A(1), A(2), etc. The elements of a matrix **B** are denoted by B(1, 2), B(7, 8), B(3, 4), etc. The first index denotes the row and the second index denotes the column of the matrix.

---

### Example 11.11

Write BASIC statements to form the scalar product of the two vectors **A** and **B**, once without using a loop and once with a loop.

---

### Solution

Without a loop

```
100 DIMENSION A(3), B(3)
110 LET C = A(1)*B(1) + A(2)*B(2) + A(3)*B(3)
```

With a loop

```
100 DIMENSION A(3), B(3)
105 LET C = 0
110 FOR I = 1 TO 3
```

```
120 LET C = C + A(I)*B(I)
130 NEXT I
```

---

The *DIMENSION statement* is required to reserve space in memory for the elements of the arrays. The number in parentheses is the number of memory units to be reserved. It can be larger than the number actually needed, but cannot be smaller. The word DIMENSION can be abbreviated to DIM. For a matrix with no more than 10 rows and 15 columns, the necessary DIMENSION statement would be

```
100 DIM A(10, 15)
```

---

### Example 11.12

Write BASIC statements that will compute the product of two 3 by 3 matrices, **A** and **B**.

---

### Solution

If **C** = **AB**, the element of **C** for the $i$th row and $j$th column is given by

$$C_{ij} = \sum_{k=1}^{3} A_{ik} B_{kj}.$$

We assume that a DIMENSION statement and statements to assign values to **A** and **B** were included in an earlier part of the program.

```
110 FOR I = 1 TO 3
120 FOR J = 1 TO 3
130 LET C(I, J) = 0
140 FOR K = 1 TO 3
150 LET C(I, J) = C(I, J) + A(I, K)*B(K, J)
160 NEXT K
170 NEXT J
180 NEXT I
```

---

### Problem 11.10

Write BASIC statements to form the trace of the matrix **A** with eight rows and eight columns.

---

Most BASIC processors have the capability to carry out matrix operations automatically. For example, if **C** is to be the matrix product of **A** and **B**, we write the statement

$$\text{MAT LET C = A*B} \tag{11.22}$$

or if we are using TrueBASIC and have included an OPTION NOLET statement

$$\text{MAT C = A*B} \tag{11.23}$$

The matrix **A** must have as many columns as **B** has rows.

For other matrix operations, we have the following statements.
To take the sum of two matrices,

$$\text{MAT LET C = A + B} \tag{11.24}$$

To take the difference of two matrices,

$$\text{MAT LET C = A - B} \tag{11.25}$$

To multiply a matrix by a scalar expression, say exp1,

$$\text{MAT LET C = (exp1)*A} \tag{11.26}$$

The scalar expression must be enclosed in parentheses. For example,

$$\text{MAT LET C = (SQR(X/U))*A}$$

or

$$\text{MATLET C = (2)*A}$$

To replace the matrix **B** by the matrix **A**,

$$\text{MAT LET B = A} \tag{11.27}$$

Some processors might not recognize this assignment statement. In that case one can use the statement

$$\text{MAT LET B= (1)*A} \tag{11.28}$$

To form the inverse of a matrix A,

$$\text{MAT LET B = INV(A)} \tag{11.29}$$

The matrix **A** must be square and nonsingular. The matrix inversion must be the only operation in a BASIC statement. The following is not permitted:

$$\text{MAT LET X = INV(A)*C} \quad \textit{invalid statement.}$$

In addition to the matrix operation statements in Eqs. (11.22) through (11.29), there are some special matrix assignment statements to produce commonly needed matrixes. To produce the identity matrix,

$$\text{MAT LET A = IDN} \tag{11.30}$$

To produce a matrix that has every element equal to unity,

$$\text{MAT LET A = CON} \tag{11.31}$$

To produce a matrix with every element equal to zero (the null matrix),

$$\text{MAT LET A = ZER} \tag{11.32}$$

**Example 11.13**

Write BASIC statements that will solve the set of three linear inhomogeneous equations

$$a_{11}x_1 + a_{12}x_2 + a_{13}x_3 = c_1 \qquad (11.33a)$$
$$a_{21}x_1 + a_{22}x_2 + a_{23}x_3 = c_2 \qquad (11.33b)$$
$$a_{31}x_1 + a_{32}x_2 + a_{33}x_3 = c_3. \qquad (11.33c)$$

**Solution**

We use Eq. (9.37):

```
100 DIM A(3, 3), C(3, 1) X(3, 1), B(3, 3)
110 MAT LET B = INV(A)
120 MAT LET X = B*C
```

You can see how easy this is, compared with manuallly solving the equations.

**Problem 11.11**

Write BASIC statements that will accomplish the following.

a. Take a given 5 by 5 matrix called **A** and find its inverse.
b. Take the resulting matrix and multiply it by another 5 by 5 matrix called **B**.

## Strings in **BASIC**

A *string* is a set of literal characters (a word, etc.). A *string variable* represents a memory unit in which a string can be stored. A string variable name ends in a dollar sign. The literal information stored in a string variable is called a *string constant*. A string constant is enclosed in quotation marks when written into a BASIC statement.

The *assignment statement* has the form

$$100 \text{ LET B\$ = "YES"} \qquad (11.34)$$

The string constant in one variable can be assigned to another, as in

$$100 \text{ LET A\$ = B\$} \qquad (11.35)$$

**Problem 11.12**

Write BASIC statements that will assign string constants of your choice to the string variables A\$, B\$, C\$, and D\$, and then put the content of A\$ into B\$, the content of B\$ into C\$, the content of C\$ into D\$, and the content of D\$ into A\$.

One common use of string variables is in IF–THEN statements. This is done in the following way:

```
100 LET B$ = "YES"
110 IF B$ = "YES" THEN 500 ELSE 400
400 executable statements
500 other executable statements
```

It is also possible to compare string variables, as in the statement

```
100 IF A$ = B$ THEN 500
```

Most BASIC processors also have *string functions*, which will do such things as discard part of the string constant in a string variable, create specific string variables, and map numbers into strings. Some also have a string addition operator (concatenation operator) which combines two strings into one string.

## Input and Output

The *INPUT statement* causes the computer to accept a value for a variable or a set of variables from the keyboard at the time of execution of the program. It has the form

```
INPUT list
```

where the word list in this statement stands for a variable or a list of the variables for which values are desired. For example, if values are needed for the variables X, Y, Z, A1, and A2, an input statement could be

```
100 INPUT X, Y, Z, A1, A2
```

When the program is executed, the computer will pause at this point in the program and prompt the user by printing a question mark on the video screen. Using the keyboard, the user then types in the value of X, a comma, the value of Y, a comma, the value of Z, a comma, and so forth. These values appear on the video screen, so that typographical errors can be detected. After typing the last value, the user then presses the carriage return on the keyboard and execution continues.

The INPUT statement can be used for arrays as well as for ordinary (scalar) variables, as in the following statements:

```
90 DIM A(10), B(10, 10)
100 INPUT A(1), A(2), A(3)
110 INPUT B(1, 1), B(1, 2), B(2, 1), B(2, 2)
```

Some of the memory units reserved by this DIMENSION statement will remain unused.

Another way to put values of arrays into the computer is by using loops:

---

### Example 11.14

Write a loop that will provide for input of a 5 by 5 matrix called C and a vector of 7 components called A.

---

### Solution

```
100 DIM C(5, 5), A(7)
```

```
110 FOR I = 1 TO 5
120 FOR J = 1 TO 5
130 INPUT C(I, J)
140 NEXT J
150 NEXT I
160 FOR I = 1 TO 7
170 INPUT A(I)
180 NEXT I
```

### Problem 11.13

Write BASIC statements that will provide for the input of a 7 by 4 matrix called **A** and a 4 by 7 matrix called **B** and will form the product **AB**.

Most BASIC processors such as TrueBASIC provide for automatic input of matrices. The required statement is of the form

```
MAT INPUT list
```

For example, the 5 by 5 matrix of Example 11.14 could be put into the computer by use of the statements

```
100 DIM C(5, 5)
110 MAT INPUT C
```

The matrix elements are assigned values one row at a time. Thus, if the following set of numbers were typed in for the matrix **C**,

1, 2, 3, 4, 5, 6, 7, 8, 9, 10, 11, 12, 13, 14, 15, 16, 17, 18, 19, 20, 21, 22, 23, 24, 25

the resulting matrix would be

$$\begin{bmatrix} 1 & 2 & 3 & 4 & 5 \\ 6 & 7 & 8 & 9 & 10 \\ 11 & 12 & 13 & 14 & 15 \\ 16 & 17 & 18 & 19 & 20 \\ 21 & 22 & 23 & 24 & 25 \end{bmatrix}.$$

### Example 11.15

Write BASIC statements that will provide for input of two 4 by 4 matrices, called **A** and **B**, and will provide for the user to type in the word "YES" if **A** is to be replaced by **B**.

### Solution

```
100 DIM A(4, 4), B(4, 4)
110 MAT INPUT A
120 MAT INPUT B
130 INPUT Y$
140 IF Y$ = "YES" THEN 150 ELSE 200
```

```
150 MAT LET A = B
200 other executable statements
```

---

### Problem 11.14

Write BASIC statements that will provide for input of 10 numbers as a one-dimensional array. Write statements that will take the arithmetic mean of the numbers if the user types in "ARITHMETIC MEAN" and take the geometric mean if the user types in "GEOMETRIC MEAN."

---

String variables can be put into the computer during execution of the program in the same way as arithmetic variables. The statement is of the form

```
100 INPUT B$
```

Quotation marks are optional when typing in the string constant.

It is possible to write a BASIC program without any input statements. The program itself can contain assignment statements that put the desired initial values into all variables. The combination of a *READ statement* and a *DATA statement* can also be used. The INPUT statement is used to put values in during the execution of the program, while the READ-DATA statements allow the numbers to be typed in as part of the program. The statements have the form

```
READ variables
DATA values
```

For example, if the variables X and Y are to be given the values 10.35 and 17, the statements would read

```
100 READ X, Y
300 DATA 10.35, 17
```

The DATA statement is an example of a *nonexecutable statement*.

DATA statements can be placed anywhere in the program. A good place for them is at the end of the program, where they can easily be located and changed when necessary. The computer will consider all the numbers in DATA statements as a block and will start reading at the beginning of the first DATA statement and will continue until the READ statements have all been executed.

---

### Example 11.16

Determine the outcome of the READ and DATA statements:

```
100 READ A, B, C, D
110 READ X
300 DATA 12, 14.1, 16
320 DATA 75, 83.57, 4, 12.001
```

---

### Solution

The variable A will be given the value 12, the variable B will be given the value 14.1, the variable C will be given the value 16, the variable D will be given the value

75, and the variable X will be given the value 83.57. The values 4 and 12.001 will not be read at this time and will remain ready to be read by the next READ statement.

---

### Problem 11.15

Write a loop containing a READ statement that will read from a DATA statement and cause the 3 by 3 array A to take the values

$$\begin{bmatrix} 12.3 & 13.7 & 19.9 \\ 75.1 & 23.9 & 37.1 \\ 14.7 & 10.7 & 7.7 \end{bmatrix}.$$

---

The *MAT READ statement* can be used to read elements of arrays. It has the form

```
MAT READ array name
```

The MAT READ statement will read in the elements of an array in the same order as the MAT INPUT statement, reading across the first row, then the second row, etc. The following statements will accomplish the same thing as a loop:

```
100 DIM A(3, 3)
110 MAT READ A
300 DATA 12.3, 13.7, 19.9, 75.1, 23 9, 37.1
310 DATA 14.7, 10.7, 7.7
```

---

### Problem 11.16

Write MAT READ and DATA statements to read in the matrices **A** and **B**. Write statements to take the product **AB = C**.

---

The principal BASIC output statement is the *PRINT statement*, which has the form

```
PRINT list
```

where the word list stands for a list of variable names, arithmetic expressions, and messages.

In its simplest version, the list part of the PRINT statement contains only the names of variables and will cause the values of these variables to be printed out on the video screen or other output device. For example, the following PRINT statement will cause the value of the variable X to be printed out on the video screen:

```
400 PRINT X
```

Several variables can have their values printed out with a single PRINT statement. The following statement will cause the current values of five variables to be printed out on the same line:

```
400 PRINT X, Y, Z, A1, A2
```

The commas in the list produce standard spacing, with 15 spaces on the line used for each variable. If the output list contains more variables than one line can hold, the computer will use more than one line. String variables can also be printed out in the same way as arithmetic variables.

In addition to the values of variables, messages can be printed with the PRINT statement. The required statement has the form

```
100 PRINT "message"
```

where the word message stands for whatever string of characters you want printed out.

---

**Example 11.17**

Modify the statements in Example 11.8 so that the computer will print a statement of what the program does and prompt the user when input is required, and will print the results.

---

**Solution**

```
100 PRINT "THIS PROGRAM FINDS THE ROOTS OF THE"
110 PRINT "QUADRATIC EQUATION AX^2 + BX + C = 0 "
115 OPTION NOLET
120 PRINT "VALUE OF A?"
130 INPUT A
140 PRINT "VALUE OF B?"
150 INPUT B
160 PRINT "VALUE OF C?"
170 INPUT C
200 D = B^2 - 4*A*C
210 IF D = 0 THEN 400
220 IF D > 0 THEN 500
230 D = -D
240 PRINT "ROOTS ARE COMPLEX"
250 R = -B/(2*A)
260 I = SQR(D)/(2*A)
270 PRINT "REAL PART OF ROOTS IS "
280 PRINT R
290 PRINT "IMAGINARY PARTS OF ROOTS ARE"
300 PRINT I, -I
310 GOTO 600
400 PRINT "ROOTS ARE REAL AND EQUAL"
410 X1 = -B/(2*A)
420 PRINT "VALUE OF ROOTS IS "
430 PRINT X1
440 GOTO 600
500 PRINT "ROOTS ARE REAL AND UNEQUAL"
510 X1 = (-B + SQR(D))/(2*A)
520 X2 = (-B - SQR(D))/(2*A)
525 PRINT "THE ROOTS ARE "
530 PRINT X1, X2
600 END
```

The END statement signals to the computer that the end of the program has been reached. It is required by all BASIC processors, and some might require a STOP statement just before the END statement.

---

### Problem 11.17

Modify the statements in Example 11.12 to provide for input of the matrices **A** and **B** and for output of the matrix **C**. Include necessary messages.

---

It is possible to include messages and variables in the same PRINT statement. Statements 525 and 530 in the previous example could be combined as

```
525 PRINT "X1 = ", X1, "X2 = ", X2
```

If the value of XI is 2.000 and the value of X2 is 4.000, the line of output would look like

```
X1 = 2.000 X2 = 4.000
```

If you wish to pack your output more closely than one item in 15 columns, you can separate the items by semicolons (;) instead of commas. The statement would then look like

```
525 PRINT "X1 = "; X1, "X2 = "; X2
```

Some processors allow close packing by omission of any separating mark between the items. The statement would then look like

```
525 PRINT "X1 = " X1, "X2 = " X2
```

The following are legitimate PRINT statements that will give closely packed output:

```
690 PRINT A$; X; Y; Z
970 MAT PRINT C;
```

The semicolon in the MAT PRINT statement allows a matrix with more than five columns to be printed with all the elements of one row on the same line. Some processors require a comma to be typed in this position in order to print the matrix in the ordinary rectangular format. If no comma is put in, each element will appear on its own line.

The PRINT statement without a list produces a blank line, and a list followed by a comma or a semicolon produces a line of output without a return to the next line, so that the next item of output will go on the same line if there is room for it.

---

### Example 11.18

Add suitable input and output statements to the statements in Example 11.13.

---

### Solution

```
100 PRINT "THIS PROGRAM SOLVES A SET OF THREE"
110 PRINT "SIMULTANEOUS LINEAR INHOMOGENEOUS"
120 PRINT "ALGEBRAIC EQUATIONS OF THE  FORM"
125 PRINT "AX = C WHERE A IS A 3 BY 3 MATRIX"
```

```
130 PRINT "AND X AND C ARE COLUMN VECTORS "
135 PRINT "X CONTAINS THE UNKNOWNS "
140 DIM A(3, 3), C(3, 1), X(3, 1), B(3, 3)
150 PRINT "TYPE IN THE ELEMENTS OF A BY ROWS"
160 MAT INPUT A
170 PRINT "TYPE IN THE ELEMENTS OF C"
180 MAT INPUT C
190 MAT LET B = INV(A)
200 MAT LET X = B*C
210 PRINT "THE ROOTS ARE:"
220 MAT PRINT X
230 END
```

Some BASIC processors have additional input/output features, such as the use of *FORMAT statements*, which determine where things are printed on the line. Some processors provide for printing a message by including it in an INPUT statement, and this is handy for reminding the user what variable is to be given a value. TrueBASIC will not allow this.

### Writing a BASIC Program

Many programmers find it convenient to construct a *flowchart* prior to writing a program. A flowchart schematically charts out the path that the computer will take through the problem. It contains variously shaped "boxes" connected by arrows. Rectangles are used to represent the calculation of values for variables, parallelograms are used for input and output, and diamonds for decision making. Figure 11.5 shows a flowchart for Example 11.17.

The second phase in constructing a computer program in BASIC is to construct all the BASIC statements that you will need in the program. With TrueBASIC on the Macintosh computer, the processor is opened in the usual way, by double-clicking on the TrueBASIC icon after opening the appropriate folder. A window appears in the video screen in which you can type the program, and a list of menus appears at the top of the screen. The program can be edited as you write it in much the same way as with a word processor, using commands in the Edit menu.

The following facts and policies should be remembered as you construct your program:

1. If you number statements, number all statements in consecutive order. Leave gaps in your numbering so that additional statements can be inserted later if needed.

2. Some processors allow for additional statements to be put on the same line if they are separated by a backslash ( \ ) or a colon (:), but do not put more than one statement on a line unless you are sure that your processor allows it and unless you know what symbol to use.

3. You can include nonexecutable statements called *REMARK statements* to remind you what various parts of the program do, etc. If you number statements, give each such statement a number. The abbreviation REM is used in the statements. For example,

```
310 REM THIS PART OF THE PROGRAM COMPUTES THE FIRST SUM
```

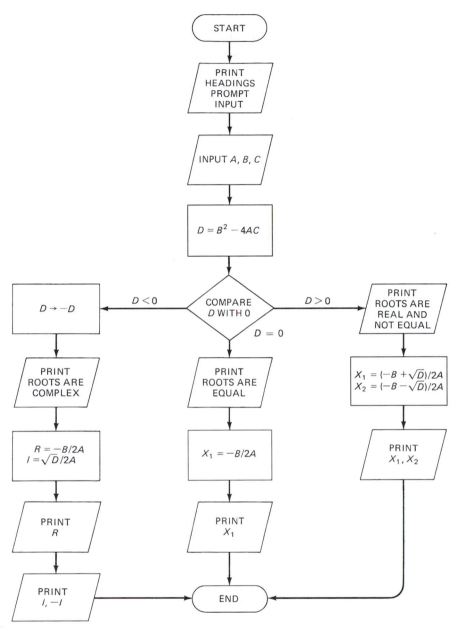

**FIGURE 11.5** Flowchart for the program of Example 11.16.

TrueBASIC also allows a remark to be added at the end of an executable statement by preceding it with an exclamation point (!). For example,

```
340 LET C = 1.000 ! THIS IS THE INITIAL CONCENTRATION
```

4. At the end of the program, a statement or statements are required to signal the computer that the program is complete. Some processors might require both a *STOP statement* and an *END statement*, but most require only an END statement.

All BASIC processors have "diagnostics" built into them, which help you to find errors. If you type in a statement that violates the BASIC rules, such as by having unpaired parentheses, two equals signs, etc., the computer will print an error message, informing you of this fact. Some messages will be printed immediately, but others will be printed when you try to run the program.

If your processor does not allow editing, and if you need to delete a statement, type in the statement number and press the carriage return. This will replace the statement by a blank. If you need to change a statement and if your processor will not allow editing, you must retype the statement, using the same number. When you press the return key, the new version will replace the old version.

If you want to keep your program on file for later use, you must save it. With TrueBASIC on the Macintosh, you save the program in the usual way, using the command "Save As" or "Save" from the "File" menu. This will store the source program on the open folder of the open disk. If you are through with the program for the time being, you select "Quit" from the "File" menu.

To recall a program from memory, you double-click on its icon in the usual way if TrueBASIC has not yet been opened. If TrueBASIC is open, you can also use the *Command Window*, by choosing "Command" from the "Window" menu. A small window will open and the letters "Ok" will appear. Letters or symbols like this which let you know that the computer is ready for an instruction are called *prompts*. You type OLD followed by a space and the name of the program and then press the return key. After a program has been recalled, you can modify or complete it in the usual way or run it.

With TrueBASIC, you can save a permanently translated version of a program. Choose the command "Compile" from the "Run" menu, and the TrueBASIC processor will compile the program and save it under a slightly different name from the name of the source program.

## Running a BASIC Program

Different processors have different ways to run a program. If the program is on the screen, some processors simply require you to type the word "run." With TrueBASIC on the Macintosh, you can run a program if it is in the screen window by using the mouse to select "Run" from the "Run" menu. You can also open the Command Window and type RUN in this window. You can run a program that is not on the video screen. To run a program called SIMEQ, open the command window and type OLD SIMEQ; RUN and then press the return key. This is a combination of two commands (separated by a semicolon): OLD SIMEQ tells the computer to retrieve a program called SIMEQ, and RUN tells the computer to run it.

If there are mistakes in the program that cause errors during execution, the computer will stop execution and print error messages. When execution is complete the computer will print the output. After you have finished with the program, you close it by choosing the command "Quit" in the "File" menu.

All processors have provision to print the output on a printer as well as on the video screen. With TrueBASIC on the Macintosh computer, you open the Command Window and type the command ECHO and press the return key. The output will be printed on the video screen in the usual way and will be printed on the printer when you close the program with the command "Quit" in the "File" menu.

If you are using TrueBASIC and want to print a list of the statements making up a program, open the Command Window, type the word LIST, and press the return key.

The printer will print out the entire program. If the program is not already in the core memory of the computer, you must first type OLD and the name of the program and press the carriage return. Other processors operate differently, but nearly all recognize the command LIST.

### Example 11.19

Type the program in Example 11.18 into the computer under the name SIMEQ and use it to solve the simultaneous equations

$$x_1 + x_2 + x_3 = 9$$
$$2x_1 + x_2 + 2x_3 = 15$$
$$3x_1 + 2x_2 + x_3 = 16.$$

### Solution

The output follows:

```
THIS PROGRAM SOLVES A SET OF THREE SIMULTANEOUS
LINEAR INHOMOGENEOUS ALGEBRAIC EQUATIONS OF THE FORM
AX = C WHERE A IS A 3 BY 3 MATRIX  AND X AND C ARE
COLUMN VECTORS. X CONTAINS THE UNKNOWNS.
TYPE IN THE ELEMENTS OF A BY ROWS.
? 1,1,1,2,1,2,3,2,1
TYPE IN THE ELEMENTS OF C.
? 9,15,16
THE ROOTS ARE:
2
3
4
```

### Problem 11.18

Type the program in Example 11.17 into the computer under the name QUADEQ and use it to solve the following two quadratic equations:

a. $3x^2 + 2x + 2 = 0$
b. $7x^2 - 5x + 1 = 0.$

### Example 11.20

Write and run a BASIC program to sort data points. That is, the program should take ordered pairs of numbers and rearrange them so that the value of the first variable, $x$, in any pair is smaller than the value of $x$ in the next pair.

### Solution

We use the "push-down" method, in which we begin by comparing the first value of $x$ with the second. If the first is larger, we exchange the data points (exchanging both the $x$ and $y$ values to keep a given value of $y$ with the same value of $x$). The second value of $x$, which is larger than the first is compared with the third value, and exchanged with it if larger. This process is continued throughout the list, which puts the largest value of $x$ in the last place in the list. The process is then repeated beginning at the first of the list but stopping one point before the end of the list, which will leave the second largest value of $x$ in the next-to-last position. The process is then repeated a third time, stopping two points before the end of the list, putting the third largest value of $x$ in the third place from the end, and so on.

The program follows:

```
100 PRINT "THIS PROGRAM SORTS DATA POINTS CONSISTING
OF "
110 PRINT "ORDERED PAIRS OF NUMBERS SO THAT THE
DATA"
120 PRINT "POINTS ARE PUT IN ORDER OF ASCENDING
VALUE"
130 PRINT "OF THE FIRST VARIABLE."
140 DIM X(100), Y(100)
150 PRINT "NUMBER OF DATA POINTS?.
160 INPUT N
170 FOR I = 1 TO N STEP 1
180 PRINT "VALUES OF X, Y FOR POINT NUMBER "; I
190 INPUT X(I), Y(I)
200 NEXT I
210 FOR I = 1 TO N - 1 STEP 1
220 FOR J = 1 TO N - I STEP 1
230 IF X(J) <= X(J + 1) THEN 300
240 T = X(J)
250 S = Y(J)
260 X(J) = X(J + 1)
270 Y(J) = Y(J + 1)
280 X(J + 1) = T
290 Y(J + 1) = S
300 NEXT J
310 NEXT I
340 FOR I = 1 TO N STEP 1
350 PRINT "X("I") =  "X(I), "Y("I") = "Y(I)
360 NEXT I
400 END
```

## Graphics in BASIC

TrueBASIC and other BASIC processors have various graphics capabilities built into them. We will describe only the procedure for plotting a simple two-dimensional graph in TrueBASIC. The first step is to specify the portion of the screen to be occupied by the graph, using the *OPEN* command. The processor uses a horizontal screen coordinate ranging from 0 to 1 and a vertical screen coordinate ranging from 0 to 1 to cover the screen from left to right and from bottom to top. If you want your graph to occupy the bottom half of the screen, you open a window as

```
OPEN #1: screen 0, 1, 0, 0.5
```

The first two numbers are the start and the end of the horizontal axis and the third and fourth numbers are the start and end of the vertical axis in screen coordinates. The #1 specifies that this window is called window #1. Several windows can be open at once.

The next step is to define the scales on the axes in the window. Let us make a graph of the sine function for arguments between 0 and $2\pi$. So that the curve will not reach the top or bottom of the window, we choose to let x (the horizontal coordinate) range from 0 to $2\pi$, and let y (the vertical coordinate) range from $-1.5$ to 1.5, using the statement

```
SET WINDOW 0, 2*PI, -1.5, 1.5
```

The next step is to write the body of the program. The *PLOT statement* is used to plot a data point. It places a point on the window at the specified coordinates. The first variable listed is the x (horizontal) coordinate, and the second is the y (vertical) coordinate. If a semicolon follows the second variable, a line segment will be drawn from one point to the next. If the semicolon is omitted, only the individual points will be plotted. A program which will plot a sine curve follows:

```
100 REM PROGRAM TO PLOT A SINE CURVE
110 OPEN "1: screen 0, 1, 0, .5
120 SET WINDOW 0, 2*PI, -1.5, 1.5
130 FOR X = 0 TO 2*PI STEP 0.01*PI
140 PLOT X, SIN(X);
150 NEXT X
160 END
```

We have given only a very cursory introduction to programming in BASIC. You can consult the manual which comes with your processor for more information. If you are a beginning programmer, you can get some help by modeling a program after one that someone else has written.

## SECTION 11.7. COMPLETE MATHEMATICS PACKAGES

There are several commonly used software packages that can carry out symbolic mathematics as well as numerical calculations and graphics. Two are MathCad and Mathematica. We provide a brief introduction to the use of Mathematica, based on Version 2.2 of Mathematica for a Macintosh computer.

There are several things that Mathematica can do that are useful in a physical chemistry course: (1) you can use Mathematica to carry out numerical calculations, including sums and numerical approximations to integrals; (2) you can use it to carry out symbolic mathematics: it will simplify an algebraic expression, provide a formula for the derivative or the indefinite integral of a function, etc.; (3) you can use it to solve equations, either numerically or symbolically, including sets of simultaneous equations and differential equations; (4) you can use it to construct graphs, including perspective views of three-dimensional surfaces; (5) you can use it as you would a word processor. All of these uses can be combined in a single document. There are many other tasks which Mathematica can carry out, such as matrix algebra, linear programming, and statistical calculations. [7]

As with other Macintosh applications, Mathematica is opened by double-clicking on the Mathematica icon after opening the folder in which the application is contained. When you do this, a blank "untitled" window is opened on the video screen. This window is called a notebook. Mathematica is now ready to accept instructions.

## Numerical Calculations with Mathematica

You can use Mathematica much as you would a calculator. For example, to obtain $(3.58731)^{56.3}$, you type in an open notebook the expression

$$3.58731^{\wedge}56.3$$

using the notation that the caret ($^{\wedge}$) stands for exponentiation. You then press the enter key (not the return key). Instead of pressing the enter key, you can press the return key while holding down the shift key (a shift-return) or you can press the return key while holding down the command key next to the space bar (a command-return).

Mathematica labels each input by a number. If you are at the beginning of a notebook, after you press the enter key you will see

*In[1]:=*
    **3.58731^56.3**

Mathematica prints input in bold-face type. It will print out the result, labeling it as output number 1:

*Out[1]=*
    $1.71194 \ 10^{31}$

In this expression, the space stands for multiplication, which is standard notation in Mathematica.

On the screen there is now a square bracket at the right of your input and another at the right of the output, as well as a larger square bracket to the right of both of these brackets and encompassing them. The smaller brackets identify cells. Mathematica assumes that any new cell is an input cell. When you press the enter key, you notify Mathematica that you are ending the input cell and that you want any expression in the input cell to be evaluated. When Mathematica printed your output, it created an output cell for the output, and also printed a larger bracket linking the input cell with its output cell. Any Mathematica notebook consists of a sequence of cells.

A cell can also be designated as a text cell, allowing Mathematica to be used like a word processor. You can convert any cell to a text cell as follows: first "select" the cell by placing the mouse cursor on the bracket to the right of the cell and pressing on the mouse button (clicking on the bracket). Then type the numeral 7 while depressing the command key next to the space bar. Mathematica will store text in a text cell, but will not perform any mathematical operations on anything in a text cell. You can delete the contents of any cell by selecting the cell and then choosing "Clear" from the "Edit" menu or typing the letter x while depressing the command key.

---

[7]See Stephen Wolfram, *Mathematica*, 2nd ed., Addison–Wesley, Redwood City, CA, 1991.

Pressing the return key does not end a cell. If a piece of input requires more than one line, you can press the return key at the end of each line, and then press the enter key to end the cell at the end of the last line. You can put several executable statements in the same cell.

The following rules apply for typing Mathematica expressions:

1. Common symbols are used for arithmetical operations:

addition: $+$
subtraction: $-$
negation: $-$
division: $/$
multiplication: blank space or $*$
exponentiation: $\wedge$
factorial: !

Be careful with multiplication. In ordinary formulas, placing two symbols together without a space between them can stand for multiplication. In Mathematica, if you write xy, the software will think you mean a variable called xy, and not the product of x and y. However, you can write either 2x or 2 x for 2 times x, but not x2.

2. Parentheses are used to determine the sequence of operations. If you aren't sure whether you need parentheses, put them in. The rule is that all operations inside a pair of parentheses will be carried out before the result is combined with anything else.

3. Complex arithmetic is done automatically, using I for $\sqrt{-1}$ .

4. Several constants are available by using symbols: Pi, E, I, Infinity, and Degree stand for $\pi, e, i = \sqrt{-1}, \infty$, and $\pi/180$ (conversion from degrees to radians). The first letter of each symbol must be capitalized.

5. Many functions are available. The names of the functions must be entered with the first letter capitalized and the other letters in lower case, and the argument of the function must be enclosed in square brackets, not parentheses. You will have to get used to this. No deviation from the capitalization rule is allowed. Some common functions are given in Table 11.3. Other functions are described in the book by Wolfram listed at the end of the chapter.

---

## Problem 11.19

Write Mathematica expressions for the following:

a. The complex conjugate of $(10)e^{2.657i}$
b. $\ln(100! ) - (100\ln(100) - 100)$
c. The complex conjugate of $(1 + 2i)^{2.5}$.

---

Mathematica will print out numerical values with any specified number of digits. For example, if you want to have the value of $(3.58731)^{56.3}$ to 15 digits, you enter

*In[1]:=*
```
N[3.58731^56.3,15]
```
*Out[1]=*
$1.71194390461979 10^{31}$

**TABLE 11.3   Mathematical Functions in Mathematica**

| Symbol | Function | Result |
|---|---|---|
| Abs[x] | Absolute value (magnitude) of x | Positive constant |
| Arg[z] | Argument $\phi$ of complex expression $|z|e^{i\phi}$ | Positive constant |
| ArcCos[x] | Inverse cosine (in radians) | Constant, $0 < c < \pi$ |
| ArcSin[x] | Inverse sine (in radians) | Constant, $-\frac{\pi}{2} < c < \frac{\pi}{2}$ |
| ArcTan[x] | Inverse tangent (in radians) | Constant, $-\frac{\pi}{2} < c < \frac{\pi}{2}$ |
| Conjugate[z] | Complex conjugate of $z = x + iy$ | $x - iy$ |
| Cos[x] | Cosine of an angle in radians | Constant, $-1 < c < 1$ |
| Exp[x] | Exponential function, $e^x$ | Positive constant |
| Im[z] | Imaginary part of complex expression $z$ | Constant |
| Log[x] | Natural logarithm (base $e$) | Constant |
| Log[b, x] | Logarithm to the base $b$ | Constant |
| n! | n factorial | Constant |
| Random[ ] | Random number generation | Constant, $0 < c < 1$ |
| Re[z] | Real part of complex expression $z$ | Constant |
| Round[x] | Closest integer to $x$ | Constant |
| Sin[x] | Sine of an angle in radians | Constant, $-1 < c < 1$ |
| Sqrt[x] | Square root | Constant |
| Tan[x] | Tangent of angle in radians | Constant |

The entire expression and the number of digits are enclosed in the square brackets following the N. If you do not specify the number of digits, Mathematica will give you a standard number of digits (usually six) for an expression that contains a decimal point. It will give all of the digits if possible for an expression that does not contain a decimal point. If you enter **30!**, it will give you the entire value, with 33 digits. If you enter **30.!**, it will give you a value with six digits. However, if you enter **Sqrt[3]**, it will not give you a value, since an exact value of $\sqrt{3}$ cannot be written with a finite number of digits. It will print Sqrt[3] as output. If you enter **Sqrt[3.]**, it will give you a value with six digits, and you can also get a six-digit answer by entering **N[Sqrt[3]]** or **Sqrt[3]//N**. If you want 20 digits, you can enter **N[Sqrt[3],20]**.

You can refer to the last output cell with a percent sign (%). For example, if you had entered **Sqrt[3]** and had obtained Sqrt[3] as your output, you could type **N[%]** and press the enter key to obtain a numerical value with six digits, or could type **N[%,15]** to obtain a numerical value with 15 digits.

Mathematica expressions can contain symbols for variables, as well as constants. A variable stands for a location in the computer memory in which a numerical value can be stored, much as in a programming language such as BASIC. Variable names can contain any number of letters and/or digits. However, they cannot begin with a digit. Begin your variable names with a lower-case letter to avoid confusion with Mathematica functions and other Mathematica objects, which always begin with a capital letter. Also remember that xy would represent a variable called xy while x y (with a space between the letters) stands for the product of the two variables $x$ and $y$.

Values are assigned to variables by using an ordinary equal sign, which stands for an assignment operator. For example, a value of 75.68 would be assigned to the variable $x$ by entering the statement

*In[1]:=*
> x = 75.68

The variable x is to be replaced by the value 75.68 whenever it occurs in a Mathematica expression, until a new value is assigned.

To remove a value from the variable $x$, type the statement **Clear[x]**. If you are not sure whether a given variable already has a value, use the **Clear** statement before using the variable.

You can also define one variable in terms of other variables. Assuming that x already has a value, the statement **y = x^3** will assign the cube of that value to y. The variable y will keep that numerical value until it is explicitly assigned a new value, even if the value of x is changed. You can also define y as a function of x equal to $x^3$, such that the value of y will change to the cube of a new value of x, if you use the second type of equal sign that is used in Mathematica, denoted by the symbol :=. The statement

*In[1]:=*
> y := x^3

will cause y to be evaluated as the cube of whatever value x has at the time of execution. You can see what the value of any variable is at the moment by typing the name of the variable and pressing the enter key. If you type a question mark followed by the name of the variable and press the enter key, you can see whether it is defined as a function of other variables.

You can also use a function notation to define a function. For example, if

$$f = abce^{-x/y}$$

you can type

*In[1]:=*
> Clear[a,b,c, f,x,y]
> f[x_] := a b c Exp[-x/y]

Note that the underline must follow the symbol for the independent variable in the function expression on the left-hand side of the statement, and that Mathematica's second type of equal sign, :=, must be used. After defining a function, you can use it in a Mathematica expression, as in the statement

$$g = x \ f[x] \ Cos[x/y]$$

The underline is not used after the symbol for the function's argument in an expression.

## Symbolic Algebra with Mathematica

Mathematica has a powerful capability to carry out symbolic mathematics on algebraic expressions and can solve equations symbolically. The principal Mathematica statements for manipulating algebraic expressions are **Expand[ ]**, **Factor[ ]**, **Simplify[ ]**, **Together[ ]**, and **Apart[ ]**. The **Expand** statement multiplies factors and powers out to give an expanded form of the expression. The following input and output illustrate this action:

*In[1]:=*

    Clear[a,x]
    Expand[(a + x)^3]

*Out[1]=*

$$a^3 + 3\,a^2\,x + 3\,a\,x^2 + x^3$$

The **Clear** statement is included in case a and x had been previously defined as variables with specific values, which would cause Mathematica to return a numerical result instead of a symbolic result.

The **Factor** statement manipulates the expression into a product of factors. The following input and output illustrate this action:

*In[2]:=*

    Clear[y]
    Factor[1 + 5 y + 6 y^2]

*Out[2]=*

$$(1 + 2\,y)\,(1 + 3\,y)$$

The **Simplify** statement manipulates an expression into the form that is considered by the rules built into Mathematica to be the simplest form (with the fewest parts). This form might be the factored form or the expanded form, depending on the expression.

The **Together** statement collects all terms of an expression together over a common denominator, while the **Apart** statement breaks the expression apart into terms with simple denominators, as in the method of partial fractions.

---

### Example 11.21

Write a Mathematica entry that will carry out the decomposition into partial fractions in Example 4.13.

---

### Solution

The input and output lines are

*In[1]:=*

    Clear[x]
    Apart[(6 x - 30)/(x^2 + 3 x + 2)]

*Out[1]=*

$$\frac{-36}{1 + x} + \frac{42}{2 + x}$$

---

### Problem 11.20

In the integration of the differential rate expression of Example 4.14

$$\int_0^{x'} \frac{1}{([A]_0 - ax)([B]_0 - bx)}\,dx,$$

where $[A]_0$ and $[B]_0$ are the initial concentrations of reactants called $A$ and $B$, $a$ and $b$ are the stoichiometric coefficients of these reactants, and $x$ is a variable specifying the extent to which the reaction has occurred. Write a Mathematica statement to decompose the denominator into partial fractions. Carry out the integration manually. (We discuss later how to let Mathematica do the integration for you.)

---

## Differential and Integral Calculus with Mathematica

Mathematica contains the rules for differentiating all of the common functions. It provides higher-order derivatives, partial derivatives, and differentials. The derivative of $f(x)$ with respect to $x$ is obtained with the entry `D[f,x]`. Any variables other than $x$ in the formula for $f(x)$ will be treated as constants. However, if a variable has been assigned a value, that value will be used rather than the symbol for the variable. For example, to obtain the derivative of $ax^2+bx+c$ with respect to $x$, the input and output are

*In[1]:=*

```
Clear[a,b,c,x]
D[ax^2 + bx + c, x]
```

*Out[1]=*

```
b + 2ax
```

However, notice the result of

*In[2]:=*

```
Clear[b,c,x]
a = 2
D[ax² + bx + c, x]
```

*Out[2]=*

```
b + 4x
```

Mathematica will also find the derivative of a previously defined function. For example, you can write

*In[1]:=*

```
Clear[a,b,c,f,x]
f[x_] :=a b c Cos[x/a]
D[ f[x],x]
```

*Out[1]=*

```
            x
-(b c Sin[-])
            a
```

You can also use the prime notation for a derivative of a previously defined function. For example, the last line of the above cell would be replaced by `f'[x]`. The second derivative is obtained with a double prime, etc.

Since all variables other than the independent variable are treated as constants, we can obtain partial derivatives.

---

### Problem 11.21

Write the Mathematica statement needed to find the partial derivative of the function

$$P = \frac{nRT}{V}$$

with respect to $V$ at constant $n$ and $T$.

---

The $n$th derivative of the function $f(x)$ is obtained by the entry

$$\mathtt{D[f,\{x,n\}]}$$

Note the use of braces (curly brackets). Any list of quantities such as a set of several equations is surrounded by braces in Mathematica.

---

### Example 11.22

Find the fifth derivative of the function $x^5 \cos(x)$.

---

### Solution

*In[1]:*

```
Clear[x]
D[x^5 Cos[x],{x,5}]
```

*Out[1]*

```
120 Cos[x] - 600 x^2 Cos[x] + 25 x^4 Cos[x] - 600 x
Sin[x] + 200 x^3 Sin[x] - x^5 Sin[x]
```

---

### Problem 11.22

Write the Mathematica entry needed to find the third derivative with respect to $x$ of $xe^{-bx^2}$.

---

Mathematica can also provide mixed partial derivatives. For example, say that

$$f(x, y) = e^{x/y}$$

and that we want the derivative $(\partial^4 f/\partial^2 x\, \partial^2 y)$. We can obtain this as follows (note the definition of a function of two variables):

*In[1]:=*

```
Clear[x,y]
f[x_,y_] := Exp[x/y]
D[ f[x,y],{x,2},{y,2}]
```

*Out[1]=*

$$\frac{E^{x/y}\, x^2}{y^6} + \frac{6\, E^{x/y}}{y^4} + \frac{6\, E^{x/y}\, x}{y^5}$$

Mathematica can also provide a formula for the differential of a function of several variables. If there are any symbols for constants in the expression for the function, Mathematica will interpret them as variables and will include a term for each such constant. Once you have the output, delete these terms. For example, if a function $f(x, y)$ is given by the formula $ax^2 + bxy + e^{x/y}$ where $a$ and $b$ are constants, Mathematica will deliver a differential with four terms instead of just two terms. The entry line to obtain the differential of this function would be

*In[1]:=*

```
Dt[a x^2 + b x y + Exp[x/y]]
```

and the output line would be

*Out[1]=*

$$x^2 \, Dt[a] \; + \; xy \, Dt[b] \; + \; 2 \, ax \, Dt[x] \; + \; by \, Dt[x] \; +$$
$$bx \, Dt[y] \; + \; E^{x/y} \left( \frac{Dt[x]}{y} \; - \; \frac{x \, Dt[y]}{y^2} \right)$$

where Dt[a] stands for *da*, Dt[x] stands for *dx*, etc. If *a* and *b* are constants, you disregard the terms in *da* and *db*, so that in normal notation

$$df = \left( 2ax + by + \frac{e^{x/y}}{y} \right) dx + \left( bx - e^{x/y} \frac{x}{y^2} \right) dy.$$

Mathematica can carry out definite and indefinite integrals. For example, the input and output statements for the indefinite integral of sin(*x*) are

*In[1]:=*

```
Clear[x]
Integrate[Sin[x],x]
```
*Out[1]=*
```
-Cos[x]
```

Mathematica appears to contain just about every indefinite integral that exists in tables. However, if you specify an integrand for which no indefinite integral exists or which is not in Mathematica's tables, Mathematica will print out what you give it.

Definite integrals are obtained by adding the limits to the input entry. To obtain the definite integral of sin(*x*) from $x = 0$ to $x = \pi$, the input and output statements are

*In[1]:=*

```
Clear[x]
Integrate[Sin[x],{x,0,Pi}]
```
*Out[1]=*
```
2
```

If you give Mathematica a definite integral with an integrand that has no indefinite integral in Mathematica's tables, Mathematica will carry out a numerical approximation to the integral and give you a numerical value.

---

### Problem 11.23

Write Mathematica entries to obtain the following integrals:

a. $\int \cos^3(x) \, dx$

b. $\int_1^2 e^{5x^2} \, dx$

c. $\int_0^\pi \sin[\cos(x)] \, dx.$

---

## Solving Equations with Mathematica

Mathematica can carry out both symbolic and numerical solutions of equations, including single algebraic equations, simultaneous algebraic equations, and differential equations. Mathematica contains the rules needed to solve polynomial equations up to the fourth degree and can solve some fifth-degree equations. The principal statements used to solve equations are **Solve, FindRoot, Eliminate**, and **Reduce.**

The **Solve** statement returns formulas for solutions, if they exist. For example, the input line to solve the equation $ax^2 + bx + c = 0$ is

*In[1]:=*

```
Solve[a x^2 + b x + c == 0,x]
```

The name of the variable to be solved for must be included at the end of the equation and separated from it by a comma. Mathematica's third kind of equal sign, a double equal sign, must be used in equations to be solved. Another use of this equal sign is in asking Mathematica to test whether an equality is true or false.

The resulting output is the standard quadratic formula:

*Out[1]=*

$$\left\{\left\{x \rightarrow \frac{-b - \text{Sqrt}[b^2 - 4\,a\,c]}{2\,a}\right\}, \left\{x \rightarrow \frac{-b + \text{Sqrt}[b^2 - 4\,a\,c]}{2\,a}\right\}\right\}$$

Note the use of the symbol $\rightarrow$, which Mathematica types as two symbols.

If no formula can be found for a solution, you can use the **FindRoot** statement to obtain a numerical result. For example, to find a root for the equation

$$e^{-x} - 0.5x = 0$$

you type the input statement,

*In[1]:*

```
FindRoot[Exp[-x] - 0.5 x == 0,{x,1}]
```

and get the output

*Out[1]=*

```
{x -> 0.852606}
```

When a numerical procedure is to be used, it is necessary to include an initial trial value of the root, which is here represented by the 1 following the x in braces. If your equation is a polynomial equation, the **NSolve** statement can be used instead of **FindRoot**. The **NSolve** statement does not require a trial root, and will find all roots, while **FindRoot**, which uses Newton's method, will generally converge to one root and then stop.

The **Solve** statement can also be used to solve simultaneous equations. The equations are typed inside curly brackets with commas between them, and the variables are listed inside curly brackets. To solve the equations

$$ax + by = c$$
$$gx + hy = k$$

we type the input entry

*In[1]:=*

```
Solve[{ax + by == c, gx + hy == k},{x,y}]
```

Notice the use of braces to notify Mathematica that we have a list of two equations and a list of two variables. The output is

*Out[1]=*

$$\left\{\left\{x \rightarrow \frac{c}{a} + \frac{b(cg - ak)}{a(-(bg) + ah)}, y \rightarrow -\left(\frac{cg - ak}{-(bg) + ah}\right)\right\}\right\}$$

To simplify the expressions for x and y, we use the fact that the percent symbol represents the last line of output and type

*In[2]:=*

```
Simplify[%]
```

We receive the output

*Out[2]=*

$$\left\{\left\{x \rightarrow \frac{-(ch) + bk}{bg - ah}, y \rightarrow \frac{-(cg) + ak}{-(bg) + ah}\right\}\right\}$$

which is the expression obtained from Cramer's rule.

The **Eliminate** statement is used to eliminate one or more of the variables in a set of simultaneous equations. For example, to obtain a single equation in x from the set of equations above, you would type the input entry (note the double equal signs)

*In[1]:=*

```
Eliminate[{ax + by == c,gx + hy == k},y]
```

and would receive the output

*Out[1]=*

```
ch == bk - bgx + ahx
```

We solve this equation for *x* by typing

*In[2]:=*

```
Solve[%,x]
```

*Out[2]=*

$$\left\{\left\{x \rightarrow -\left(\frac{ch - bk}{bg - ah}\right)\right\}\right\}$$

## Graphics with Mathematica

Mathematica can produce sophisticated graphics, including two-dimensional graphs and perspective views of three-dimensional graphs. For example, to make a graph of $\sin(x)$ from $x = 0$ to $x = 2\pi$, one enters the input

*In[1]:=*

```
Plot[Sin[x],{x,0,2Pi}]
```

Figure 11.6 shows the output graph.

A graph of a function can be obtained more easily with Mathematica than with CricketGraph, since CricketGraph requires you to construct a table of values while Mathematica can work from the formula for the function. The graphic capabilities of Mathematica are very extensive. It makes very nice perspective views of

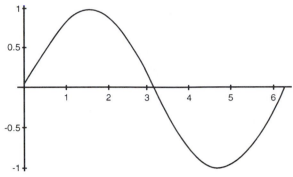

**FIGURE 11.6**   Graph of the sine function prepared with mathematica.

three-dimensional graphs. You can read more about this in the book by Wolfram and in the manual that is supplied with Mathematica.

## SUMMARY OF THE CHAPTER

In this chapter we have presented a brief introduction to the use of several kinds of computer software. We have included Microsoft Word, a word processing program, Excel, a spreadsheet, CricketGraph, a graphics program for constructing graphs, BASIC, a programming language, and Mathematica, a complete mathematics package. Only enough information is presented to allow a beginner to get started.

## ADDITIONAL READING

E. J. Billo, *Microsoft Excel for Chemists: A Comprehensive Guide*, Wiley/VCH, New York, 1997. This is a much more useful guide to Excel than the manual provided by the manufacturer.

Richard E. Crandall, *Mathematica for the Sciences*, Addison–Wesley, Reading, MA, 1991. This is a small book that contains a variety of Mathematica applications to physics problems, including some related to physical chemistry.

William H. Cropper, *Mathematica Computer Programs for Physical Chemistry*, Springer, New York, 1998. This is a book of Mathematica programs specifically designed for physical chemistry. It includes a CD-ROM containing the programs.

William S. Davis, *BASIC: Getting Started*, Addison–Wesley, Reading, MA, 1981. This is an elementary introduction to BASIC.

DeLos F. DeTar, *Computer Programs for Chemistry*, Vols. 1 and 2, Benjamin, New York, 1968. These two volumes contain a large number of ready-to-use programs for problems in chemistry, including analysis of NMR spectra, analysis of chemical reaction rate data, etc. The programs are written in FORTRAN.

Dermot Diamond and Venita C. A. Hanratty, *Spreadsheet Applications in Chemistry Using Microsoft Excel*, Wiley–Interscience, New York, 1997. This is a comprehensive introduction to the use of Excel for chemists.

T. R. Dickson, *The Computer and Chemistry*, Freeman, San Francisco, 1968. This is a fairly complete introduction to programming in the FORTRAN language, with applications to chemistry problems.

Richard M. Jones, *Introduction to Computer Applications Using BASIC*, Allyn & Bacon, Newton, MA, 1981. This is another of the books recommended by the creators of TrueBASIC.

Joseph H. Noggle, *Physical Chemistry Using MathCad*, Pike Creek Publishing, Newark, Delaware, 1998. This book is a supplement to standard physical chemistry course. It introduces MathCad, a competitor to Mathematica, and presents a number of physical chemistry problems that can be solved using this software. As of late 1998, information about the book can be obtained at the website http://www.udel.edu/noggle/mcadcad.html.

Rama N. Reddy and Carol A. Ziegler, *FORTRAN 77 with Applications for Scientists and Engineers*, West Publishing Co., St. Paul, MN, 1981. This book describes a version of the FORTRAN language that is still used.

Leonard Soltzberg, Arvind A. Shah, John C. Saber, and Edgar T. Canty, *BASIC and Chemistry*, Houghton Mifflin, Boston, 1975. This book is an introduction to the BASIC language and a manual for its use in chemistry problems. It contains a number of ready-to-use programs. It is available as a fairly inexpensive paperback book.

Stephen Wolfram, *Mathematica*, 2nd ed., Addison–Wesley, Reading, MA, 1991. This is a textbook which provides a complete introduction to the use of Mathematica, written by the developer of the software.

## ADDITIONAL PROBLEMS

### 11.24

Set up a workbook in Excel that will calculate the mean and standard deviation of ten values placed in column A. Use the workbook to calculate the mean and standard deviation of the values: 1.25, 2.37, 5.43, 6.84, 7.32, 8.14, 9.31, 10.42, 11.71, and 12.00.

### 11.25

a. Using CricketGraph, make a properly labeled graph of the function $y(x) = \ln(x) + \cos(x)$ for values of $x$ from 0 to $2\pi$, at intervals of $\pi/100$. Use Excel to create the table of values of the function.

b. Repeat part a using Mathematica.

### 11.26

a. Using CricketGraph, construct a graph of the pressure of a gas obeying the van der Waals equation containing three curves of $P$ as a function of $V$: one for T $= 10a/27Rb$, one for T $= 8a/27Rb$, and one for T $= 6a/27Rb$. Use the Excel spreadsheet to construct the table of values and import this table into CricketGraph to construct the graph.

b. Repeat part a using Mathematica.

### 11.27

The following data were taken for the thermal decomposition of $N_2O_3$:

| t/s | 0 | 184 | 426 | 867 | 1877 |
|---|---|---|---|---|---|
| $[N_2O_3]$/mol L$^{-1}$ | 2.33 | 2.08 | 1.67 | 1.36 | 0.72 |

a. Make three graphs, using CricketGraph: one with $\ln([N_2O_3]/\text{mol L}^{-1})$ as a function of $t$, one with $1/[N_2O_3]$ as a function of $t$, and one with $1/[N_2O_3]^2$ as a function of $t$. Use the simple math and data transform functions to generate the data table. Determine which graph is most nearly linear. If the first graph is most nearly linear, the reaction is first order; if the second graph is most nearly linear, the reaction is second order; and if the third graph is most nearly linear, the reaction is third order.

b. Repeat part a using Excel to create the data table.

c. Repeat part a using Mathematica.

## 11.28

Each of the parts of this problem consists of a formula and a BASIC statement that is supposed to calculate a value for the dependent variable in that formula. Find and correct any mistakes in the BASIC statements.

| Formula | BASIC statement |
|---------|-----------------|
| a. $z = \dfrac{xy}{ac}$ | Z = X*Y/A*C |
| b. $y = mx + b$ | Y = MX + B |
| c. $y = ax^2 + bx + c$ | Y = A*X^2 + B*X + C |
| d. $d = b^2 - 4ac$ | D = B^2 - 4A*C |
| e. $z = (ax^2 + b)^n/an$ | Z = (A*X^2 + B)*N/A/N |
| f. $y = e^{e^x}$ | Y = EXP(EXP(X)) |
| g. $y = \log_{10}(\cos(x))$ | Y = LOG(COS(X)) |
| h. $y = (a + x)^{-1}$ | Y = 1/(A + X) |

## 11.29

Write BASIC statements that will calculate a value for each of the dependent variables in the formulas:

a. $y = (A + B)/(C + D)$
b. $z = (17y^2 + 3x^2)^{1/2}/14ab$
c. $f = (a + b)^{3/2}/(c + d)$
d. $s = (a_1 + a_2 + a_3 + a_4 + \cdots + a_n)/n!$
e. $g = (b \sin^2(ax) + a \sin^2(bx))^{1/2}$
f. $h = (ax^2 + bx + c)^{-1}$
g. $s = 1/(n - 1) \left[ \displaystyle\sum_{i=1}^{n} (x_i - m)^2 \right]^{1/2}.$

## 11.30

Write a BASIC program that will convert a list of distance measurements in meters to miles, feet, and inches. Include input and output statements and label your output.

## 11.31

Write a BASIC program to calculate the pH of an aqueous solution of a weak monobasic acid. Neglect the hydrogen ions from the water so that you can use the formula

$$\frac{[H^+][A^-]}{[HA]} = K_a,$$

where $[H^+]$ is the hydrogen-ion concentration, $[A^-]$ the acid-anion concentration, and $[HA]$ the concentration of undissociated acid. This formula assumes that all activity coefficients are equal to unity. Include INPUT statements for the value of $K_a$, the acid dissociation constant, and for the value of $[HA] + [A^-]$, the gross concentration of acid. Write the program to test the discriminant $b^2 - 4ac$ (where the quadratic equation is written $ax^2 + bx + c = 0$) to make sure the roots are real and to reject any negative roots. Label your output, which should include $[H^+]$ as well as pH, which is equal to $-\log_{10}([H^+])$ with our assumptions.

## 11.32

a. Modify the program that you wrote for Problem 11.31 so that it will produce a table of values of pH for various values of the gross acid concentration.

b. Modify the program so that will produce a table of values of the percent ionization for various values of the gross acid concentration.

## 11.33

Write a BASIC program that will carry out all the calculations for a gravimetric determination of sodium chloride in an unknown mixture. Write the program so that it will carry out the calculation for any number of repetitions of the determination on the same mixture. For each sample, arrange for the mass of the dried sample and its container and the mass of the empty container to be read in, as well as the mass of the precipitated silver chloride plus its container and the mass of the empty container. Have the program calculate the percentage of each sample that was chloride ion and the percentage that was sodium chloride and have it calculate the mean and the standard deviation. Label all output.

## 11.34

Write a BASIC program to calculate the pressure of a nonideal gas described by the van der Waals equation of state

$$\left( P + \frac{an^2}{V^2} \right)(V - nb) = nRT,$$

where $R$ is the ideal gas constant and $a$ and $b$ are constants whose values are to be read into the computer for the particular gas. Give the user the option of receiving a table of values of the pressure for a set of values of $n$, $V$, and $T$ instead of a single value. The equation is not valid if $V \leq nb$. Include a test to ensure that this condition is met.

## 11.35

Rearrange the van der Waals equation of state into a cubic equation in $V$ as a function of $T$, $P$, and $n$. Write a BASIC program to carry out a numerical solution for the value of $V$, using either the method of bisection or Newton's method. Run the program for several sets of values of $T$, $P$, and $n$.

## 11.36

When $T$ is smaller than $8a/27Rb$ ( critical value of the temperature) there is a relative maximum in the curve given by the van der Waals equation for the pressure.

Write a BASIC program to determine the value of $V$ and the value of $P$ at the relative maximum. Include a test to make sure that $T$ is smaller than $8a/27Rb$ and a test to ensure that the program has found the maximum and not a minimum. A relative minimum also occurs. Modify the program to find both the maximum and the minimum.

### 11.37

The van der Waals equation of state is

$$\left(P + \frac{n^2 a}{V}\right)(V - nb) = nRT,$$

where $a$ and $b$ are temperature-independent parameters that have different values for each gas. For carbon dioxide, $a = 0.3640 \, \text{Pa m}^6 \, \text{mol}^{-2}$ and $b = 4.267 \times 10^{-5} \, \text{m}^3 \text{mol}^{-1}$.

a. Write this equation as a cubic equation in $V$.

b. Use the **NSolve** statement to find the volume of 1.000 mol of carbon dioxide at $P = 1.000$ bar (100000 Pa) and $T = 298.15$ K. Define a function of $a, b, P, V,$ and $T$ which is to be set equal to zero and use assignment statements to provide values of $a, b, P,$ and $T$. Notice that two of the three roots are complex and must be ignored.

c. Repeat part b for $P = 10.000$ bar ($1.0000 \times 10^6$ Pa) and $T = 298.15$ K.

d. Repeat part b for $P = 100.00$ bar ($1.0000 \times 10^7$ Pa) and $T = 298.15$ K.

e. Make graphs showing the pressure of 1.000 mol of carbon dioxide as a function of the volume for several different temperatures. You can use the statement

$$\textbf{PlotRange-> \{ymin,ymax\}}$$

as part of your **Plot** statement (separate it by commas) to adjust the vertical axis.

f. Change variables to measure volumes in liters and pressures in atmospheres. Repeat parts a–e. Remember to change the value of $R$ to 0.08206 L atm K$^{-1}$ mol$^{-1}$.

### 11.38

Use Mathematica to solve the set of simultaneous equations

$$3x + 4y + 5z = 1$$
$$4x - 3y + 6z = 3$$
$$7x + 2y - 6z = 2.$$

### 11.39

Stirling's approximation for $\ln(N!)$ is

$$\ln(N!) \approx \frac{1}{2}\ln(2\pi N) + N\ln(N) - N.$$

Use Mathematica to determine the validity of this approximation and of the less accurate version

$$\ln(N!) \approx N\ln(N) - N$$

for values of $N$ up to $10^5$.

### 11.40

Use Mathematica to generate a short table of values of the error function.

# VALUES OF PHYSICAL CONSTANTS[1]

Avogadro's constant,

$$N_{Av} = 6.02214 \times 10^{23} \, \text{mol}^{-1}.$$

Molar ideal gas constant,

$$R = 8.3145 \, \text{J K}^{-1} \, \text{mol}^{-1} = 0.082056 \, \text{liter atm K}^{-1} \, \text{mol}^{-1}$$
$$= 1.9872 \, \text{cal K}^{-1} \, \text{mol}^{-1}.$$

The magnitude of an electron's charge,

$$e = 1.602177 \times 10^{-19} \, \text{C}.$$

Planck's constant,

$$h = 6.62608 \times 10^{-34} \, \text{J s}.$$

Boltzmann's constant,

$$k_B = 1.38066 \times 10^{-23} \, \text{J K}^{-1}.$$

The rest-mass of an electron,

$$m_e = 9.10939 \times 10^{-31} \, \text{kg}.$$

---

[1] From E. G. Cohen and B. N Taylor, "The 1986 Adjustment of the Fundamental Physical Constants," CODATA Bulletin Number 63, November 1986.

The rest-mass of a proton,

$$m_p = 1.672623 \times 10^{-27} \text{ kg}.$$

The rest-mass of a neutron,

$$m_n = 1.674929 \times 10^{-27} \text{ kg}.$$

The speed of light (exact value, used to define the standard meter),

$$c = 2.99792458 \times 10^8 \text{ m s}^{-1} = 2.99792458 \times 10^{10} \text{ cm s}^{-1}.$$

The acceleration due to gravity near the earth's surface (varies slightly with latitude; this value applies near the latitude of Washington, DC, or San Francisco),

$$g = 9.80 \text{ m s}^{-7}.$$

The gravitational constant,

$$G = 6.673 \times 10^{-11} \text{ m}^3 \text{ s}^{-2} \text{ kg}^{-1}.$$

The permittivity of a vacuum,

$$\varepsilon_0 = 8.8545187817 \times 10^{-12} \text{ C}^2 \text{ N}^{-1} \text{ m}^{-2}.$$

The permeability of a vacuum (exact value, by definition),

$$\mu_0 = 4\pi \times 10^{-7} \text{ NA}^{-2}.$$

## Some Conversion Factors

1 pound $= 0.4535924$ kg
1 inch $= 2.54$ cm (exact value by definition)
1 calorie $= 4.184$ J (exact value by definition)
1 electron volt $= 1.60219 \times 10^{-19}$ J
1 erg $= 10^{-7}$ J (exact value by definition)
1 atm $= 760$ torr $= 101,325$ N m$^{-2}$ $= 101,325$ pascal (Pa) (exact values by definition)
1 atomic mass unit (u) $= 1.66054 \times 10^{-27}$ kg
1 horsepower $= 745.700$ watt $= 745.700$ J s$^{-1}$

# SOME MATHEMATICAL FORMULAS AND IDENTITIES

1. The arithmetic progression of the first order to $n$ terms,

$$a + (a + d) + (a + 2d) + \cdots + [a + (n - 1)d] = na + \frac{1}{2}n(n - 1)d$$
$$= \frac{n}{2}(\text{1st term} + n\text{th term}).$$

2. The geometric progression to $n$ terms,

$$a + ar + ar^2 + \cdots + ar^{n-1} = \frac{a(1 - r^n)}{1 - r}.$$

3. The definition of the arithmetic mean of $a_1, a_2, \ldots, a_n$,

$$\frac{1}{n}(a_1 + a_2 + \cdots + a_n).$$

4. The definition of the geometric mean of $a_1, a_2, \ldots, a_n$,

$$(a_1, a_2, \ldots, a_n)^{1/n}.$$

5. The definition of the harmonic mean of $a_1, a_2, \ldots, a_n$: If $\bar{a}_H$ is the harmonic mean, then

$$\frac{1}{\bar{a}_H} = \frac{1}{n}\left(\frac{1}{a_1} + \frac{1}{a_2} + \frac{1}{a_3} + \cdots + \frac{1}{a_n}\right).$$

6. If

$$a_0 + a_1 x + a_2 x^2 + a_3 x^3 + \cdots + a_n x^n = b_0 + b_1 x + b_2 x^2 + b_3 x^3 + \cdots + b_n x^n$$

for all values of $x$, then

$$a_0 = b_0, \qquad a_1 = b_1, \qquad a_2 = b_2, \ldots, a_n = b_n.$$

## Trigonometric Identities

7. $\sin^2(x) + \cos^2(x) = 1.$

8. $\tan(x) = \dfrac{\sin(x)}{\cos(x)}.$

9. $\operatorname{ctn}(x) = \dfrac{1}{\tan(x)}.$

10. $\sec(x) = \dfrac{1}{\cos(x)}.$

11. $\csc(x) = \dfrac{1}{\sin(x)}.$

12. $\sec^2(x) - \tan^2(x) = 1.$

13. $\csc^2(x) - \operatorname{ctn}^2(x) = 1.$

14. $\sin(x + y) = \sin(x)\cos(y) + \cos(x)\sin(y).$

15. $\cos(x + y) = \cos(x)\cos(y) - \sin(x)\sin(y).$

16. $\sin(2x) = 2\sin(x)\cos(x).$

17. $\cos(2x) = \cos^2(x) - \sin^2(x) = 1 - 2\sin^2(x).$

18. $\tan(x + y) = \dfrac{\tan(x) + \tan(y)}{1 - \tan(x)\tan(y)}.$

19. $\tan(2x) = \dfrac{2\tan(x)}{1 - \tan^2(x)}.$

20. $\sin(x) = \dfrac{1}{2i}(e^{ix} - e^{-ix}).$

21. $\cos(x) = \dfrac{1}{2}(e^{ix} + e^{-ix}).$

22. $\sin(x) = -\sin(-x).$

23. $\cos(x) = \cos(-x).$

24. $\tan(x) = -\tan(-x).$

25. $\sin(ix) = i\sinh(x).$

26. $\cos(ix) = \cosh(x).$

27. $\tan(ix) = i\tanh(x).$

28. $\sin(x \pm iy) = \sin(x)\cosh(y) \pm i\cos(x)\sinh(y).$

29. $\cos(x \pm iy) = \cos(x)\cosh(y) \mp i\sin(x)\sinh(y).$

30. $\cosh(x) = \dfrac{1}{2}(e^x + \bar{e}^x).$

31. $\sinh(x) = \dfrac{1}{2}(e^x - \bar{e}^x).$

32. $\tanh(x) = \dfrac{\sinh(x)}{\cosh(x)}.$

33. $\operatorname{sech}(x) = \dfrac{1}{\cosh(x)}.$

34. $\operatorname{csch}(x) = \dfrac{1}{\sinh(x)}.$

35. $\text{ctnh}(x) = \dfrac{1}{\tanh(x)}.$
36. $\cosh^2(x) - \sinh^2(x) = 1.$
37. $\tanh^2(x) + \text{sech}^2(x) = 1.$
38. $\text{ctnh}^2(x) - \text{scsh}^2(x) = 1.$
39. $\sinh(x) = -\sinh(-x).$
40. $\cosh(x) = \cosh(-x).$
41. $\tanh(-x) = -\tanh(-x).$
42. Relations obeyed by any triangle with angle $A$ opposite side $a$, angle $B$ opposite side $b$, and angle $C$ opposite side $c$:

a. $A + B + C = 180° = \pi$ radians
b. $c^2 = a^2 + b^2 - 2ab\cos(C)$
c. $\dfrac{a}{\sin(A)} = \dfrac{b}{\sin(B)} = \dfrac{c}{\sin(C)}.$

# INFINITE SERIES

## Series with Constant Terms

1. $1 + \dfrac{1}{2} + \dfrac{1}{3} + \dfrac{1}{4} + \cdots = \infty.$

2. $1 + \dfrac{1}{2^2} + \dfrac{1}{3^2} + \dfrac{1}{4^2} + \cdots = \dfrac{\pi^2}{6}.$

3. $1 + \dfrac{1}{2^4} + \dfrac{1}{3^4} + \dfrac{1}{4^4} + \cdots = \dfrac{\pi^4}{90}.$

4. $1 + \dfrac{1}{2^p} + \dfrac{1}{3^p} + \dfrac{1}{4^p} + \cdots = \zeta(p).$

   The function $\zeta(p)$ is called the *Riemann zeta function.* See H. B. Dwight, *Tables of Elementary and Some Higher Mathematical Functions*, 2nd ed., Dover, New York, 1958, for tables of values of this function.

5. $1 - \dfrac{1}{2} + \dfrac{1}{3} - \dfrac{1}{4} + \cdots = \ln(2).$

6. $1 - \dfrac{1}{2^p} + \dfrac{1}{3^p} + \dfrac{1}{4^p} + \cdots = \left(1 - \dfrac{2}{2^p}\right)\zeta(p).$

## Power Series

7. Maclaurin's series. If there is a power series in $x$ for $f(x)$, it is

$$f(x) = f(0) + \frac{df}{dx}\bigg|_{x=0} x = \frac{1}{2!}\frac{d^2 f}{dx^2}\bigg|_{x=0} x^2 = \frac{1}{3!}\frac{d^3 f}{dx^3}\bigg|_{x=0} x^3 + \cdots.$$

8. Taylor's series. If there is a power series in $x - a$ for $f(x)$, it is

$$f(x) = f(a) + \frac{df}{dx}\Big|_{x=a}(x - a) + \frac{1}{2!}\frac{d^2f}{dx^2}\Big|_{x=a}(x - a^2) + \cdots.$$

In Eqs. (7) and (8), $\frac{df}{dx}\big|_{x=a}$ means the value of the derivative at $x = a$.

9. If, for all values of $x$,

$$a_0 + a_1 x + a_2 x^2 + a_3 x^3 + \cdots = b_0 + b_1 x + b_2 x^2 + b_3 x^3 + \cdots$$

then

$$a_0 = b_0, \qquad a_1 = b_1, \qquad a_2 = b_2, \qquad \text{etc.}$$

10. The reversion of a series. If

$$y = ax + bx^2 + cx^3 + \cdots$$

and

$$x = Ay + By^2 + Cy^3 + \cdots$$

then

$$A = \frac{1}{a}, \qquad B = -\frac{b}{a^3}, \qquad C = \frac{1}{a^5}(2b^2 - ac),$$

$$D = \frac{1}{a^7}(5abc - a^2d - 5b^3), \qquad \text{etc.}$$

See Dwight, *Table of Integrals and Other Mathematical Data* (cited above), for more coefficients.

11. Powers of a series. If

$$S = a + bx + cx^2 + dx^3 + \cdots$$

then

$$S^2 = a^2 + 2abx + (b^2 + 2ac)x^2 + 2(ad + bc)x^3$$
$$+ (c^2 + 2ae + 2bd)x^4 + 2(af + be + cd)x^5 + \cdots$$

$$S^{1/2} = a^{1/2}\left[1 + \frac{b}{2a}x + \left(\frac{2}{2a} - \frac{b^2}{8a^2}\right)x^2 + \cdots\right]$$

$$S^{-1} = a^{-1}\left[1 - \frac{b}{a}x + \left(\frac{b^2}{a^2} - \frac{c}{a}\right)x^2 + \left(\frac{2bc}{a^2} - \frac{d}{a} - \frac{b^3}{a^3}\right)x^3 + \cdots\right].$$

12. $\sin(x) = x - \dfrac{x^3}{3!} + \dfrac{x^5}{5!} - \dfrac{x^7}{7!} + \cdots.$

13. $\cos(x) = 1 - \dfrac{x^2}{2!} + \dfrac{x^4}{4!} - \dfrac{4^6}{6!} + \cdots.$

14. $\sin(\theta + x) = \sin(\theta) + x\cos(\theta) - \dfrac{x^2}{2!}\sin(\theta) - \dfrac{x^3}{3!}\cos(\theta) + \cdots.$

15. $\cos(\theta + x) = \cos(\theta) - x\sin(\theta) - \dfrac{x^2}{2!}\cos(\theta) + \dfrac{x^3}{3!}\sin(\theta) + \cdots.$

16. $\sin^{-1}(x) = x + \dfrac{x^3}{2 \cdot 3} + \dfrac{1 \cdot 3x^5}{2 \cdot 4 \cdot 5} + \dfrac{1 \cdot 3 \cdot 5x^7}{2 \cdot 4 \cdot 6 \cdot 7} + \cdots,$

where $x^2 < 1$. The series gives the principal value, $-\pi/2 < \sin^{-1}(x) < \pi/2$.

17. $\cos^{-1}(x) = \dfrac{\pi}{2} - \left( x + \dfrac{x^3}{2 \cdot 3} + \dfrac{1 \cdot 3x^5}{2 \cdot 4 \cdot 5} + \cdots \right),$

where $x^2 < 1$. The series gives the principal values, $0 < \cos^{-1}(x) < \pi$.

18. $e^x = 1 + x + \dfrac{x^2}{2!} + \dfrac{x^3}{3!} + \dfrac{x^4}{4!} + \cdots \ (x^2 < \infty).$

19. $a^x = e^{x \ln(a)} = 1 + x \ln(a) + \dfrac{(x \ln(a))^2}{2!} + \cdots.$

20. $\ln(1 + x) = x - \dfrac{x^2}{2} + \dfrac{x^3}{3} - \dfrac{x^4}{4} \cdots \ (x^2 < 1 \text{ and } x = 1).$

21. $\ln(1 - x) = -\left( x + \dfrac{x^2}{2} + \dfrac{x^3}{3} + \dfrac{x^4}{4} + \cdots \right)(x^2 < 1 \text{ and } x = -1).$

22. $\sinh(x) = x + \dfrac{x^3}{3!} + \dfrac{x^5}{5!} + \dfrac{x^7}{7!} + \cdots \ (x^2 < \infty).$

23. $\cosh(x) = 1 + \dfrac{x^2}{2!} + \dfrac{x^4}{4!} + \dfrac{x^6}{6!} + \cdots \ (x^2 < \infty).$

# A SHORT TABLE OF DERIVATIVES[1]

In the following list, $a$, $b$, and $c$ are constants, and $e$ is the base of natural logarithms.

1. $\dfrac{d}{dx}(au) = a\dfrac{du}{dx}$.

2. $\dfrac{d}{dx}(uv) = u\dfrac{dv}{dx} + v\dfrac{du}{dx}$.

3. $\dfrac{d}{dx}(uvw) = uv\dfrac{dw}{dx} + uw\dfrac{dv}{dx} + vw\dfrac{du}{dx}$.

4. $\dfrac{d(x^n)}{dx} = nx^{n-1}$.

5. $\dfrac{d}{dx}\left(\dfrac{u}{v}\right) = \dfrac{1}{v}\dfrac{du}{dx} - \dfrac{u}{v^2}\dfrac{dv}{dx} = \dfrac{1}{v^2}\left(v\dfrac{du}{dx} - u\dfrac{dv}{dx}\right)$.

6. $\dfrac{d}{dx}f(u) = \dfrac{df}{du}\dfrac{du}{dx}$, where $f$ is some differentiable function of $u$ and $u$ is some differentiable function of $x$ (the *chain rule*).

7. $\dfrac{d^2}{dx^2}f(u) = \dfrac{df}{du}\dfrac{d^2u}{dx^2} + \dfrac{d^2f}{du^2}\left(\dfrac{du}{dx}\right)^2$.

8. $\dfrac{d}{dx}\sin(ax) = a\cos(ax)$.

---

[1] These formulas, and the other material in Appendices 3 through 7, are from H. B. Dwight, *Tables of Integrals and other Mathematical Data,* 4th ed., Macmillan, New York, 1961.

9. $\dfrac{d}{dx}\cos(ax) = -a\sin(ax).$

10. $\dfrac{d}{dx}\tan(ax) = a\sec^2(ax).$

11. $\dfrac{d}{dx}\text{ctn}(ax) = -a\csc^2(ax).$

12. $\dfrac{d}{dx}\sec(ax) = a\sec(ax)\tan(ax).$

13. $\dfrac{d}{dx}\csc(ax) = -a\csc(ax)\text{ctn}(ax).$

14. $\dfrac{d}{dx}\sin^{-1}\left(\dfrac{x}{a}\right) = \dfrac{1}{\sqrt{a^2 - x^2}}$      if $x/a$ is in the first or fourth quadrant

$\phantom{14. \dfrac{d}{dx}\sin^{-1}\left(\dfrac{x}{a}\right)} = \dfrac{-1}{\sqrt{a^2 - x^2}}$      if $x/a$ is in the second or third quadrant.

15. $\dfrac{d}{dx}\cos^{-1}\left(\dfrac{x}{a}\right) = \dfrac{-1}{\sqrt{a^2 - x^2}}$      if $x/a$ is in the first or second quadrant

$\phantom{15. \dfrac{d}{dx}\cos^{-1}\left(\dfrac{x}{a}\right)} = \dfrac{1}{\sqrt{a^2 - x^2}}$      if $x/a$ is in the third or fourth quadrant.

16. $\dfrac{d}{dx}\tan^{-1}\left(\dfrac{x}{a}\right) = \dfrac{a}{a^2 + x^2}.$

17. $\dfrac{d}{dx}\text{ctn}^{-1}\left(\dfrac{x}{a}\right) = \dfrac{-a}{a^2 + x^2}.$

18. $\dfrac{d}{dx}e^{ax} = ae^{ax}.$

19. $\dfrac{d}{dx}a^x = a^x\ln(a).$

20. $\dfrac{d}{dx}a^{cx} = ca^{cx}\ln(a).$

21. $\dfrac{d}{dx}u^y = yu^{y-1}\dfrac{du}{dx} + u^y\ln(u)\dfrac{dy}{dx}.$

22. $\dfrac{d}{dx}x^x = x^x[1 + \ln(x)].$

23. $\dfrac{d}{dx}\ln(ax) = \dfrac{1}{x}.$

24. $\dfrac{d}{dx}\log_a(x) = \dfrac{\log_a(a)}{x}.$

25. $\dfrac{d}{dq}\displaystyle\int_p^q f(x)\,dx = f(q)$      if $p$ is independent of $q$.

26. $\dfrac{d}{dq}\displaystyle\int_p^q f(x)\,dx = -f(p)$      if $q$ is independent of $p$.

# A SHORT TABLE OF INDEFINITE INTEGRALS

In the following, an arbitrary constant of integration is to be added to each equation. $a, b, c, g,$ and $n$ are constants.

1. $\int dx = x.$

2. $\int x \, dx = \dfrac{x^2}{2}.$

3. $\int \dfrac{1}{x} \, dx = \ln(|x|)$ do not integrate from negative to positive values of $x$.

4. $\int x^n \, dx = \dfrac{x^{n+1}}{n+1}$, where $n \neq -1$.

5. $\int (a+bx)^n dx = \dfrac{(a+bx)^{n+1}}{b(n+1)}.$

6. $\int \dfrac{1}{(a+bx)} \, dx = \dfrac{1}{b} \ln(|a+bx|).$

7. $\int \dfrac{1}{(a+bx)^n} \, dx = \dfrac{-1}{(n-1)b(a+bx)^{n-1}}.$

8. $\int \dfrac{x}{(a+bx)} \, dx = \dfrac{1}{b^2}[(a+bx) - a \ln(|a+bx|)].$

9. $\int \dfrac{a+bx}{c+gx} \, dx = \dfrac{bx}{g} + \dfrac{ag-bc}{g^2} \ln(|c+gx|).$

10. $\int \dfrac{1}{(a + bx)(c + gx)}\,dx = \dfrac{1}{ag - bc}\ln\left(\left|\dfrac{c + gx}{a + bx}\right|\right).$

11. $\int \dfrac{1}{a^2 + x^2}\,dx = \dfrac{1}{a}\tan^{-1}\left(\dfrac{x}{a}\right).$

12. $\int \dfrac{x}{(a^2 + x^2)^2}\,dx = \dfrac{-1}{2(a^2 + x^2)}.$

13. $\int \dfrac{x}{(a^2 + x^2)}\,dx = \dfrac{1}{2}\ln(a^2 + x^2).$

14. $\int \dfrac{1}{(a^2 - b^2x^2)}\,dx = \dfrac{1}{2ab}\ln\left(\left|\dfrac{a + bx}{a - bx}\right|\right).$

15. $\int \dfrac{x}{(a^2 - x^2)}\,dx = -\dfrac{1}{2}\ln(|a^2 - x^2|).$

16. $\int \dfrac{x^{1/2}}{(a^2 + b^2x)}\,dx = \dfrac{2x^{1/2}}{b^2} - \dfrac{2a}{b^3}\tan^{-1}\left(\dfrac{bx^{1/2}}{a}\right).$

17. $\int \dfrac{1}{(a + bx^2)^{p/2}}\,dx = \dfrac{-2}{(p - 2)b(a + bx)^{(p-2)/2}}.$

18. $\int \dfrac{1}{(x^2 + a^2)^{1/2}}\,dx = \ln(x + (x^2 + a^2)^{1/2}).$

19. $\int \dfrac{x}{(x^2 + a^2)^{1/2}}\,dx = (x^2 + a^2)^{1/2}.$

20. $\int \dfrac{1}{(x^2 - a^2)^{1/2}}\,dx = \ln(x + (x^2 - a^2)^{1/2}).$

21. $\int \dfrac{x}{(x^2 - a^2)^{1/2}}\,dx = (x^2 - a^2)^{1/2}.$

22. $\int \sin(ax)\,dx = -\dfrac{1}{a}\cos(ax).$

23. $\int \sin(a + bx)\,dx = -\dfrac{1}{b}\cos(a + bx).$

24. $\int x \sin(x)\,dx = \sin(x) - x\cos(x).$

25. $\int x^2 \sin(x)\,dx = 2x\sin(x) - (x^2 - 2)\cos(x).$

26. $\int \sin^2(x)\,dx = \dfrac{x}{2} - \dfrac{\sin(2x)}{4} = \dfrac{x}{2} - \dfrac{\sin(x)\cos(x)}{2}.$

27. $\int x \sin^2(x)\,dx = \dfrac{x^2}{4} - \dfrac{x\sin(2x)}{4} - \dfrac{\cos(2x)}{8}.$

28. $\int \dfrac{1}{1 + \sin(x)}\,dx = -\tan\left(\dfrac{\pi}{4} - \dfrac{x}{2}\right).$

29. $\int \cos(ax)\,dx = \dfrac{1}{a}\sin(ax).$

30. $\int \cos(a + bx)dx = \dfrac{1}{b} \sin(a + bx).$

31. $\int x \cos(x)dx = \cos(x) + x \sin(x).$

32. $\int x^2 \cos(x)dx = 2x \cos(x) + (x^2 - 2) \sin(x).$

33. $\int \cos^2(x)dx = \dfrac{x}{2} + \dfrac{\sin(2x)}{4} = \dfrac{x}{2} + \dfrac{\sin(x)\cos(x)}{2}.$

34. $\int x \cos^2(x)dx = \dfrac{x^2}{4} + \dfrac{x \sin(2x)}{4} + \dfrac{\cos(2x)}{8}.$

35. $\int \dfrac{1}{1 + \cos(x)} dx = \tan\left(\dfrac{x}{2}\right).$

36. $\int \sin(x)\cos(x)dx = \dfrac{\sin^2(x)}{2}.$

37. $\int \sin^2(x)\cos^2(x)dx = \dfrac{1}{8}\left[x - \dfrac{\sin(4x)}{4}\right].$

38. $\int \sin^{-1}\left(\dfrac{x}{a}\right) dx = x \sin^{-1}\left(\dfrac{x}{a}\right) + (a^2 - x^2)^{1/2}.$

39. $\int \left[\sin^{-1}\left(\dfrac{x}{a}\right)\right]^2 dx = x\left[\sin^{-1}\left(\dfrac{x}{a}\right)\right]^2 - 2x + 2(a^2 - x^2)^{1/2} \sin^{-1}\left(\dfrac{x}{a}\right).$

40. $\int \cos^{-1}\left(\dfrac{x}{a}\right) dx = x \cos^{-1}\left(\dfrac{x}{a}\right) - (a^2 - x^2)^{1/2}.$

41. $\int \left[\cos^{-1}\left(\dfrac{x}{a}\right)\right]^2 dx = x\left[\cos^{-1}\left(\dfrac{x}{a}\right)\right]^2 - 2x - 2(a^2 - x^2)^{1/2} \cos^{-1}\left(\dfrac{x}{a}\right).$

42. $\int \tan^{-1}\left(\dfrac{x}{a}\right) dx = x \tan^{-1}\left(\dfrac{x}{a}\right) - \dfrac{a}{2} \ln(a^2 + x^2).$

43. $\int x \tan^{-1}\left(\dfrac{x}{a}\right) dx = \dfrac{1}{2}(x^2 + a^2)\tan^{-1}\left(\dfrac{x}{a}\right) - \dfrac{ax}{2}.$

44. $\int e^{ax} dx = \dfrac{1}{a}e^{ax}.$

45. $\int a^x dx = \dfrac{a^x}{\ln(a)}.$

46. $\int xe^{ax} dx = e^{ax}\left(\dfrac{x}{a} - \dfrac{1}{a^2}\right).$

47. $\int x^2 e^{ax} dx = e^{ax}\left[\dfrac{x^2}{a} - \dfrac{2x}{a^2} + \dfrac{2}{a^3}\right].$

48. $\int e^{ax} \sin(x)dx = \dfrac{e^{ax}}{a^2 + 1}[a \sin(x) - \cos(x)].$

49. $\int e^{ax} \cos(x)dx = \dfrac{e^{ax}}{a^2 + 1}[a \sin(x) + \sin(x)].$

50. $\displaystyle\int e^{ax}\sin^2(x)\,dx = \frac{e^{ax}}{a^2+4}\left[a\sin^2(x) - 2\sin(x)\cos(x) + \frac{2}{a}\right].$

51. $\displaystyle\int \ln(ax)\,dx = x\ln(ax) - x.$

52. $\displaystyle\int x\ln(x)\,dx = \frac{x^2}{2}\ln(x) - \frac{x^2}{4}.$

53. $\displaystyle\int \frac{\ln(ax)}{x}\,dx = \frac{1}{2}[\ln(ax)]^2.$

54. $\displaystyle\int \frac{1}{x\ln(x)}\,dx = \ln(|\ln(x)|).$

55. $\displaystyle\int \tan(ax)\,dx = \frac{1}{a}\ln(|\sec(ax)|) = -\frac{1}{a}\ln(|\cos(ax)|).$

56. $\displaystyle\int \cot(ax)\,dx = \frac{1}{a}\ln(|\sin(ax)|).$

# A SHORT TABLE OF
# DEFINITE INTEGRALS

In the following list, $a$, $b$, $m$, $n$, $p$, and $r$, are constants.

1. $\displaystyle\int_0^\infty x^{n-1}e^{-x}\,dx = \int_0^1 \left[\ln\left(\frac{1}{x}\right)\right]^{-1} dx = \Gamma(n)\ (n > 0).$

    The function $\Gamma(n)$ is called the gamma function. It has the following properties:

    $$\Gamma(n+1) = n\Gamma(n) \text{ for any } n > 0$$
    $$\Gamma(n) = (n-1)! \text{ for any integral value of } n > 0$$
    $$\Gamma(n)\Gamma(1-n) = \frac{\pi}{\sin(n\pi)} \text{ for } n \text{ not an integer}$$
    $$\Gamma\left(\frac{1}{2}\right) = \sqrt{\pi}.$$

2. $\displaystyle\int_0^\infty \frac{1}{1+x+x^2}\,dx = \frac{\pi}{3\sqrt{3}}.$

3. $\displaystyle\int_0^\infty \frac{x^{p-1}}{(1+x)^p}\,dx = \frac{\pi}{\sin(p\pi)}\ (0 < p < 1).$

4. $\displaystyle\int_0^\infty \frac{x^{p-1}}{a+x}\,dx = \frac{\pi a^{p-1}}{\sin(p\pi)}\ (0 < p < 1).$

5. $\displaystyle\int_0^\infty \frac{x^p}{(1+ax)^2}\,dx = \frac{p\pi}{a^{p+1}\sin(p\pi)}.$

6. $\displaystyle\int_0^\infty \frac{1}{1+x^p}\,dx = \frac{\pi}{p\sin(\pi/p)}.$

7. $\displaystyle\int_0^{\pi/2} \sin^2(mx)\,dx = \int_0^{\pi/2} \cos^2(mx)\,dx = \frac{\pi}{4}\quad (m=1,2,\ldots).$

8. $\displaystyle\int_0^{\pi} \sin^2(mx)\,dx = \int_0^{\pi} \cos^2(mx)\,dx = \frac{\pi}{2}\quad (m=1,2,\ldots).$

9. $\displaystyle\int_0^{\pi/2} \tan^p(x)\,dx = \int_0^{\pi/2} ctn^p(x)\,dx = \frac{\pi}{2\cos(p\pi/2)}\quad (p^2<1).$

10. $\displaystyle\int_0^{\pi/2} \frac{x}{\tan(x)}\,dx = \frac{\pi}{2}\ln(2).$

11. $\displaystyle\int_0^{\pi/2} \sin^p(x)\cos^p(x)\,dx$

$$= \frac{\Gamma((p+1)/2)\Gamma((q+1)/2)}{2\Gamma((p+q)/2+1)}\quad (p+1>0, q+1>0).$$

12. $\displaystyle\int_0^{\pi} \sin(mx)\sin(nx)\,dx = \begin{cases} 0 & \text{if } m \neq n \\ \dfrac{\pi}{2} & \text{if } m = n \end{cases}\quad (m,n \text{ integers}).$

13. $\displaystyle\int_0^{\pi} \cos(mx)\cos(nx)\,dx = \begin{cases} 0 & \text{if } m \neq n \\ \dfrac{\pi}{2} & \text{if } m = n \end{cases}\quad (m,n \text{ integers}).$

14. $\displaystyle\int_0^{\pi} \sin(mx)\sin(nx)\,dx$

$$= \begin{cases} 0 & \text{if } m = n \\ 0 & \text{if } m \neq n \text{ and } m+n \text{ is even} \\ \dfrac{2m}{m^2-n^2} & \text{if } m \neq n \text{ and } m+n \text{ is odd} \\ & (m,n \text{ integers}). \end{cases}$$

15. $\displaystyle\int_0^\infty \sin\left(\frac{\pi x^2}{2}\right)dx = \int_0^\infty \cos\left(\frac{\pi x^2}{2}\right)dx = 1/2.$

16. $\displaystyle\int_0^\infty \sin(x^p)\,dx = \Gamma\left(1+\frac{1}{p}\right)\sin\left(\frac{\pi}{2p}\right)\quad (p>1).$

17. $\displaystyle\int_0^\infty \cos(x^p)\,dx = \Gamma\left(1+\frac{1}{p}\right)\cos\left(\frac{\pi}{2p}\right)\quad (p>1).$

18. $\displaystyle\int_0^\infty \frac{\sin(mx)}{x}\,dx = \begin{cases} \dfrac{\pi}{2} & \text{if } m>0 \\ 0 & \text{if } m=0 \\ -\dfrac{\pi}{2} & \text{if } m<0. \end{cases}$

19. $\displaystyle\int_0^\infty \frac{\sin(mx)}{x^p}\,dx = \frac{\pi m^{p-1}}{2\sin(p\pi/2)\Gamma(p)}\quad (0<p<2, m>0).$

20. $\displaystyle\int_0^\infty e^{-ax}\,dx = \frac{1}{a}\ (a>0).$

21. $\displaystyle\int_0^\infty xe^{-ax}\,dx = \frac{1}{a^2}\ (a>0).$

22. $\displaystyle\int_0^\infty x^2 e^{-ax}\,dx = \frac{2}{a^3}\ (a>0).$

23. $\displaystyle\int_0^\infty x^{1/2} e^{-ax}\,dx = \frac{\sqrt{\pi}}{2a^{3/2}}\ (a>0).$

24. $\displaystyle\int_0^\infty e^{-r^2 x^2}\,dx = \frac{\sqrt{\pi}}{2r}\ (r>0).$

25. $\displaystyle\int_0^\infty xe^{-r^2 x^2}\,dx = \frac{1}{2r^2}\ (r>0).$

26. $\displaystyle\int_0^\infty x^2 e^{-r^2 x^2}\,dx = \frac{\sqrt{\pi}}{4r^3}\ (r>0).$

27. $\displaystyle\int_0^\infty r^{2n+1} e^{-r^2 x^2}\,dx = \frac{n!}{2r^{2n+2}}\ (r>0, n=1,2,\ldots).$

28. $\displaystyle\int_0^\infty x^{2n} e^{-r^2 x^2}\,dx = \frac{(1)(3)(5)\cdots(2n-1)}{2^{n+1}r^{2n+1}}\sqrt{\pi}\ (r>0, n=1,2,\ldots).$

29. $\displaystyle\int_0^\infty x^a e^{-(rx)^b}\,dx = \frac{1}{br^{a+1}}\Gamma\left(\frac{a+1}{b}\right)\ (a+1>0, r>0, b>0).$

30. $\displaystyle\int_0^\infty \frac{e^{-ax}-e^{-bx}}{x}\,dx = \ln\left(\frac{b}{a}\right).$

31. $\displaystyle\int_0^\infty e^{-ax}\sin(mx)\,dx = \frac{m}{a^2+m^2}\ (a>0).$

32. $\displaystyle\int_0^\infty xe^{-ax}\sin(mx)\,dx = \frac{2am}{(a^2+m^2)^2}\ (a>0).$

33. $\displaystyle\int_0^\infty x^{p-1} e^{-ax}\sin(mx)\,dx = \frac{\Gamma(p)\sin(p\theta)}{(a^2+m^2)^{p/2}}\ (a>0, p>0, m>0),$

where $\sin(\theta)=m/r$, $\cos(\theta)=a/r$, $r=(a^2+m^2)^{1/2}$.

34. $\displaystyle\int_0^\infty e^{-ax}\cos(mx)\,dx = \frac{a}{a^2+m^2}\ (a>0).$

35. $\displaystyle\int_0^\infty xe^{-ax}\cos(mx)\,dx = \frac{a^2-m^2}{(a^2+m^2)^2}\ (a>0).$

36. $\displaystyle\int_0^\infty x^{p-1} e^{-ax}\cos(mx)\,dx = \frac{\Gamma(p)\cos(p\theta)}{(a^2+m^2)^{p/2}}\ (a>0, p>0),$

where $\theta$ is the same as given in Eq. (33).

37. $\displaystyle\int_0^\infty \frac{e^{-ax}}{x}\sin(mx)\,dx = \tan^{-1}\left(\frac{m}{a}\right)\ (a>0).$

38. $\displaystyle\int_0^\infty \frac{e^{-ax}}{x}[\cos(mx)-\cos(nx)]\,dx = \frac{1}{2}\ln\left(\frac{a^2+n^2}{a^2+m^2}\right)\ (a>0).$

39. $\displaystyle\int_0^\infty e^{-ax}\cos^2(mx)\,dx = \frac{a^2 + 2m^2}{a(a^2 + 4m^2)}$  $(a > 0)$.

40. $\displaystyle\int_0^\infty e^{-ax}\sin^2(mx)\,dx = \frac{2m^2}{a(a^2 + 4m^2)}$  $(a > 0)$.

41. $\displaystyle\int_0^1 \left[\ln\left(\frac{1}{x}\right)\right]^q dx = \Gamma(q + 1)\ (q + 1 > 0)$.

42. $\displaystyle\int_0^1 x^p \ln\left(\frac{1}{x}\right) dx = \frac{1}{(p + 1)^2}$  $(p + 1 > 0)$.

43. $\displaystyle\int_0^1 x^p \left[\ln\left(\frac{1}{x}\right)\right]^q dx = \frac{\Gamma(q + 1)}{(p + 1)^{q+1}}$  $(p + 1 > 0, q + 1 > 0)$.

44. $\displaystyle\int_0^1 \ln(1 - x)\,dx = -1$.

45. $\displaystyle\int_0^1 x\ln(1 - x)\,dx = \frac{-3}{4}$.

46. $\displaystyle\int_0^1 \ln(1 + x)\,dx = 2\ln(2) - 1$.

47. $\displaystyle\int_0^\infty e^{-ax^2}\cos(kx)\,dx = \frac{\sqrt{\pi}}{2\sqrt{a}}e^{-k^2/(4a)}$.

# SOME INTEGRALS WITH EXPONENTIALS IN THE INTEGRANDS: THE ERROR FUNCTION

We begin with the integral

$$\int_0^\infty e^{-x^2} dx = 1.$$

We compute the value of this integral by a trick,

$$I^2 = \left[\int_0^\infty e^{-x^2} dx\right]^2 = \int_0^\infty e^{-x^2} dx \int_0^\infty e^{-y^2} dy \int_0^\infty \int_0^\infty e^{-(x^2+y^2)} dx\, dy.$$

We now change to polar coordinates,

$$I^2 = \int_0^{\pi/2} \int_0^\infty e^{-p^2} \rho\, d\rho\, d\phi = \frac{\pi}{2} \int_0^\infty e^{-p^2} \rho\, d\rho$$

$$= \frac{\pi}{2} \int_0^\infty \frac{1}{2} e^{-z}\, dz = \frac{\pi}{4}.$$

Therefore,

$$I = \int_0^\infty e^{-x^2} dx = \frac{\sqrt{\pi}}{2}$$

and

$$\boxed{\int_0^\infty e^{-ax^2} dx = \frac{1}{2}\sqrt{\frac{\pi}{a}}}.$$

(A.1)

Another trick can be used to obtain the integral,

$$\int_0^\infty x^{2n} e^{-ax^2} dx,$$

where $n$ is an integer. For $n = 1$,

$$\int_0^\infty x^2 e^{-ax^2} dx = -\int_0^\infty \frac{d}{da}\left[e^{-ax^2}\right] dx = -\frac{d}{da}\int_0^\infty e^{-ax^2} dx$$

$$= -\frac{d}{da}\left[\frac{1}{2}\sqrt{\frac{\pi}{a}}\right] = \frac{1}{4a}\sqrt{\frac{\pi}{a}} = \frac{\pi^{1/2}}{4a^{3/2}}. \tag{A.2}$$

For $n$ an integer greater than unity,

$$\int_0^\infty x^{2n} e^{-ax^2} dx = (-1)^n - \frac{d^n}{da^n}\left[\frac{1}{2}\sqrt{\frac{\pi}{a}}\right]. \tag{A.3}$$

Equations (A.2) and (A.3) depend on the interchange of the order of differentiation and integration. This can be done if an improper integral is uniformly convergent. The integral in Eq. (A.1) is uniformly convergent for all real values of $a$ greater than zero. Similar integrals with odd powers of $x$ are easier. By the method of substitution,

$$\int_0^\infty x e^{-ax^2} dx = \frac{1}{2a}\int_0^\infty e^{-y} dy = \frac{1}{2a}. \tag{A.4}$$

We can apply the trick of differentiating under the integral sign just as in Eq. (A.3) to obtain

$$\int_0^\infty x^{2n+1} e^{-ax^2} dx = (-1)^n \frac{d^n}{da^n}\left(\frac{1}{2a}\right). \tag{A.5}$$

The integrals with odd powers of x are related to the gamma function, defined in Appendix 6. For example,

$$\int_0^\infty x^{2n+1} e^{-x^2} dx = \frac{1}{2}\int_0^\infty y^n e^{-y} dy = \frac{1}{2}\Gamma(n+1). \tag{A.6}$$

## The Error Function

The indefinite integral

$$\int e^{-x^2} dx$$

has never been expressed as a closed form (a formula not involving an infinite series or something equivalent). The definite integral for limits other than 0 and $\infty$ is therefore not obtainable in closed form. Because of the frequent occurrence of such definite integrals, tables of numerical approximations have been generated.[1] One form in which the tabulation is done is as the *error function*, denoted by erf($x$) and defined by

$$\text{erf}(x) = \frac{2}{\sqrt{\pi}}\int_0^x e^{-t^2} dt.$$

---

[1] Two commonly available sources are Eugene Jahnke and Fritz Emde, *Tables of Functions*. Dover, New York, 1945, and Milton Abramowitz and Irene A. Stegun, Eds., *Handbook of Mathematical Functions with Formulas, Graphs and Mathematical Tables*, U.S. Government Printing Office, Washington, DC, 1964.

As you can see from Eq. (A.1),

$$\lim_{x \to \infty} \text{erf}(x) = 1.$$

The name "error function" is chosen because of its frequent use in probability calculations involving the Gaussian probability distribution.

Another form giving the same information is the *normal probability integral*[2]

$$\frac{1}{\sqrt{2\pi}} \int_{-x}^{x} e^{-t^2/2}\, dt.$$

## Values of the Error Function*

$$\text{erf}(x) = \frac{2}{\sqrt{\pi}} \int_{0}^{x} e^{-t^2}\, dt$$

| x | | 0 | 1 | 2 | 3 | 4 | 5 | 6 | 7 | 8 | 9 |
|---|---|---|---|---|---|---|---|---|---|---|---|
| 0.0 | 0.0 | 000 | 113 | 226 | 338 | 451 | 564 | 676 | 789 | 901 | *013 |
| 0.1 | 0.1 | 125 | 236 | 348 | 459 | 569 | 680 | 790 | 900 | *009 | *118 |
| 0.2 | 0.2 | 227 | 335 | 443 | 550 | 657 | 763 | 869 | 974 | *079 | *183 |
| 0.3 | 0.3 | 286 | 389 | 491 | 593 | 694 | 794 | 893 | 992 | *090 | *187 |
| 0.4 | 0.4 | 284 | 380 | 475 | 569 | 662 | 755 | 847 | 937 | *027 | *117 |
| 0.5 | 0.5 | 205 | 292 | 379 | 465 | 549 | 633 | 716 | 798 | 879 | 959 |
| 0.6 | 0.6 | 039 | 117 | 194 | 270 | 346 | 420 | 494 | 566 | 638 | 708 |
| 0.7 | | 778 | 847 | 914 | 981 | *047 | *112 | *175 | *238 | *300 | *361 |
| 0.8 | 0.7 | 421 | 480 | 538 | 595 | 651 | 707 | 761 | 814 | 867 | 918 |
| 0.9 | | 969 | *0.19 | *068 | *116 | *163 | *209 | *254 | *299 | *342 | *385 |
| 1.0 | 0.8 | 427 | 468 | 508 | 548 | 586 | 624 | 661 | 698 | 733 | 768 |
| 1.1 | | 802 | 835 | 868 | 900 | 931 | 961 | 991 | *020 | *048 | *076 |
| 1.2 | 0.9 | 103 | 130 | 155 | 181 | 205 | 229 | 252 | 275 | 297 | 319 |
| 1.3 | | 340 | 361 | 381 | 400 | 419 | 438 | 456 | 473 | 490 | 507 |
| 1.4 | 0.95 | 23 | 39 | 54 | 69 | 83 | 97 | *11 | *24 | *37 | *49 |
| 1.5 | 0.96 | 61 | 73 | 84 | 95 | *06 | *16 | *26 | *36 | *45 | *55 |
| 1.6 | 0.97 | 63 | 72 | 80 | 88 | 96 | *04 | *11 | *18 | *25 | *32 |
| 1.7 | 0.98 | 38 | 44 | 50 | 56 | 61 | 67 | 72 | 77 | 82 | 86 |
| 1.8 | | 91 | 95 | 99 | *03 | *07 | *11 | *15 | *18 | *22 | *25 |
| 1.9 | 0.99 | 28 | 31 | 34 | 37 | 39 | 42 | 44 | 47 | 49 | 51 |
| 2.0 | 0.995 | 32 | 52 | 72 | 91 | *09 | *26 | *42 | *58 | *73 | *88 |
| 2.1 | 0.997 | 02 | 15 | 28 | 41 | 53 | 64 | 75 | 85 | 95 | *05 |
| 2.2 | 0.998 | 14 | 22 | 31 | 39 | 46 | 54 | 61 | 67 | 74 | 80 |
| 2.3 | | 86 | 91 | 97 | *02 | *06 | *11 | *15 | *20 | *24 | *28 |
| 2.4 | 0.999 | 31 | 35 | 38 | 41 | 44 | 47 | 50 | 52 | 55 | 57 |
| 2.5 | | 59 | 61 | 63 | 65 | 67 | 69 | 71 | 72 | 74 | 75 |
| 2.6 | | 76 | 78 | 79 | 80 | 81 | 82 | 83 | 84 | 85 | 86 |
| 2.7 | | 87 | 87 | 88 | 89 | 89 | 90 | 91 | 91 | 92 | 92 |
| 2.8 | 0.9999 | 25 | 29 | 33 | 37 | 41 | 44 | 48 | 51 | 54 | 56 |
| 2.9 | | 59 | 61 | 64 | 66 | 68 | 70 | 72 | 73 | 75 | 77 |

* From Eugene Jahnke and Fritz Emde, *Tables of Functions*, p. 24, Dover, New York, 1945.

[2] See, for example, Herbert B. Dwight, *Tables of Integrals and Other Mathematical Data*, 4th ed., Macmillan, New York, 1961.

# A PROGRAM FOR NUMERICAL INTEGRATION: NUMINT

This program implements Simpson's rule or its equivalent for equally spaced data points, for unequally spaced data points, or for an integrand function, which the user can specify when running the program.

For the equally spaced data points or the integrand function, Simpson's rule in its usual form is used. For unequally spaced data points, the program fits parabolas to data points three at a time but uses the parabola to represent the integrand only in the first panel. Therefore, the number of panels can be even or odd for this option. Since the usual form of Simpson's rule requires the number of panels to be even, the unequally spaced point option can be used for equally spaced points if you have an odd number of panels.

```
100 OPTION NOLET
105 PRINT "THIS PROGRAM CALCULATES AN INTEGRAL BY
SIMPSON'S ONE-THIRD RULE."
110 PRINT "IT GIVES YOU A CHOICE OF ENTERING A LIST
OF VALUES FOR THE."
120 PRINT "INTEGRAND, OR ENTERING AN INTEGRAND
FUNCTION."
130 DIM X(101), Y(101), A(101), B(101), C(101)
140 PRINT "DO YOU WANT TO USE A TABLE OF EQUALLY
SPACED DATA POINTS?"
145 INPUT A$
150 IF A$[1:1] = "Y" THEN 500 ELSE 160
```

```
160 PRINT "DO YOU WANT TO USE A TABLE OF DATA POINTS
WHICH ARE NOT"
170 PRINT "NECESSARILY EQUALLY SPACED?"
175 INPUT C$
180 IF C$[1:1] = "Y" THEN 800 ELSE 200
200 PRINT "DO YOU HAVE AN INTEGRAND FUNCTION?"
205 INPUT B$
210 IF B$[1:1] = "Y" THEN 220 ELSE 1500
220 PRINT "NUMBER OF POINTS (MUST BE ODD, LESS THAN
OR EQUAL TO 101)?"
225 INPUT N
230 PRINT "LOWER LIMIT OF INTEGRATION?"
235 INPUT X1
240 PRINT "UPPER LIMIT OF INTEGRATION?"
245 INPUT X9
250 W = (X9 - X1)/(N -1)
260 X(1) = X1
270 FOR I = 2 TO N STEP 1
280 X(I) = X(I -1) + W
290 NEXT I
292 PRINT "RETURN TO THE PROGRAM BY TYPING
COMMAND-PERIOD."
293 PRINT "TYPE IN BASIC STATEMENTS WHICH WILL
EVALUATE YOUR INTEGRAND"
294 PRINT "FUNCTION. NUMBER THE STATEMENTS WITH
NUMBERS BETWEEN 300"
295 PRINT "AND 400. CALL YOUR INTEGRAND Y(I) AND
YOUR INDEPENDENT"
296 PRINT "VARIABLE X(I). THEN START THE PROGRAM
AGAIN, AND ANSWER"
297 PRINT "YES WHEN ASKED IF YOU WANT AN INTEGRAND
FUNCTION."
299 FOR I = 1 TO N STEP 1
410 NEXT I
420 S = Y(1)
430 FOR I = 2 TO N -1 STEP 2
440 S = S + 4*Y(I) + 2*Y(I+1)
450 NEXT I
460 S = (S - Y(N))*W/3
470 PRINT "INTEGRAL FROM";X1;" TO ";X9;" IS
EQUAL TO";S
480 GOTO 32767
500 PRINT "NUMBER OF POINTS (MUST BE ODD)?"
505 INPUT N
510 PRINT "INTERVAL BETWEEN THE VALUES OF THE
INDEPENDENT VARIABLE?"
515 INPUT W
520 FOR I = 1 TO N STEP 1
```

```
530 PRINT "VALUE OF INTEGRAND FOR POINT NUMBER"; I
540 INPUT Y(I)
550 NEXT I
560 S = Y(1)
570 FOR I = 2 TO N -1 STEP 2
580 S = S + 4*Y(I) + 2*Y(I+1)
590 NEXT I
600 FOR I = 1 TO N -1 STEP 1
610 BI = BI + Y(I)
620 T = T + (Y(I) + Y(I+1))/2
630 NEXT I
640 S = (S - Y(N))*W/3
650 BI = BI*W
660 T = T*W
670 PRINT "INTEGRAL BY BAR GRAPH AREA =";BI
680 PRINT "INTEGRAL BY TRAPEZOIDAL RULE =";T
690 PRINT "INTEGRAL BY SIMPSON'S RULE =";S
700 GOTO 32767
800 PRINT "YOU HAVE CHOSEN TO USE A LIST OF VALUES
FOR THE INTEGRAND"
810 PRINT "AND THE INDEPENDENT VARIABLE WHICH ARE
NOT NECESSARILY"
820 PRINT "EQUALLY SPACED. THIS PROGRAM FITS
PARABOLAS TO THE POINTS"
824 PRINT "THREE AT A TIME, MUCH LIKE SIMPSON'S
RULE. THE PARABOLA IS"
826 PRINT "USED TO REPRESENT THE INTEGRAND BETWEEN
THE FIRST AND SECOND"
828 PRINT "POINTS, EXCEPT FOR THE LAST THREE POINTS,
FOR WHICH THE"
829 PRINT "PARABOLA IS USED FOR BOTH PANELS."
830 PRINT "NUMBER OF DATA POINTS? (CAN BE EVEN OR
ODD)"
835 INPUT N
840 PRINT "DO YOU WANT TO DO ANY OPERATIONS ON THE
VARIABLES?"
845 INPUT D$
850 IF D$[1:1] = "Y" THEN 855 ELSE 880
855 PRINT "RETURN TO THE PROGRAM BY TYPING
COMMAND-PERIOD."
860 PRINT "TYPE IN BASIC STATEMENTS TO OPERATE ON
YOUR VARIABLES."
865 PRINT "USE STATEMENT NUMBERS BETWEEN 900 AND
940. CALL YOUR"
870 PRINT "INDEPENDENT VARIABLE X(I) AND YOUR
INTEGRAND Y(I)."
872 PRINT "THEN START THE PROGRAM AGAIN BUT DON'T
TYPE YES"
```

```
874 PRINT "WHEN ASKED IF YOU WANT TO OPERATE ON YOUR
VARIABLES."
880 PRINT "X(1) MUST BE LOWER LIMIT OF INTEGRATION."
885 PRINT "X(";N;") MUST BE UPPER LIMIT OF
INTEGRATION."
890 FOR I = 1 TO N
892 PRINT "VALUE OF X(";I;"), Y(";I;")"
894 INPUT X(I), Y(I)
945 NEXT I
950 FOR I = 1 TO N -2
960 E = Y(I)- Y(I+2)-(Y(I+1)- Y(I+2))*(X(I)- X(I+2))/
(X(I+1)- X(I+2))
970 G = X(I)*X(I)- X(I+2)*X(I+2)
980 H = (X(I+1)*X(I+1)- X(I+2)*X(I+2))*(X(I)-X(I+2))
990 G = G - H/(X(I+1)- X(I+2))
1000 F = Y(I+1)- Y(I+2)- E*(X(I+1)*X(I+1)-
X(I+2)*X(I+2))/G
1010 A(I) = E/G
1020 B(I) = F/(X(I+1)- X(I+2))
1030 C(I) = Y(I+2)-E*X(I+2)*X(I+2)/G - F*X(I+2)/
(X(I+1) - X(I+2))
1040 NEXT I
1050 Z = 0
1060 FOR I = 1 TO N -3
1070 Z = Z + A(I)*(X(I+1)^3- X(I)^3)/
3+B(I)*(X(I+1)^2- X(I)^2)/2
1080 Z = Z + C(I)*(X(I+1)- X(I))
1090 NEXT I
1100 Z = Z + A(N -2)*(X(N)^3- X(N -2)^3)/3
1110 Z = Z + B(N -2)*(X(N)^2- X(N -2)^2)/2
+ C(N -2)*(X(N) - X(N -2))
1120 PRINT "INTEGRAL FROM";X(1);"TO";X(N);"=";Z
1130 GOTO 32767
1500 PRINT "SORRY. THERE ARE NO OTHER OPTIONS IN
THIS PROGRAM."
32767 END
```

# ■■■■ INDEX